Selene's Two Faces

Nuncius Series

Studies and Sources in the Material and Visual History of Science

Series Editors

Marco Beretta (*University of Bologna*)
Sven Dupré (*Utrecht University / University of Amsterdam*)

VOLUME 3

The titles published in this series are listed at *brill.com/nuns*

Selene's Two Faces

From 17th Century Drawings to Spacecraft Imaging

Edited by

Carmen Pérez González

BRILL

LEIDEN | BOSTON

Cover illustration: *Man with a telescope*, Japanese Ukiyo-e, mid-19th century, Louwman Collection of Historical Telescopes, The Hague, The Netherlands (Also included as illustration 0.7).

Library of Congress Cataloging-in-Publication Data

Names: Pérez González, Carmen, 1969- editor.
Title: Selene's two faces : from 17th century drawings to spacecraft imaging / edited by Carmen Perez Gonzalez.
Description: Leiden ; Boston : Brill, [2018] | Series: Nuncius series. Studies and sources in the material and visual history of science ; volume 3 | Includes bibliographical references and index.
Identifiers: LCCN 2018033518 (print) | LCCN 2018035698 (ebook) | ISBN 9789004298873 (ebook) | ISBN 9789004298866 (hardback : alk. paper)
Subjects: LCSH: Moon--Pictorial works. | Lunar geography--History. | Astronomy--History.
Classification: LCC QB595 (ebook) | LCC QB595 .S45 2018 (print) | DDC 523.3--dc23
LC record available at https://lccn.loc.gov/2018033518

Typeface for the Latin, Greek, and Cyrillic scripts: "Brill". See and download: brill.com/brill-typeface.

ISSN 2405-5077
ISBN 978-90-04-29886-6 (hardback)
ISBN 978-90-04-29887-3 (e-book)

This book is printed on acid-free paper and produced in a sustainable manner.

Contents

Acknowledgments

First and foremost, I would like to thank all co-authors in this collaborative undertaking for their hard work, patience and collegial good nature throughout the gestation process of this book. I am indebted to all of them.

I shall like to thank the unknown peer-reviewer of this book, who made valuable suggestions on a previous draft of the book. I would like to thank also the editors of the Nuncius Series at Brill, Marco Beretta and Sven Duprè for their help and support with this book project. At Brill I would like to thank as well Wendel Scholma, Isadora Lyra, Rosanna Woensdregt, and Pieter te Velde, who took care of the editorial tasks related to the book at different stages and who were always helpful during the production of the book.

I would like to thank as well a few scholars who read different parts of the book and made valuable comments on them: Volker Remmert; Tofigh Heidarzadeh; and Edgar Scholz. Joseph Swann did excellent English copy-edit work on most of the chapters of this book.

We would like to thank the following institutions and private collectors for letting us print the images selected for this book, kept in their collections and archives: Hamburger Sternwarte; Argelander-Institut der Universität Bonn; Sammlung Historischer Himmelsaufnahmen der Universität Bonn; Birgminghan City Library; Adler Planetarium (Chicago); Museum of the History of Science, University of Oxford; JAXA/SELENE (Japan); Ueda City Museum (Japan); Kimia Foundation (Los Angeles); Staatsbibliothek zu Berlin, Preußischer Kulturbesitz; Bibliothèque Nationale de France; National Media Museum/Science & Society Picture Library; Museum für Kunst und Gewerbe Hamburg; ESA/NASA/JPL/University of Arizona; Science Museum in London; University of Manchester; Royal Geographical Society; Royal Observatory of Edinburg; NASA; Hasselblad Foundation (Göteborg); CNSA; Marco Antonello Collection (Italy) and Maria Francesca Bonetti; Louwman Collection of Historical Telescopes; Houman Sarshar Kimia Foundation (Los Angeles); Azita and Elmar Seibel Collection (Boston); S. Shimada (Japan); and Christopher B. Steiner.

Special thanks are due to Volker Remmert, director of the IZWT (Interdisciplinary Centre for Science and Technology Studies) at Bergische Universität Wuppertal, for his support and encouragement of this project.

Reggio Emilia
June 2018

List of Figures and Tables

Figures

Tables

Notes on Contributors

Charlotte Bigg

is an historian of science at the Centre National de la Recherche Scientifique/ Centre Alexandre Koyré and she teaches at the EHESS, Paris. She has published widely on the social and cultural history of the physical and astronomical sciences in the nineteenth and twentieth centuries. Her work focuses especially on the elaboration of optical instruments and visual cultures in scientific practice and their circulation to a range of scientific and popular audiences. She is currently preparing a short book entitled Photography and Astronomy for the Exposure series at Reaktion books.

Michael Geffert

1953 born in Bonn, studied at the Rheinische Friedrich-Wilhelms-Universität in Bonn (Diplom 1978, PhD 1986). He worked from 1978–1986 at the Geodätisches Institut der Universität Bonn and became 1987 staff member of Hoher List Observatory of University of Bonn. He worked on astrometry, star clusters and structure of the Milky Way. Beside his scientific interest, Geffert was engaged in public outreach, didactics and history of astronomy. In 1991 he founded the primary school project "Astronomie / vor Ort" and became 2007 the coordinator of the German activities of the IYA2009.

Geffert is also member of the german artist community (BBK) and participated in more than 30 art exhibitions (e.g. in Bexbach, Bonn, Pulheim, Glauchau, Halberstadt, Jülich, Schleiden).

After the closure of Hoher List Observatory in 2013, Geffert is the curator of the "Sammlung Historischer Himmelsaufnahmen" of university of Bonn.

Detlef Groote

author was born in 1949, went to school and studied physics in Berlin. At the Technische Universität he earned his doctorate 1982 in astronomy for his work on infrared measurements of peculiar stars. He moved to Hamburger Sternwarte in the same year and served there in research and as head of the IT-group until his retirement in 2014. He was deeply engaged in large scientific projects as the Hamburg Quasar Surveys (HQS and HES) and started his digitization project in 2010 which he will lead until its end in 2019 and which is supported by the Deutsche Forschungsgemeinschaft (grants DFG LIS(25) GR 969/4-1 and SCM 1032/552).

Tsuko Nakamura

got in 1975 a Ph.D. in astronomy from the University of Tokyo. At the Tokyo Astronomical Observatory, he studied solar system small bodies like asteroids and comets till 2006. In parallel with such modern astronomy, he has written several tens of research papers on the history of astronomy both in Japanese and English. He also worked as one of the editors for *Highlighting the History of Astronomy in the Asia-Pacific Region* (2011) and *The Emergence of Astrophysics in Asia* (2017 in press), both published from Springer. He has been the president of the International Conference on Oriental Astronomy (ICOA) since 2008.

Omar W. Nasim

is Professor for the History of Science at the University of Regensburg, Germany. Nasim has worked at in Zurich, Oxford, Canterbury, Florence, Berlin and was recently a senior fellow at the Vossius Center for History of Humanities and Science at the University of Amsterdam. He is the author of numerous articles and two awarding winning monographs, including Observing by Hand: Sketching the Nebulae in the Nineteenth Century (2013), which won the History of Science Society's prestigious Pfizer Book Award for most outstanding scholarly book in the field for 2016. Nasim is currently working on two book projects, one dealing with the history of photography and another with the role of furniture in the history of science.

Carmen Pérez González

holds a master's degree in Astrophysics (University of Barcelona, 1993), and a doctorate in Art History (Leiden University, 2010), where she graduated with a dissertation awarded with the ICAS Best PhD Thesis Prize in Humanities in 2011. A revised and augmented version of her dissertation was published with the title *Local Portraiture: Through the Lens of the 19thCentury Iranian Photographers* (Leiden University Press, Iranian Studies Series, 2012). She wrote the exhibition's catalogue *From Istanbul to Yokohama: The Camera Meets Asia, 1839-1900* (Walther König Verlag, 2014); co-edited two special issues on 19th century photography: "The First Hundred Years of Iranian Photography" (*History of Photography*); and "175 Years of Photography in Spain" (*PhotoResearcher*). She is currently working at the IZWT (Interdisciplinary Centre for Science and Technology Studies), Bergische Universität Wuppertal.

Pedro M.P. Raposo, DPhil

is a curator at Adler Planetarium, Chicago, where he oversees a world-class collection of scientific instruments and rare books. His research interests include the history of celestial cartography, the history of modern planetaria, and the material culture of timekeeping. He served as content lead in the award-winning exhibition 'What is a Planet?'. Raposo has published on topics such as the history of nineteenth-century observatories, astronomy and empire, and the history of modern planetaria. His current research interests include the history of visual and mechanical representations of the solar system, and the notions of size and scale in astronomical imagery. He has served in the Committee of the Historical Astronomy Division of the American Astronomical Society.

Pedro Ré
is currently an Associate Professor at the University of Lisbon (Faculty of Sciences) with a PhD in Animal Ecology (1984). His current research interests deal mainly with the study of Plankton and Benthos Ecology. Pedro Ré has also been an amateur astronomer for the last 30 years mainly interested in imaging the deep-sky, the solar system. His images were published in several astronomy magazines: *Sky and Telescope*; CCD; *Astronomy*; *Astronomy Now*; *Astronomy Magazine*; *Ciel et Espace*. He is also the author of more than 150 scientific papers (54 peer-reviewed) and 18 books (author, co-author or editor), and about 125 outreach papers dealing with amateur astronomy, history of astrophotograhy and history of the telescope.

INTRODUCTION

On the Visible, the Invisible, Presences and Absences in Lunar Portraiture

Carmen Pérez González

For Luna, Mitra and Mani, my own celestial bodies

∴

If any scientific object has over the course of human history aroused the fascination of both scientists and artists worldwide, it is beyond doubt the moon.* Scientific curiosity, poetic and visual fascination, aesthetic infatuation and attraction – all of these it has attracted, and done so the more intensely because it is so out of reach. The moon is also by far the most interesting celestial body for reflecting on the dualistic nature of photography as applied to the study of the universe. Against this background *Selene's Two Faces* sets out to look at the scientific purposes, the aesthetic expression, and the influence of early lunar drawings, maps and photographs. Nineteenth century writing on photography contains frequent references to the medium as "beautiful art-science", "the true retina of the scientist"[1], and even today photography continues to play a fundamental role in bridging the imaginary gap between art and science – the volume of discussion currently devoted to objectivity and the role of *first* pho-

* Thanks are due to Volker Remmert; Erhard Scholz; Franziska Ehmcke, Maria Francesca Bonetti, Henk Zoomers, Peter Louwman, Christopher Steiner, and Huib J. Zuidervaart.

1 "The sensitive photographic film is the true retina of the scientists ... for it possesses all the properties which science could want; it faithfully preserves images which deposit themselves upon it, and reproduces and multiplies them indefinitely on request; in the radioactive spectrum (electromagnetic spectrum) it covers a range more than double that which the eye can perceive and soon perhaps will cover it all; finally it takes advantage of the admirable property which allows the accumulation of events, and whereas our retina erases all impressions more than a tenth of a second old, the photographic retina preserves them and accumulates them over a practically limitless time": cited in Jon Darius. Beyond Vision. Oxford: Oxford University Press, 1984, 11. As pointed out by Kelley Wilder, this description of photography ("true retina of the scientists") was originally from Jean Baptiste Biot, but was often appropriated by other scientists, such as Jules Janssen: see note 1 in Kelley Wilder, "Photography and the Art of Science", in *Visual Studies* 24/2 (2009): 163-168, 167.

 | DOI 10.1163/9789004298873_002

tographs by historians of photography is a good indicator of this.[2] Nineteenth century astronomers experimented with optics and the action of light and chemistry, as well as with photography's visual effects and reproduction methods. But the opening pages in the history of lunar photography tell of astronomers who were artists in their own right, already skilled in drawing through the telescope well before photography came on the scene. Accordingly, what we present in this book is not a tale of competition between the two media, drawing *versus* photography, but rather a history of how each adapted to and complemented the other through several decades.

In its approach, ***Selene's Two Faces*** is intermedial, intercultural and interdisciplinary. It brings together various media (photography, maps, engravings, lithographs, globes, texts), cultures, and theoretical perspectives (visual analysis, history of science and technology, art history and history of photography). The contributors to this book focus their attention especially on *how* the particular physical conditions of the moon (distance, lighting and motion) have conditioned the development of photographic processes and telescope technology, from the first lunar photographs to spacecraft imaging, and also on *how* astronomy and photography shaped each other. In order to fulfil this main objective, the book takes an interdisciplinary approach, blending the history of science and technology with that of art.

Thus a number of authors elaborate, from a history of science and technology standpoint, how the early history of moon photography was built on the parallel refinement of photographic technique (aimed at improving both exposure time and image definition) and telescope technology, and outline the principal scientific achievements attained with lunar photography. Among the issues addressed are:

1. Development of photographic technique, which by the end of the 19th century had solved two main problems of lunar photography: long exposure time and low definition of the photographic image. The problem of *seeing*, however, remained unsolved.[3]
2. Development of telescopes suitable for lunar photography, in particular the clock mechanism used to drive the equatorial. In order to record a

2 Tanya Sheehan and Andrés Mario Zervigón, ed., *Photography and Its Origin* (New York and London: Routledge, 2015); Symposium *Rethinking Early Photography*, University of Lincoln, Lincoln, 16th and 17th June 2015.

3 See chapter 4 by Charlotte Bigg, and chapter 8 by Pedro Ré and Carmen Pérez González in this book on the definition and evolution of the phenomenon of *seeing*.

perfectly sharp photographic moon image, it was essential that the instrument should accurately follow the motion of the moon throughout the exposure.

3. *What* scientific achievements came about as a result of the application of photography to the study of the moon, and *how* were they achieved? When did the first useful results of the application of photography to astronomy occur? How is photography used for lunar research today?

Other contributors to this book elaborate from an art history standpoint how early scientific photographs have been read by scientists and artists at different times from 1840 to 2015. Were these images seen as mechanically objective or rather as aesthetically pleasing? How and when (if at all) did the aesthetic pleasure of viewing lunar photographs yield to a sense of their scientific usefulness?

Finally, we will explore why it was that lunar cartography and photography was basically a European concern. At the same time we will present two non-European contributions to these two fields.

The first part of this book (chapters 1, 2 and 3) concentrates on early drawings and maps produced from 1610 to the 1890s. Several issues are treated here:

1. Exploration of the act of drawing through a telescope, focusing on the process of drawing *through* visual memory, due to the gap between the moment of observing with the eye and that of transferring the seen object via the hand to paper. Concepts such as familiarization of the astronomer with the observed object will be explored (as introduced by Omar W. Nasim, Chapter 5).
2. As mentioned above, lunar cartography and photography was basically a European concern. Why was this so? A possible answer would be that the required technology (the telescope) did not arrive early enough in non-western countries. But this is not correct: the telescope spread around the world within a few decades of its invention in 1608. As will be shown later in this chapter, the reason is subtler than mere technological limitation, for non-western (or, more precisely, Asian) astronomers lacked the obsession of their western counterparts with producing realistic drawings of the moon. We will explore this phenomenon in two case studies, that of Japan and that of Iran.

The second part of the book (chapters 4, 5, 6, and 7) explores the impact of the introduction of photography to moon cartography. Several concepts are studied:

1. The (at the time) dichotomous discourse of the *beautiful* and the *scientifically useful*.
2. Analysis of photographs of lunar models by placing them within a larger context of modeling and imagining practices, as proposed by Omar Nasim.
3. The concept of *blur* in photography as defined by Charlotte Bigg and her argument that early lunar photography was part of the widespread interest in photographic portraiture in the 19th century.

The third part of the book (chapter 8) explores another phase in the history of lunar visualization: that of images produced from space. The history of the visualization of our satellite opens a new chapter with photographs of the (to us) invisible face of the moon.

1 Selene's Visible Face: the Moon from Our Planet

Between the invention of the telescope in 1608 and the invention of photography in 1839 the moon was studied and mapped by numerous observers using different techniques. The astronomer and specialist in lunar studies Ewen A. Whitaker has proposed four periods in the study of our satellite[4]: 1. 1610 to 1650: the first observers used the newly invented telescope; the moon was recognized as a new world to explore. This period marked the birth of selenography. 2. After a long period of inactivity a second phase began in the early nineteenth century. Observers used achromatic refractors and filar micrometers to study and map the moon with outstanding precision. Scientific cartography of the moon was born. 3. The third period started in 1890 when the large telescopes at the Lick and Paris observatories began a systematic photographic survey of the moon. 4. The fourth period started in 1959 with close-up detailed observations performed by spacecraft, and manned exploration of the moon.

Galileo Galilei was the first astronomer to publish drawings and descriptions of the moon's surface made directly from telescopic observation. He published these sketches in his book *Sidereus Nuncius* (Sidereal Messenger) in 1610.[5] But before Galileo, Thomas Harriot actually made the first known tele-

4 Ewen A. Whitaker, *Mapping and Naming the Moon. A History of Lunar Cartography and Nomenclature* (Cambridge: Cambridge University Press, 1999).

5 Galileo Galilei, *Siderevs// Nvncivs: Magna, Longeqve Admirabilia Spectacula pandens, suspiciendáque proponens vnicuique, praesertim verò Philosophis, atq[ue] Astronomis* (Venetiis: Baglionus, 1610).

scopic sketch of the moon on 5th August 1609.[6] Pedro Raposo provides an overview of selenographic activity between the invention of the telescope and the rise of photography. The first part of his chapter addresses the early production of observational and visual evidence for a moon with its own topography and surface features, as opposed to the old concept of the moon as an immaculate body. The second part describes how the moon was increasingly approached as a cartographic object to be thoroughly charted and named, highlighting some fundamental aspects in the emergence and development of the lunar atlas. The third and final section brings under focus the early attempts to represent the moon as a tridimensional object through the production of lunar globes.[7]

The first real map of the moon was published by the Dutch cartographer Michile Van Langren in 1645, and the last map before the introduction of photography was made by the German astronomer Johann Heinrich Mädler: *Mappa Selenographica* published in four volumes between 1834 and 1836. Only two years after Mädler's lunar map, the introduction of the daguerreotype allowed faithful reproductions of our satellite to be made without the need for time-consuming topographical drawing based only on telescopic observation. Nevertheless, early attempts at photographing the moon through a telescope were especially difficult, because of the inherent limitations of the photographic process, which still required very long exposure times. In the case of the moon, an added problem was the need to guide the telescope at lunar speed throughout the exposure. No matter how great the interest commanded by these early specimens, the fact remained that the accuracy and precision offered by the daguerreotype (or later processes) could not replace the precision of the astronomer's eye and hand.

The advent of the new medium marked a pivotal point in the history of scientific representation. Several scholars have pointed out, however, that technical limitations obfuscated the actual entry of photography into the different scientific disciplines, and how this time-lag has repercussions on cross-disciplinary claims about the transition from a "truth-to-nature" to a "mechanical" mode of

---

6 Whitaker, *Mapping and Naming the Moon*, 17.

7 The idea of making the moon tangible through globes is a fascinating topic within the history of the visualization of the moon. Specially interesting is the attempts done so far to make the moon tangible and *visible* to the visually impaired. For instance, a great project is the one done at Valencia Observatory: <http://www.scienceknowledge.org/2011/06/13/moon-created-the-first-touch-for-the-blind/>. Before that, a pioneering project on this field is the book: David Hurd and John Matelok, *Getting a Feel for Lunar Craters* (Collingdalle: Diane Publishing Co., 2011).

representation.[8] Especially important in this argument is the contribution of Kelley Wilder, who states that

> visualizing with photography is not some kind of mechanical observation akin to the mechanical objectivity so ably defined by Lorraine Daston and Peter Galison. Since cameras, lenses, emulsions, and other photographic materials register images that lie so far outside of the scope of the human senses, it is also never merely a matter of making visible the previous invisible. To make a thing visible to the unaided eye out of objects which are too small, too large, too close, too fast or too slow, and too far outside of the spectrum of human vision is only the precondition for observation, and dozens of small decisions go into the process of giving these ephemeral phenomena form.[9]

As Wilder goes on to argue, none of these small decisions is absolutely objective, even less so the summed result of all of them. Photographs, as visualizations, are a formal (or pictorial manifestation) of what already existed in the astronomer's mind, as well as being material objects (daguerreotypes, wet collodion plates, salt prints, lithographic plates, glass, or silver gelatin prints) that react to the moon's reflected light in very different ways. Visualizing our natural satellite is therefore a combination of the setting up of the conditions under which the moon can be observed (mechanical skills in the use of the telescope) with the learned practice of imaginative photographic manipulation. Photographic methods employed by scientists are multiple and complex, as Wilder has argued, and hence there has never been *a* photographic method, but many individual methods, each particular to a time, a scientist, and most importantly, to a phase in the history of photography.[10]

---

8 Klaus Hentschel, *Mapping the Spectrum. Techniques of Visual Representation in Research and Teaching* (Oxford and New York: Oxford University Press, 2002); Wilder, "Photography and the Art of Science," 163-168; Kelley Wilder, "Visualizing Radiation: The Photographs of Henri Becquerel," in *Histories of Scientific Observation,* ed. Lorraine Daston et al. (Chicago: University of Chicago Press, 2011), 349-368.

9 Wilder, "Visualizing Radiation," 351. See also: Lorraine Daston, and Peter Galison, eds., *Objectivity* (New York: Zone Books, 2010); and Joel Snyder, "Res Ipsa Loquitur," in *Things That Talk. Object Lessons from Art and Science,* ed. Lorraine Daston (New York: Zone Books, 1998), 195-221.

10 Wilder, "Visualizing Radiation", 352. See also: Jennifer Tucker, *Nature Exposed: Photography as Eyewitness in Victorian Science* (Baltimore: Johns Hopkins University Press, 2005); and Kelley Wilder, *Photography and Science* (London: Reaktion Books, 2011).

In fact, the number of early photographic techniques is much greater than we might imagine: between 1822 and 1879 there were twenty such processes, starting from the niépceotype heliography (asphaltum-coated glass plate, 6 hours exposure time) to the improved gelatin emulsions (silver nitrate added to gelatinous solution of ammonium bromide, needing less than 1 second exposure time).[11] Each scientific discipline gave birth to a different mode of photography, developed to meet its particular needs and conditions. In astronomical photography, all the celestial bodies shared the difficult conditions of distance, lighting and motion. Every scientific object has an inherent *resistance* to photography[12], caused by its particular physical characteristics. In the case of the moon, we can define five different types of *resistance*: three caused by the special conditions of the object (distance, lighting and motion) and two caused by the inherent limitations of the early photographic medium (lack of definition and long exposure time). Parallel to the improvement of the telescope, it was the improvement of the camera and the development of photographic techniques adapted to imaging the moon that marked the different periods in the history of early lunar photography.

As argued by John Lankford, the 1840s were a decade of *firsts* in the history of astronomical photography: the first successful daguerreotype of the moon in March 1840, by the Anglo-American chemist J.W. Draper; the first daguerreotype of a solar eclipse in 1842 by G.A. Majocchi; the first daguerreotype of the solar spectrum in 1843 by Draper; and the first daguerreotype of the sun in 1845 by J.B.L. Foucault and A.H.L. Fizeau.[13]

Many scientists involved in the discovery and dissemination of the daguerreotype process were astronomers. From the moment of its inception, the daguerreotype was seen as a significant scientific tool for recording monuments, as well as for making photographic maps of the moon. This latter purpose was remarked upon by D.F.J. Arago, director of the Paris Observatory, when he informed the French Chamber of Deputies of the remarkable invention by the French artist Louis-Jacques Mandé Daguerre:

---

11 William Crawford, *The Keepers of Light. A History of Working Guide to Early Photographic Processes* (New York: Dobbs Ferry Morgan, 1979).

12 I mean here: an inherent resistance (or difficulty) to being photographed. Eleanor Robson has used this term, *resistance,* as well, but applied to cuneiform tablets: Eleanor Robson, "Bel and the Dragons: Deciphering Cuneiform after Decipherment," in *William Henry Fox Talbot: Beyond Photography,* ed. Mirjam Brusius, Katrina Dean, and Chitra Ramalingam (New Haven and London: Yale University Press, 2013), 193-218, 212.

13 John Lankford, "The impact of photography on astronomy," in *Astrophysics and twentieth-century astronomy to 1950,* part A, ed. Owen Gingerich (Cambridge: Cambridge University Press, 1984), 16-39.

> The preparation on which Mr. Daguerre operates is a reactive, much more liable to the effects of light than any that has hitherto been made use of. The rays of the moon, we do not say naturally but condensed in the focus of a lens of the largest size, never produced any physical effect. The sheets of plated metal prepared by Mr. Daguerre on the contrary become so white, when exposed to the same light and to the subsequent operations, that we may really hope to make a photographic map of our satellite. That is to say that in a few minutes, one of the longest, most minute and delicate labours of astronomy may be effected.[14]

Arago's expectations about the use of the daguerreotype as a tool to master the pictorial representation of a scientific object were achieved as early as March 1840, although the results were not of the kind he had envisioned.[15] This was when the American physiologist John William Draper succeeded in making the first daguerreotype of the moon after a 20-minute exposure. Daguerre himself is believed to have made the very first photographic images of the moon, together with a daguerreotype of the solar spectrum. Undertaken at the request of Arago prior to the announcement of Daguerre's process to the Académie, as "proof that *scientific work* could be conducted using this process" (original emphasis), this original daguerreotype has regrettably been lost.[16] In fact, even if Arago was the scholar who gained most credit when the invention of the new medium was made public, the fundamental role played by another French physicist, Jean Baptiste Biot, in the dissemination of a parallel photographic process to the daguerreotype must also be stressed, namely the first paper photographic process, the talbotype or calotype (introduced in 1841, using paper coated with silver iodide), invented by William Henry Fox Talbot. As pointed out by Theresa Levitt, Arago and Biot were well-known rival scholars, and each of them threw his support behind one of the competing processes: Arago supported the daguerrian silver plates and Biot favoured paper soaked in silver solution.[17] For example, in a letter written to Talbot on 13th February 1839 Biot wrote:

---

14 Beaumount Newhall and Louis Jacques Mandé Daguerre, *Daguerre* (*An historical and descriptive account of the various processes of the Daguerreotype and the Diorama, by Daguerre*) (New York: Winter House, first edition, 1971), 26.

15 See Charlotte Bigg' chapter in this book for an insightful exploration of this topic.

16 Susan M. Barger, *The Daguerreotype: Nineteenth-Century Technology and Modern Science* (Washington and London: Smithsonian Institution Press, 1991).

17 For an enlightening in-depth insight into this rivalry see: Theresa Levitt, "Biot's Paper and Arago's Plates: Photographic Practice and the Transparency of Representation," *Isis* 94 (2003): 456-476.

> Biot urges Talbot to publish. It might be necessary to wait several months for Daguerre's process and Biot would like to try Talbot's. Since Daguerre's process has artistic advantages that one might expect from a painter's, Talbot's should be published as an experimental process, useful to those exploring the physics of light [...]. Biot assures Talbot of his high respect and that he is concerned solely with the advancement of science.[18]

Another enthusiastic supporter of Talbot's invention was the great British pioneer of photography John Herschel, as we can read in the many letters exchanged between them during the years known as the "pre-history" of photography – generally considered to be the first five years after its public announcement. For example, in a letter dated 7th December 1839 Talbot agrees with Herschel that

> although the perfection of the French method of Photography cannot be surpassed in some respects; yet in others the English is decidedly superior. For instance in the capability of multiplication of copies and therefore publishing a work with photographic plates.[19]

In the very early years of photography, as Ann Thomas has remarked, contemporary astronomers and astronomer-photographers viewed lunar photographs produced with the first two processes, the daguerreotype and the wet collodion plate, as basically *beautiful images of the moon.* Thus Herschel wrote to Talbot that "to do it *beautifully* is an art to be learned, and many curious minutiae will have to be discovered and reduced into practice before either of us can arrive at that perfection which I am confident the thing is capable of" (1839, original emphasis).[20] William Crookes[21], a great advocate of the collodion wet plate process, sent examples of some of his photographs taken in 1857 to Talbot. Talbot reacted by thanking Crookes for the "beautiful photographs of the

18 Larry J. Schaaf, *Selected Correspondence of William Henry Fox Talbot, 1823-1874* (Bradford: Science Museum and National Museum of Photography, Film & Television 1994), 19.

19 Talbot to Herschel, December 7, 1839, London Royal Society, Herschel Collection, HS 17:299.

20 Quoted in: Ann Thomas, "Capturing Light: Photographing the Universe," in *Beauty of Another Order: Photography in Science*, ed. Ann Thomas et al. (Yale: Yale University Press, 1997), 186-219.

21 For detailed information on the work of Crookes, see: William H. Brock, *William Crookes (1832-1919) and the Commercialization of Science* (Hampshire: Asghate, 2008); and William Crookes, "On the Photography of the Moon," in *Proceedings of the Royal Society of London* 8 (1856-1857): 363-371.

moon", but noted that they had not yet been perfected and suggested that they be looked upon as the "precursor of still more important results".[22] As noted by Ann Thomas, this comment reflects the lack of consensus at the end of the 1850s on the effectiveness of photography in recording the moon's surface, the results of which were still perceived as scientifically useless. Hence many disappointed viewers of moon photographs were prone to dismiss them as "pretty pictures".[23] For example, the Anglo-Irish astronomer Lord Rosse from the Photographic Society of Liverpool complained in a comment to the Astronomer Royal in 1854, that although a "very pretty picture of the moon can be obtained", there was no known photographic process "sufficiently sensitive to give details in the least degree approaching the way in which they are brought out to the eye".[24] Other astronomers expressed doubts about the "scientific value of any photographs that are likely to be taken of the moon", and proposed that photography's real contribution to astronomy lay in its ability to record rapid change, rather than in its delivery of detail. A couple of decades later, the French astronomer Hervé Faye presented a selection of the American astronomer Lewis Morris Rutherford's beautiful views of the moon's phases to the Académie des Sciences in late 1872. As well as being admired for their aesthetic qualities, Faye remarked that Rutherford's photographs "should be recognized for their contribution to lunar geology".[25]

Charlotte Bigg states in *The Science of Lunar Portraiture* that the moon figures prominently as an object of early astronomical photography. She argues that the reasons for this are not only technical. Bringing in new work in the cultural history of photography and the history of science, she seeks to contextualize early lunar photography as part of the widespread interest in photographic portraiture in the mid-nineteenth century, as well as within the context of the emergent geographical physiognomies of landscape. Bigg explains that lost daguerreotypes figuring the moon as a small white blur usually play the role of the first "firsts" in the history of astrophotography, the prelude to a succession of increasingly sharp, detailed and reliable representations of the moon and other celestial bodies. In her chapter, she considers blur as a materialization of the underdetermined nature of both photography and astronomy: a surface of projection for competing conceptions of what astrophotography

22 Larry J. Schaaf, *Selected Correspondence of William Henry Fox Talbot* (London: Science Museum and National Museum of Photography, Film and Television, 1994), 69.

23 Thomas, "Capturing Light: Photographing the Universe," 202: Angelo Pietro Secchi, "Lunar Photography," *The Daguerrian Journal* 8/23 (1857): 354.

24 Thomas, "Capturing Light: Photographing the Universe," 202: LPS 1854, 127.

25 Thomas, "Capturing Light: Photographing the Universe," 205.

was about, was good for and was to become. Whether in the making of lunar cartographies or the emergent geographical physiognomies of landscape, blurry photographs are central to a history of astrophotography understood as a history of experimentation involving a wide range of actors, as Bigg states in her chapter.

The moon was among the first astronomical objects to be photographed. It is strange, therefore, as Omar W. Nasim remarks, to find that during and well after these firsts, one continues to encounter photographs made not of the moon but of plaster and wood models of the moon. Nasim's chapter in the present volume seeks to make sense of photographs of lunar models by placing them within a larger context of modelling and imagining practices. So for instance, the essay relates these activities to plaster and wood models made in the geological sciences and medicine. It also explores the ways in which photography was used to enhance the availability of what was depicted, and investigates ways in which scientific phenomena like the moon were, in the process, made more photogenic.

An important concept introduced by Nasim in his previous work is the **familiarity** with the object developed through a long period of observation, often years or even decades.[26] Nasim returns to this concept in the present volume. Detailed descriptions of the process of drawing are, in fact, quite common in the work of 19th century astronomer-artists, where we also read of their frustration at having to deal with the difficult technical conditions of drawing through a telescope. An especially fascinating case is that of the geologist John Phillips, who described in minute detail the drawings he made of the Gassendi Crater[27] over the decade 1852-1862, using different telescopes and under varying conditions of observation:

> My first serious attempts to portray the aspect of the moon were made with the noblest instrument of modern times, the great telescope of Lord Rosse, in 1852. The mirror was not in adjustment, so that the axes of the incident and convergent pencils of light were inclined at a very sensible angle. This being met by a large reduction of the working area of the mirror, the performance was found to be excellent. I have never seen some parts of the moon as well as on that occasion. But when I came to represent what was seen, the difficulty of transferring from the blaze of the picture to the dimly lighted paper, on a high exposed station, with little

26 Omar W. Nasim, *Observing by Hand: Sketching the Nebulae in the Nineteenth Century* (Chicago: Chicago University Press, 2013).

27 Coordinates: 17.55°S, 39.96°W, diameter: 110 km; depth: 1.9 km

> power of arranging the drawing-apparatus, was found to be insuperable, and the effect was altogether disheartening. It was like setting down things *ex memoria,* to give a rude general meaning, not an accurate and critical copy. I present as a specimen of this memorial a sketch of the great crater of Gassendi, marked No.1.[28]

Phillips was reduced to drawing *ex memoria* as the only solution possible after gaining the required familiarity with the observed object, in this case, the Gassendi Crater. He further describes his second attempt:

> I next mounted, in my garden at York, a small but fine telescope of Cooke, only 2*4 inches in the aperture; and, aware of the nature of the difficulty which beset me at Birr Castle, I gave it an equatorial mounting, without, however, a clock movement. The whole was adapted to a large solid stone pillar in the open air. It was not possible with 1/60 of the (reduced) light of the Rosse mirror to *see* so well; but it was easy to represent far better what one saw. A conveniently placed board to hold the drawing-paper, a well-arranged light, no necessity of changing position, I made in this manner the drawing of Gassendi which is marked No. 2.[29]

Finally, the intrepid observer gives us the details of his third and last attempt:

> My next attempt was made in the same situation with a fine telescope by Cooke, of 6¼ inch aperture and 11 feet sidereal focus, mounted equatorially, in the old English mode, and carried by clockwork. With this excellent arrangement I was enabled to use photography very successfully, and to obtain selenographs 2 inches across in 5 s of time. The drawing of Gassendi was made with this instrument (Figure 0.1).

Phillips concludes from these descriptions that, for obtaining good drawings of the moon, convenient mounting was actually more important than great optical power; and that for that purpose it was desirable to increase in every way the comfort of the observer, and furnish him with suitable arrangements for his own position, for the placing of his drawing-board, and for the necessary light.[30]

28 John Phillips, "Notices of Some Parts of the Surface of the Moon," *Philosophical Transactions of the Royal Society of London* 158 (1868): 333.

29 Phillips, "Surface of the Moon," 333.

30 Phillips, "Surface of the Moon," 333.

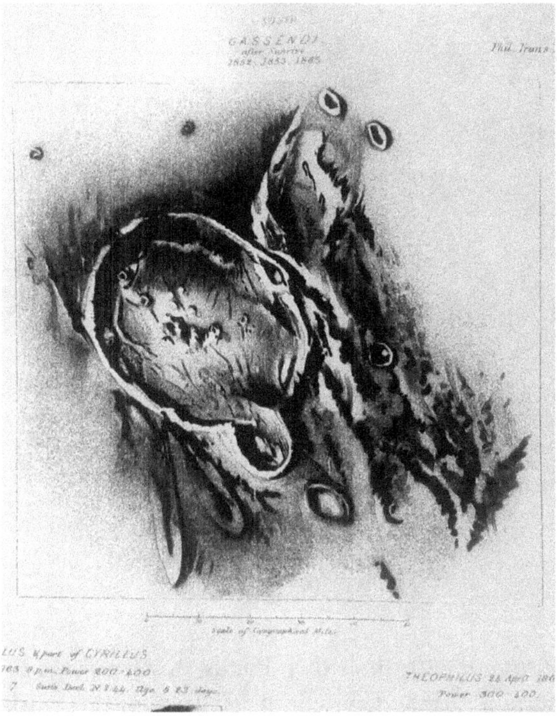

FIGURE 0.1
John Phillips, Gassendi Crater, 1853, made with a Cooke telescope of 6,25 inch aperture and 11 feet sidereal focus, mounted equatorially and carried by clockwork. Published in John Phillips, "Surface of the Moon", 1868.

Another interesting article published by Phillips on the topic of drawing through the telescope was entitled "Notes on the Drawing of "Copernicus" presented to the Royal Society by P.A. Secchi".[31] Here he again explains the process of drawing and compares the outstanding drawing of the crater Copernicus done by the Italian astronomer Fr. Pietro Angelo Sechhi at the Roman Observatory (for a photograph of one of Secchi's drawings, see Figure 0.2) with his own work:

> Of the few attempts which have been made of late years to prepare drawings,[32] on a large scale, of selected lunar mountains, this contribution from the Roman Observatory appears to be one of the most successful. It is on a scale of magnitude (about 10 geographical miles to one inch) such as only the larger modern telescopes can command, and characterized by such firmness of definition as to honour alike to the maker of the instrument and to the artist engaged in the delineation. It may assist

31 John Phillips, "Notes on the Drawing of 'Copernicus' presented to the Royal Society by P.A. Secchi," *Proceedings of the Royal Society of London* 8 (1856-57): 73-75.

32 Original footnote from Phillips' article: Phillips, "Notes on the Drawing of 'Copernicus'," 73: "As distinct from mere *plans*. The drawings must however be based on exact plans".

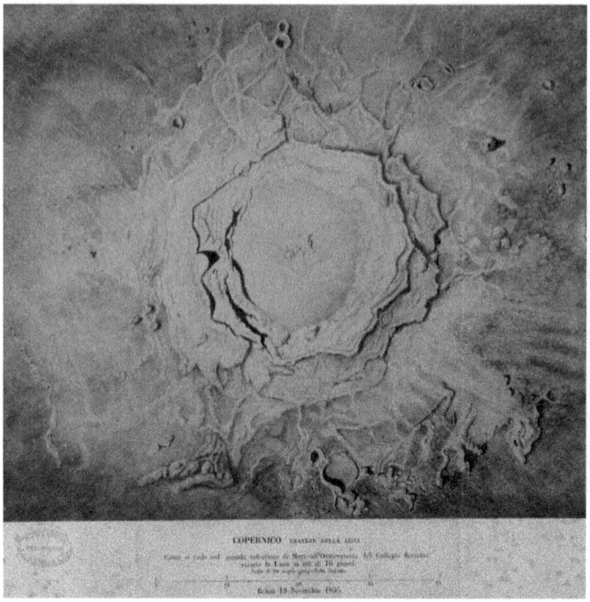

FIGURE 0.2
James Anderson, Crater Copernicus, 19.11.1855, as seen through the big refracting Metz telescope at the Observatory of the Colegio Romano, when the moon was 10 days old; albumen print, 300 × 336 mm, MARCO ANTONETTO COLLECTION, INV. NR. 1591

> those who have not attempted, with their own hands, any drawings of this kind, and desire to form a right judgement of the value of this work of P. Secchi, if I send for comparison a drawing of Gassendi, executed from my object-glass of 6¼ (Cook), with a focal length of 11 feet.[33] The drawing is on a scale of 20 geographical miles to an inch, and Gassendi thus appears of half the linear dimension of Copernicus, being really almost of the same diameter.
>
> Placing together the two drawings, and remembering the appearance of Copernicus, as I have seen through the telescopes,[34] some reflections arise which it may be permitted me to express, in the hope that we are now fairly entered on the long career of discoveries in the moon, to which the attention of astronomers has been of late systematically drawn by the Earl of Rosse and the Committee of the British Association.[35]

When these astronomers embarked on the new medium of photography, they were, then, already extremely familiar with lunar topography, and could

33 He refers here to the drawing explained on the previous article.

34 This sentence "and remembering the appearance of Copernicus, as I have seen through the telescopes", very much is an example of the familiarity with the observed object that will be introduced in Omar W. Nasim's chapter later on this book.

35 Phillips, "Notes on the Drawing of "Copernicus"," 73.

devote their efforts entirely to overcoming the technical difficulties of linking telescope and camera.

Towards the end of the nineteenth century, the final scientific and aesthetic *tour de force* of lunar photography, in the words of Ann Thomas in *Beauty of Another Order,* came with the translation of a number of photographic negatives of the moon into large, astonishingly detailed and visually stunning photogravures published in the *Atlas photographique de la Lune 1896-1910* by Pierre Puiseux and Maurice Loewy.[36] These results were achieved, as Bigg shows in this volume, after many failed attempts and *uselessly* blurred photos. The two French astronomers had, in fact, systematically photographed our satellite from 1896 to 1909 with the large equatorial coudé, working on an atlas of the moon composed of 100,000 photographs.

Many other observatories around the world had large telescopes suitable for photographing the moon, and produced images of the kind described above. In 1886, the librarian of the Observatoire royal de Bruxelles, Albert Benoit Marie Lancaster (1849-1908), published a *Liste Générale des Observatoires et des Astronomes des Sociétés et des Revues Astronomiques.* The success of the book can be gauged from the fact that by 1890 a third edition had been published. Lancaster provided detailed information about more than 200 observatories that existed at that time around the world, most of them in Europe and America: Europe 150, North America 42, Latin America 15, Asia 6 (Constantinople Royal Observatory, Tokyo Royal Observatory, and further observatories at Bombay, Calcutta, Madras, and Hong Kong), Africa 5 and Australia 5.[37]

Thus in the course of the nineteenth century the number of astronomical observatories worldwide had vastly increased.[38] Much of the history of these individual observatories remains unknown, but Michael Geffert informs us in the present volume about one of them. As he explains in his chapter on "Karl Friedrich Küstners' Observations of the Moon in Bonn", when Karl Friedrich Küstner became director of the Bonn observatory, he introduced photography as a new technique for studying astronomical objects. Between 1899 and 1922 Küstner took more than 500 photographic plates of star clusters and other star fields with a new 5.1m focal length telescope. These plates were used after 1950

---

36 Thomas, "Capturing Light," 186-219.

37 Albert Lancaster, *Liste Générale des Observatoires et des Astronomes des Sociétés et des Revues Astronomiques,* 3rd edition (Harvard: Harvard University, 1890).

38 For a study of the exponential growth of observatories around the world in the nineteenth century, see: Dieter B. Hermann, "An Exponential Law for the Establishment of Observatories in the Nineteenth-Century," *Journal for the History of Astronomy* 4 (1973): 57-58.

for the determination of proper and space motions of star clusters in the Milky Way. For a long time it was forgotten that the first observations with the Bonn refractor were made of the moon. While some 60 plates of the moon were kept in Bonn (and are still available today), other observations were sent to Julius Franz in Breslau. The author discusses these observations with reference to others from the Potsdam and Lick observatories that Julius Franz used for his work.

Today old photographic plates constitute a scientific heritage providing unique information for many astronomical tasks; hence the importance of preserving them as a treasure-trove for astronomers and historians of astronomy worldwide. The WFPDB (Wide Field Plate Database: <http://www.wfpdb.org>) project, for example, contains 2.4 million wide-field (>1°) plates from 476 archives in astronomical observatories and institutions all over the world. The GAVO (German Astrophysical Virtual Observatory: <http://www.g-vo.org/>) project is also a fundamental resource for astronomical plates. Using these facilities, everyone can, with time and patience, enjoy amazing images of our satellite.

More locally, other joint projects aim at digitizing and cataloguing astronomical plates. One is the DFG-funded project APPLAUSE (<https://www.plate-archive.org/applause/>), which is open to integrate any other plate collection. In his chapter "The Digital Sky of Hamburg Observatory: How to bring astro-photographic plates from the 20th into the 21st century", Detlef Groote (the leader of the APPLAUSE project), reflects below on the fact that millions of photographic plates made in the course of astronomical and astrophysical research remain hidden from modern research in plate archives worldwide. In 2010 he started a project to digitize their more than 40,000 plates and make their images available on the Internet. In cooperation with Bamberg and Potsdam Observatories the wider APPLAUSE project was then launched with the additional goal of calibrating most of the plates astrometrically and photometrically and presenting the data from nearly 100,000 plates for worldwide research in virtual observatories like the German GAVO.

Some early plates taken by Bernhard Schmidt of the moon led finally to his coming to Hamburg, where he later invented the Schmidt telescope. Such types of telescopes are nowadays used worldwide in observatories, are also mounted on satellites, and are even available for modern cameras. About two thirds of Hamburg Observatory's plate archive is already digitized and published on dedicated web pages. An easy-to-use web interface with a search engine was designed to serve astronomers, historians and members of the public without any knowledge of SQL databases. Each plate can be found on its own web page, with low-resolution images of the plate, plate envelope and corresponding pages of logbooks and observer-notes taken during the night of exposure.

The moon fascinated not only early astronomer-photographers, but also early studio photographers, who created the very popular "Moon Paper" portraits. Among the thousands of extant portraits of this kind, there is a special class, which seems to recreate the century-long wish of humanity to land on the moon (Figure 0.3).

Georges Méliès' film *Le Voyage dans la Lune* features a woman sitting on a crescent moon, and the image became popular as a background in early 20th century studios. The film scene may well have inspired high-street portrait photographers, who sensed huge marketing possibilities and offered "Moon Paper" portraits at very affordable prices. The popular lunar backdrop became a widespread studio fixture – also perhaps, as suggested by Hal Morgan and Andreas Brown, inspired by such popular songs as "Moonbeams" (1906), "Shine on Harvest Moon" (1908), and "By the Light of the Silvery Moon" (1909).[39] But the portraits were not confined to the studio: the new stop-action photographs were also offered at carnivals and amusement parks (Figure 0.4).

2 Selene's Local Histories: Non-western Lunar Cartography and Photography

Lunar cartography and photography were initially developed in Europe. But what happened outside Europe, and more precisely in Asia? What was the situation in countries like Iran, which were not exposed to modern astronomy until the 19th century? Did Persian astronomers have an interest in making drawings of the moon at all, and, if so, for what purpose? Did they have the technology required to do so, and when was it developed? Are there any early Persian lunar photographs?

Little is known about non-Western contributions to the history of lunar cartography. The field is still in its infancy, but it is a fascinating one that deserves attention. A most welcome exception is that of Japan, due to the research (published in English) by the Japanese historian of astronomy Tsuko Nakamura, who also contributes to this book.[40] He has shared with us his discovery

39 Hal Morgan and Andreas Brwon, *Prairie Fires and Paper Moons. The American Photographic Postcard: 1900-1920* (Boston: David R. Godine Publisher, 1981), 3.

40 Tsuko Nakamura, "A Lunar Map Made in 1813 by a Japanese Feudal Warlord," *Mapping the Oriental Sky. Proceedings of the Seventh International Conference on Oriental Astronomy-ICOA*, 82-87 (Tokyo: National Astronomical Observatory of Japan, 2011). See also his chapter in this book.

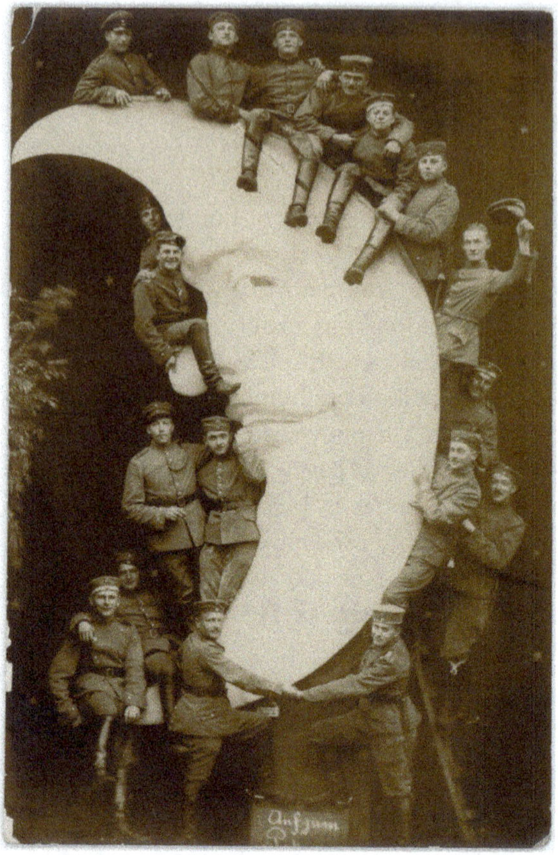

FIGURE 0.3
Moon Paper studio postcard, c.1920, COLLECTION CHRISTOPHER STEINER.

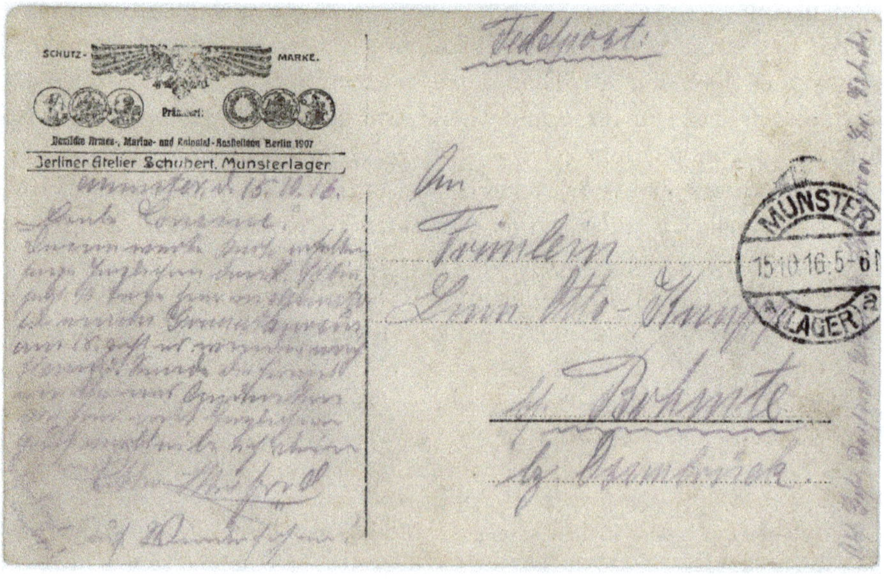

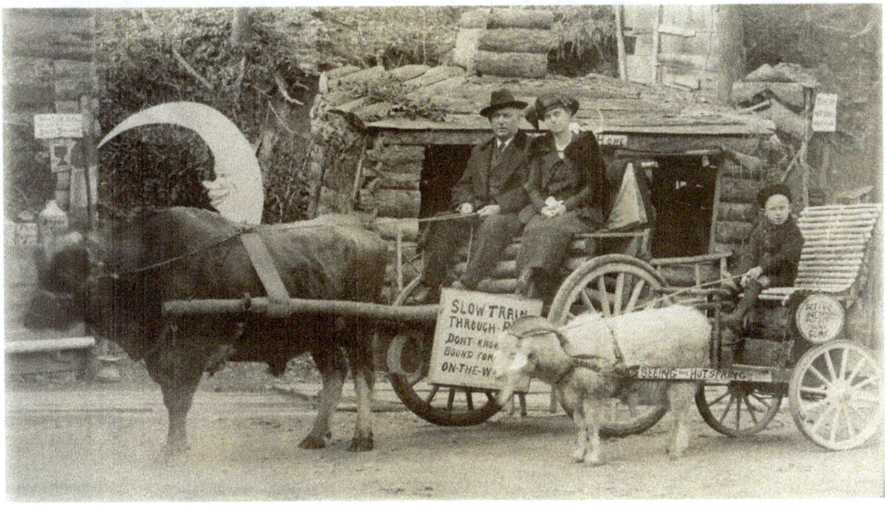

FIGURE 0.4 Moon Studio, c.1920, COLLECTION CHRISTOPHER STEINER.

of a lunar map entitled *Taiin-no-Zu* ("Figure of the moon") prepared by Hotta Masatami, a feudal warlord, on the basis of observations made in 1813.[41] Also remarkable is the only Chinese drawing of the moon known to date.[42]

Thanks are due to the leading historian of Indian astronomy, S.M. Razaullah Ansari, who brought to my attention the only known lunar drawings found on the Indian subcontinent. In his insightful article "The Observatories Movement in India during the 17-18th Centuries", he describes the great effort of the Maharaja Jai Singh to establish observatories across the north Indian plain (Delhi, Jaipur, Banares, Mathura and Ujjain). Ansari explains his motivation on the basis of the Maharajah's own words in his astronomical tables *Zij-I Muhammad Shahi:*

> Since the computations of planetary positions and lunar visibility, as calculated on the basis of various *zijes* (both Persian and those written in India, and also of LA HIRE), were not in agreement with the observed

41 Tsuko Nakamura, "A Lunar Map made in 1813 by a Japanese Feudal Warlord", *Mapping the Oriental Sky. Proceedings of the Seventh International Conference on Oriental Astronomy* (*ICOA-7*), ed. Tsuko Nakamura at al. (Tokyo: National Astronomical Observatory of Japan, 2011).

42 Shi Yunli, "The First Telescopic Map of the Moon's Surface Drawn by a Chinese Astronomer", *China Historical Materials of Science and Technology* 12/4, 1991, 88-92.

> values, he (Jai Singh) made representations to emperor Muhammad Shah for permission to build an observatory in order that the prevalent *zijes* could be rectified.[43]

Jai Singh had a close association with Jesuit missionaries working in India, many of whom were cartographer-geographers or astronomer-mathematicians. Ansari relates that the Jesuits gave Jai Singh a telescope, which he employed for observations, including that of the moon.[44]

But, *why* there are so few drawings and maps of the moon in these Asian cultures? Records inform us that the telescope arrived in many Asian countries within the first two decades after its invention in Holland. Even if the telescope was soon transmitted to China, Mughal India and the Ottoman Empire, those cultures did not respond in the way Europeans did. The telescope is a powerful, emblematic symbol of the scientific revolution. Toby E. Huff postulates three contexts as necessary to understand its significance:

> First, it is the emblematic instrument of the modern scientific revolution. It is *the* instrument of empirical observation most associated with the scientific revolution of the seventeenth century.[45]
>
> Second, and most practically, the telescope transformed the practice of astronomy in the seventeenth century: it transformed astronomy from a plodding science into an active, exploratory inquiry, that constantly looks for new discoveries. In that sense, the telescope was a newly invented "discovery machine". [...] And, at the same time, the telescope is a "portable laboratory" that could be taken anywhere in the world and used to explore the heavens.[46]
>
> The third context concerns the telescope as a precision instrument.[47]

So, at least technically, Asian astronomers would also have been able to undertake telescopic observations of the moon and to make telescopic drawings. There will undoubtedly be unknown lunar drawings and maps produced in non-western countries by non-western astronomers, but the number will predictably not be high. The reasons are manifold, and more rooted in the

43 S.M. Razaulah Ansari, "The Observatories Movement in India During the 17-18th Centuries," *Vistas in Astronomy* 28 (1985): 380.

44 Ansari, "The Observatories Movement in India," 380.

45 Toby E. Huff, *Intellectual Curiosity and the Scientific Revolution. A Global Perspective* (Cambridge: Cambridge University Press, 2011), 18.

46 Huff, *Intellectual Curiosity*, 18.

47 Huff, *Intellectual Curiosity*, 19.

cultural reception (or rejection) of modern western astronomy, which often encountered resistance from the ancient local astronomy traditions, than in any lack of adequate technology for the practice of observational astronomy. In fact, the transmission of the telescope to Asia in particular was rapid and had been achieved by the mid 17th century. The Dutch V.O.C. and their expeditions to East Asia; the French king Louis XIV, the Jesuit missions to China, the Capuchin missions to Iran, etc, were the paths through which the telescope was introduced into many Asian countries.[48] By the end of the 17th century, all these countries had the technology required to make lunar drawings and maps (Figure 0.5). In Japan, for instance, as Franziska Ehmcke explained to me in e-mail communication, already during the 18th century, telescopes had been placed to help visitors enjoy viewing far away places in popular shrines such as the Shinto Shrine Kôzu-no-Yashiro (Figure 0.6).[49]

In the present volume Carmen Pérez González explores the slow and difficult introduction of modern astronomy to mid 19th century Iran. Her chapter

---

48 For the introduction of the telescope in different Asian countries, see: Henk Zoomers, "The Netherlands, Siam and the Telescope. The first Asian encounter with a Dutch invention," in *The Origins of the Telescope,* ed. Albert Van Helden et al. (Amsterdam: Knaw Press, 2010), 301-320; Kameswara N. Rao, "Father J. Richard and Early Telescope Observations in India," *Bulletin of the Astronomical Society of India* 12 (1984): 81-85; Engel Sluiter, "The First Known Telescopes carried to America, Asia and the Arctic, 1614-39," *Journal for the History of Photography* 28 (1997): 141-145; and for a good reference to the history of the telescope with a chapter on historical Japanese telescopes, see: Peter J.K. Louwman and Huib J. Zuidervaart, *A Certain Instrument for Seeing Far* (Wassenaar: Louwman Collection of Historical Telescopes, 2013) and Ramesh Kapoor, "Nur ud-Din and Father Kirwitzer: the independent discovery of the Great Comets of November 1618 and the first astronomical use of the telescope in India", *Journal of Astronomical History and Heritage* 19 (2016): 264-297. This book presents an interesting study on the striking local differences found for the various configurations designs related to the diverse functions the instrument acquire over time.

49 Franziska Ehmcke kindly shared with me this never-published image (Figure 0.6.), giving me permission to print it and sharing with me all the information that she herself researched for her own ongoing project "The Modernity of Urban Life in Osaka in the 1830s and 1840s as depicted by a "pair of Folding Screens depicting Famous Places (*meisho*) in the region of Osaka". As Franziska Ehmcke explained to me by e-mail: Akisato Ritô mentions in his *Settsumeisho zu'e* ("Illustrated Collected Edition of Famous Places in Settsu", 1796-78) that telescopes had been placed here long ago to provide visitors of the Shinto shrine with a look at far distant places for some money. The probably oldest depiction of telescopes at this Shinto Shrine can be found in the illustration *Kôzu-no-kei* ("Scenery at the Kôzu Shrine") by Hasegawa Mitsunobu in the book *Ehon o-togi shinakagami* ("Illustrated Book Listing Enjoyable Things", 3 Volumes, 1. Volume, left side) <http://dl.ndl.go.jp/info:ndljp/pid/2534308> (two-sided picture, 21; 29.08.2018) (E-mail communication: 20.08.2017).

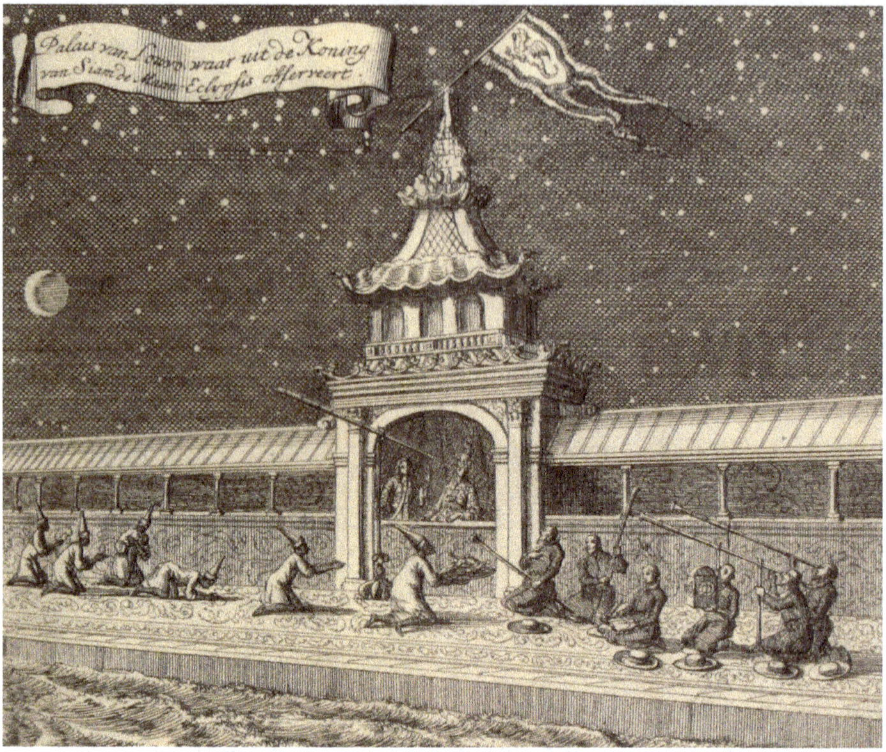

FIGURE 0.5 Observation of the Siamese king Narai and French Jesuits ("the mathematicians") of a lunar eclipse in Louvo 1685. Printed in Henk Zoomers, "The Netherlands, Siam and the Telescope. The first Asian encounter with a Dutch invention", in *The Origins of the Telescope*, edited by Albert Van Helden, Sven Dupré, Rob Van Gent, Huib J. Zuidervaart, 301-320. Amsterdam: Knaw Press, 2010: Figure III. 4 (page 318).

"Mapping Time Rather Than Mapping Space: the Introduction of Modern Astronomy in Iran during the Naseri Period (1846-1896)" outlines the role that modern astronomy played in lunar studies in that country.

In "On Japanese Lunar Maps, Drawings and Photographs in the 19th Century", Tsuko Nakamura explains that before the first Japanese encounter with Europeans (Portuguese) in 1543, Japan was under the exclusive influence of Chinese astronomy. Even after the introduction of the first telescope to the country in 1613, Japanese astronomers long showed no interest in directing telescopes at celestial bodies. The reasons were twofold: one was that the duty of *Shogunal* astronomers consisted in the compiling of calendars, a long-standing tradition of Chinese astronomy; and another is probably that most telescopes of the period had neither high magnification powers nor good

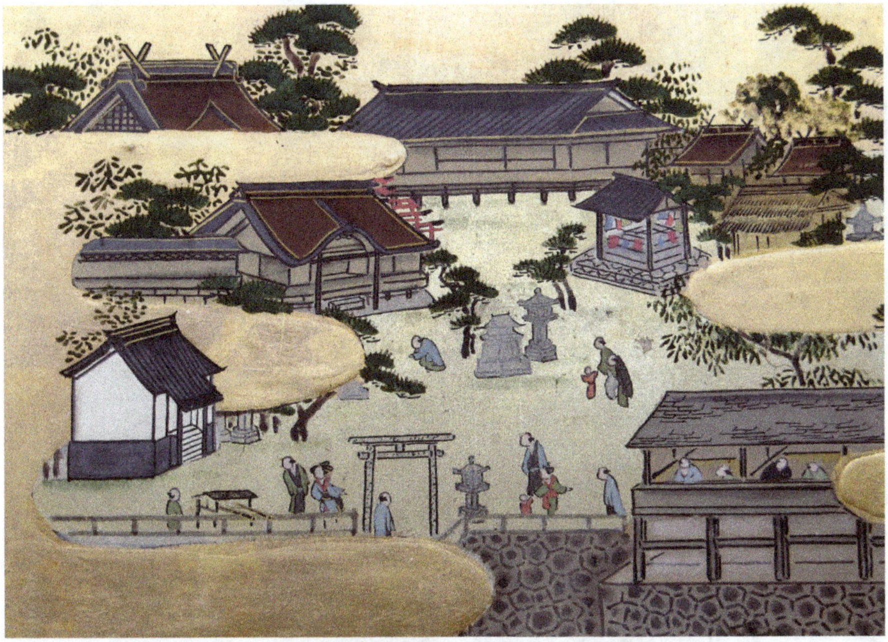

FIGURE 0.6 Unknown artist, Pair of Folding Screens "Famous Places (*meisho*) in the Region of Osaka", around 1840. Detail: Telescopes at the Shinto Shrine Kôzu-no-Yashiro (Left Screen, Panel 3). Personal / Private Property. As argued by Franziska Ehmcke, "Akisato Ritô mentions in his *Settsumeisho zu'e* ("Illustrated Collected Edition of Famous Places in Settsu", 1796-78) that telescopes had been placed here long ago to provide visitors of the Shinto shrine with a look at far distant places for some money".

enough glass quality to scrutinize the surfaces of the moon and planets.[50] In the late eighteenth century, however, amateur astronomers began to use telescopes to observe the moon mainly out of curiosity (Figure 0.7). In 1813, a Samurai warlord and his son made two fairly detailed lunar maps with imported telescopes. In the 1830s, after a few years' of painstaking effort, a gunsmith named Kunitomo developed on his own at least four high-magnification Gregorian reflector telescopes and left several detailed sketches of lunar surface topography as performance tests of his instruments. But Japanese interest in the moon was limited to mere curiosity until the 1870s, and did not attain the level of scientific astronomy.

50 On this topic, see: Tsuko Nakamura, "The Earliest Telescope Preserved in Japan," *Journal of Astronomical History and Heritage* 11/3 (2008): 203-212.

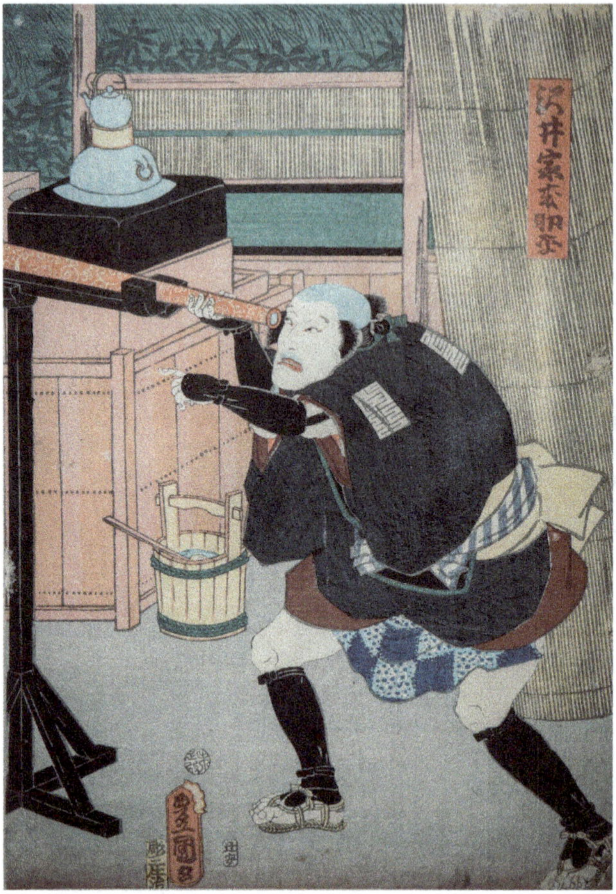

FIGURE 0.7
Man with a telescope, Japanese Ukiyo-e, mid 19th century, LOUWMAN COLLECTION OF HISTORICAL TELESCOPES, The Hague, The Netherlands.

As Albert van Helden has remarked[51], the analysis of lunar crater nomenclature reveals a similarly Eurocentric approach: only 24 craters – just to give one example to be explored in this book – are named after Islamic scientists. And with only 28 craters named after women, the nomenclature could also be called gender biased – a discrimination intensified by the fact that women, like non-Western scientists, are often represented spatially on the lunar periphery. One of them, Williamina Fleming, was a member of the educated group of women working for Pickering at the Harvard Observatory who were known at the time as the "Harvard computers" – or alternatively as "Pickering's harem". Williamina Fleming's role in moon cartography mirrors the position she had

51 Albert van Helden, "Moon maps in History (review of E. Whitaker's Mapping and Naming the Moon", *Journal for the History of Astronomy* 33 (2002): 285-289.

on earth: she shares a crater name, Fleming, with the scientist Alexander Fleming, and the crater is located on the far side of the moon, invisible from earth.[52]

3 9% of the Lunar Surface between Visibility and Invisibility, and Stereoscopic Lunar Photography

Even though the same face of the moon is always presented to the earth, 59% of its surface is visible at one time or another. We can see 9% of the far side of the moon thanks to lunar libration.[53]

As explained by Jaroslav Wlodarczyk, even if the optical libration of the moon is a phenomenon that can be seen with the naked eye, its discovery was strictly tied to the development of telescopic observations of the Moon's surface in the first half of the seventeenth century, as well as to the emergence of selenography, a new discipline in the field of celestial cartography. He argues further that,

> [t]he oldest preserved source that most likely refers to the phenomenon of libration is a note from Dec. 1611, made by Thomas Harriot in the course of one of the three lunar observations he made that month. The English mathematician noticed that two features of the Moon's surface (today called Mare Frigoris and Sinus Moris) appeared closer to the edge of the lunar disk than was the case on the map he had drawn sometime earlier.[54]

However, the handwritten notes left by Harriot were not discovered until the end of the eighteenth century, so Michael Florentius van Langren and Galileo Galilei are generally given credit for the first mention of lunar libration.[55]

The first relatively complete description of lunar libration derives from observations conducted in the 1630s and early 1640s. Between autumn 1643 and spring 1646 Johannes Hevelius observed the lunar surface systematically with a telescope, publishing his results in a book entitled *Selenographia: sive, Lunae descriptio* (1647). This consisted of three large maps (see Figure 1.3 in the next

52 Coordinates: 15°N, 109,6°E, diameter: 106 km, depth: unknown.

53 Libration is a perceived oscillating motion of orbiting bodies relative to each other, such as the motion of the Moon relative to the Earth. The principal libration, or apparent rocking back and forth of the moon, results from the fact that the moon travels in an elliptical orbit around the earth while spinning at a constant rate about its axis.

54 Jaroslav Wlodarczyk, "Libration of the Moon, Hevelius Theory and its early Reception in England," *Journal for the history of astronomy* 42 (2011): 496.

55 Wlodarczyk, "Libration of the Moon," 496.

chapter) and 40 prints of the moon in different chronological phases, starting from the first visible crescent after the new moon. Each map covers the entire section of the lunar surface visible from the Earth. As note by Wlodarczyk, in the history of astronomy these are the first full renditions of the regions close to the edge of the lunar disk that periodically appear and disappear due to the combined effect of the principal types of optical libration.[56]

One of the most interesting physical characteristics of the moon for the student of drawings and photography is the phenomenon of libration, which enables astronomers to observe for short periods of time a very small part of the hidden face of our satellite and permitted two leading astronomer-photographers to produce astounding stereographs of the moon. Using only two photographs taken at different times in order to register maximum libration, these aimed to give the moon a 3D-dimensional likeness.

The aim of a stereograph is to produce an effect of 3-dimensionality. This depends on the fact that the eyes see a scene from slightly different angles. The average distance between adult eyes is about 7.5 cm, and objects as far away as 300 m can be viewed stereoscopically in pictures taken with a stereo camera whose two lenses are separated by this distance. However, in the case of the moon, the two pictures must be taken with two lenses some 100,000 kms apart. Charles Wheatstone, who wrote the first paper on the stereoscopic effect in 1838, suggested that the libration of the moon be used to obtain the necessary baseline.[57]

The great amateur astronomer Warren de la Rue[58] used lunar photographs to produce extraordinary stereoscopic pictures by grouping pairs of photographs taken at different stages of lunar libration, as he explained in detail in his article "The Present State of Celestial Photography in England" published in 1859 in *The Report of the 29th Meeting of the British Association for the Advancement of Science:*

> Taking advantage of the libration, we may, by combining two views taken at sufficiently distant periods, produce stereoscopic pictures, which present to the eyes the moon as a sphere. It has been remarked by the Astronomer Royal, that such a result is an experimental proof of the rotundity of our satellite. A dispute has been going on between photographers as to

56 Wlodarczyk, "Libration of the Moon," 503.

57 For a study on the earliest stereoscopic images of the Moon, see: Thomas B. Greenslade Jr., "The First Stereoscopic Pictures of the Moon," *Am. J. Phys.* 40 (1972): 536-40.

58 For a detailed study of his live and work, see: David Le Conte, "Warren De La Rue – Pioneer astronomical photographer," *The Antiquarian Astronomer* 5 (2011): 14-35.

> the proper angle for taking terrestrial stereoscopic pictures, and I infer that one side of the disputants would consider my arrangement of moon-pictures to produce stereographs unnatural, because under no circumstances could the moon itself be so seen by human eyes; but, to use Sir John Herschel's words, the view is such as would be seen by a giant with eyes thousands of miles apart: after all, the stereoscope affords such a view as we should get if we possessed a perfect model of the moon and placed it at a suitable distance from the eyes, and we may be well satisfied to possess such means of extending our knowledge respecting the moon, by thus availing ourselves of the giant eyes of science. It does not follow as a matter of course that any two pictures of the moon taken under different conditions of libration will make a true stereoscopic picture; so far from this being the case, a most distorted image would result, unless attention he paid first to the selection of the lunar pictures, and then to their position on the stereoscopic slide. It is possible to determine before hand, by calculation, the epochs at which the two photographs must be taken in order to produce a stereoscopic picture; but so many circumstances stand in the way of celestial photography, that the better course is to take the lunar photographs on every favourable occasion, and afterwards to group such pictures as are known to be suitable.[59]

De la Rue provides further details in his description of the exhibited stereographs:

> At the meeting at Leeds last year, there were exhibited some of my stereoscopic lunar pictures 8 inches in diameter, and an apparatus constructed expressly for viewing them. The instrument is of similar construction to Wheatstone's reflecting stereoscope; but, the objects being transparent, the usual arrangements and adjustments are considerably modified. Prisms with slight curvatures worked on their surfaces are employed, instead of mirrors, for combining the pictures, which can be revolved and moved horizontally and vertically in order to place them in the true position. The effect of rotundity is perfect over the whole surface; and parts, which appear like plane surfaces in a single photograph, in the stereoscope, present the most remarkable undulations and irregularities.[60]

---

59 Warren de la Rue, "The Present State of Celestial Photography in England," *The Report of the 29th Meeting of the British Association for the Advancement of Science* (1859): 130.

60 Warren de la Rue, "The Present State of Celestial Photography in England," 130.

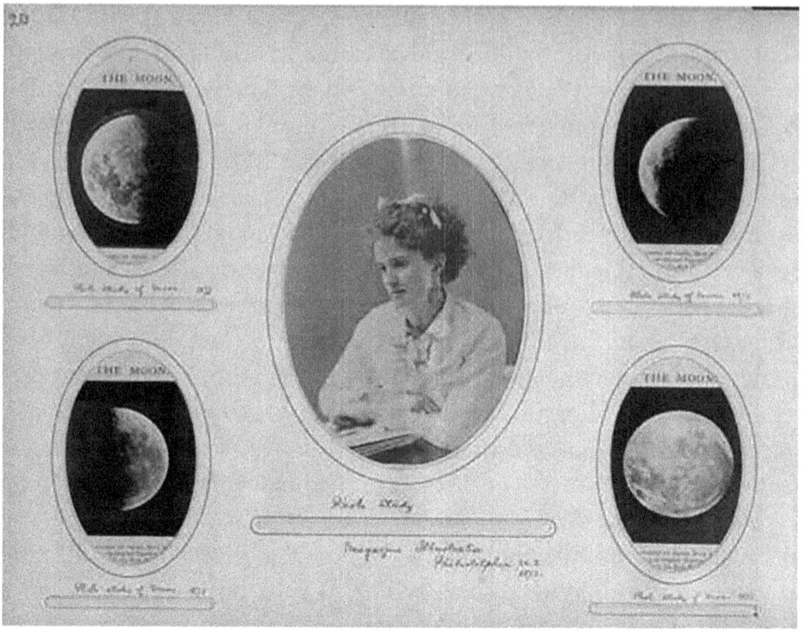

FIGURE 0.8 Album page from Sir John Benjamin Stone's album 31, Miscellaneous (1868-1879), showing four Warren de la Rue's moon photographs (1873), REPRODUCED BY PERMISSION OF THE LIBRARY OF BIRMINGHAM, MS 3196 ALBUM 31, LL/29705

One of those astonishing lunar stereographs, published by Smith, Beck & Beck, illustrates the aforementioned method: the photograph on the left was taken on February 27, 1858, at 1.50 a.m. when the moon was 14.2 days old, that on the right was taken on September 11, 1859 at 11.20 a.m. when the moon was 14.8 days old.[61] In this stereograph, we can observe that the Mare Crisium (on the right) is closer to the edge of the moon in the left than in the right-hand image.

Sir John Herschel, among many other astronomers, expressed his wonder and admiration at their effect:

> It is a step in *nature* but beyond *human* nature as if a giant with eyes some thousands of miles apart look at the Moon through a binocular. What surprises me most is the extraordinary difference in the two pictures as seen by either of the eyes separately not only in form but in shadow & light & the way in which they blend into one is something quite astonishing.[62]

61 To see the set of images published by Smith, Beck & Beck online, please go to: <http://www.londonstereo.com/modern_stereos_moons.html (Accessed August 26, 2017).

62 Letter from Herschel to de la Rue, 10 October 1858 (Royal Society HS.6.D.143)

De la Rue's lunar portraits became so popular that they are even found in typical Victorian albums – for their aesthetic rather than scientific value (Figure 0.8).[63]

Smith, Beck & Beck published De la Rue's lunar stereographs, with detailed observational information of the two photographs used for the stereograph on their back. The images of the moon published in the Keystone Publishing Company sets (*600 SET* and *1200 SET*) were done using the same method that De la Rue used some decades before, but the text on their backs was of another nature, written in a precise and clear way to match the student's educational needs.

The two images of the moon (Figures 0.9 and 0.10) present in the *600 SET* are not attributed to any astronomer-photographer, nor to any observatory. But we could attribute them to William Wallace Campbell after going through the list of stereographs published in *An Illustrated Catalogue of Astronomical Photographs* (1923), who used lunar libration, as explained above, to get the two photographs needed for the stereograph.[64]

The astronomical stereographs produced by pioneer astronomer photographers during the 1850s and 1860s remained rather unknown to a non-scientific audience, circulating mostly among astronomers, even if they were printed in very reduced sets by Smith, Beck & Beck. Towards the end of the century, the photographic dry plate enabled systematic photographic surveys of the Moon and beyond. The American publishing companies, most notably Keystone, and other companies, played a fundamental role on bringing these astronomical images to a wider audience. They enabled a democratization of astronomical portraiture thanks to including astronomical images in their educational set of stereographs, which were used as leading visual tools to help students learn about our earth and its neighbours.[65] The astronomical images of the Keystone 600 SET and 1200 SET were produced by professional astronomers working on leading observatories, most remarkably the Yerkes Observatory, but the texts printed on their backs were written to match the student's educational needs, very much in contrast to the texts on the backs of De la Rue's lunar stereo-

63 I came across this image through Sophy Rickett and Bettina von Zewh's project "Album 31": <http://www.album31.com/research/4f89trzimo4jmusobzi2mitxsqi2jp>. Thanks are due to Sophy Rickett for help on getting the image.

64 *An Illustrated Catalogue of Astronomical Photographs including Latern Slides, Transparencies, and Prints from Negatives Made at Yerkes Observatory*, 3rd edition (Chicago: Chicago University Press, 1923), 29.

65 For a study of this topic, see: Carmen Pérez González, "From the Observatory to the Classroom: space images in the Keystone *600 SET* and *1200 SET*", *International Journal on Stereo & Immersive Media* 1/1 (2017): 40-63.

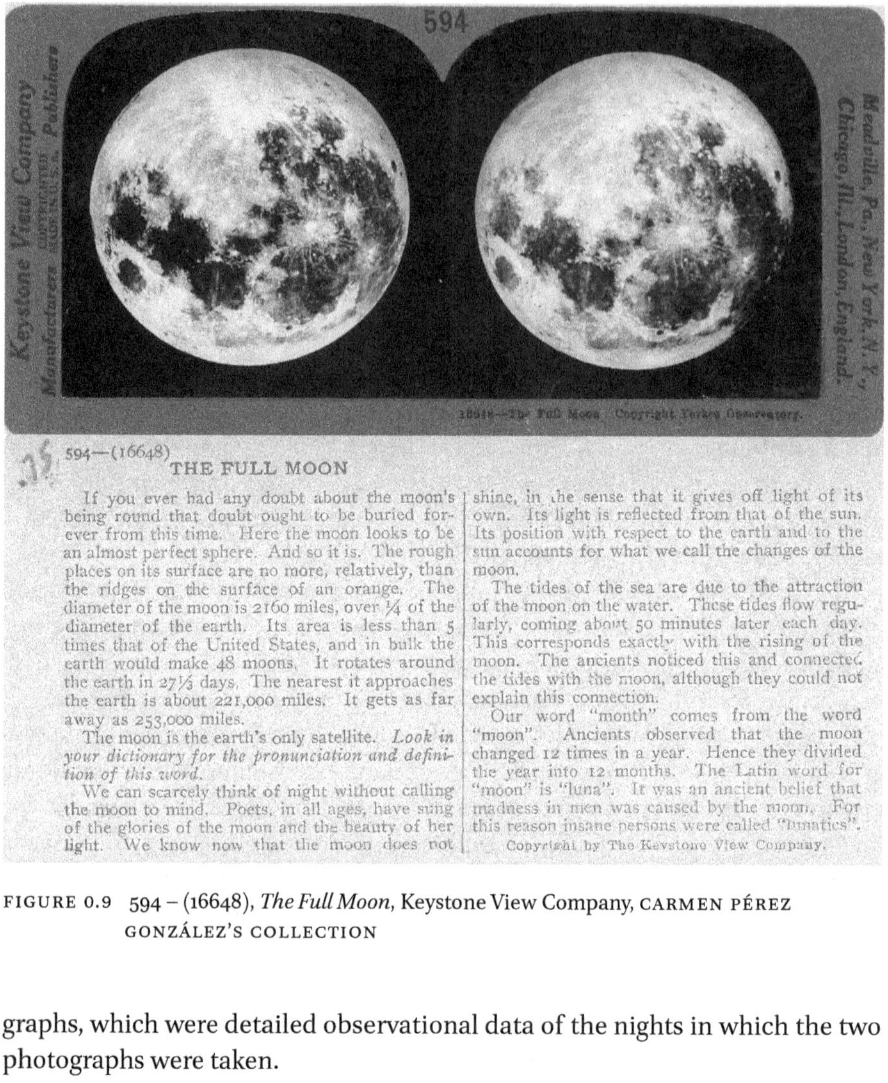

594—(16648)
THE FULL MOON

If you ever had any doubt about the moon's being round that doubt ought to be buried forever from this time. Here the moon looks to be an almost perfect sphere. And so it is. The rough places on its surface are no more, relatively, than the ridges on the surface of an orange. The diameter of the moon is 2160 miles, over ¼ of the diameter of the earth. Its area is less than 5 times that of the United States, and in bulk the earth would make 48 moons. It rotates around the earth in 27⅓ days. The nearest it approaches the earth is about 221,000 miles. It gets as far away as 253,000 miles.

The moon is the earth's only satellite. *Look in your dictionary for the pronunciation and definition of this word.*

We can scarcely think of night without calling the moon to mind. Poets, in all ages, have sung of the glories of the moon and the beauty of her light. We know now that the moon does not shine, in the sense that it gives off light of its own. Its light is reflected from that of the sun. Its position with respect to the earth and to the sun accounts for what we call the changes of the moon.

The tides of the sea are due to the attraction of the moon on the water. These tides flow regularly, coming about 50 minutes later each day. This corresponds exactly with the rising of the moon. The ancients noticed this and connected the tides with the moon, although they could not explain this connection.

Our word "month" comes from the word "moon". Ancients observed that the moon changed 12 times in a year. Hence they divided the year into 12 months. The Latin word for "moon" is "luna". It was an ancient belief that madness in men was caused by the moon. For this reason insane persons were called "lunatics".

Copyright by The Keystone View Company.

FIGURE 0.9 594 – (16648), *The Full Moon*, Keystone View Company, CARMEN PÉREZ GONZÁLEZ'S COLLECTION

graphs, which were detailed observational data of the nights in which the two photographs were taken.

4 Selene's Two Faces: the Moon from Space

In "From pioneer astronomer-photographers to astronaut-photographers: Close-up detailed observations performed by spacecraft and manned exploration of the moon", Pedro Ré and Carmen Pérez González present an historical overview of the space missions launched with the aim of studying the Moon. With the beginning of space exploration, the moon was extensively photographed by unmanned (e.g. Luna, Ranger, Surveyor, and Lunar Orbiter) and manned space vehicles (e.g. Apollo). Photographing from spacecraft changed

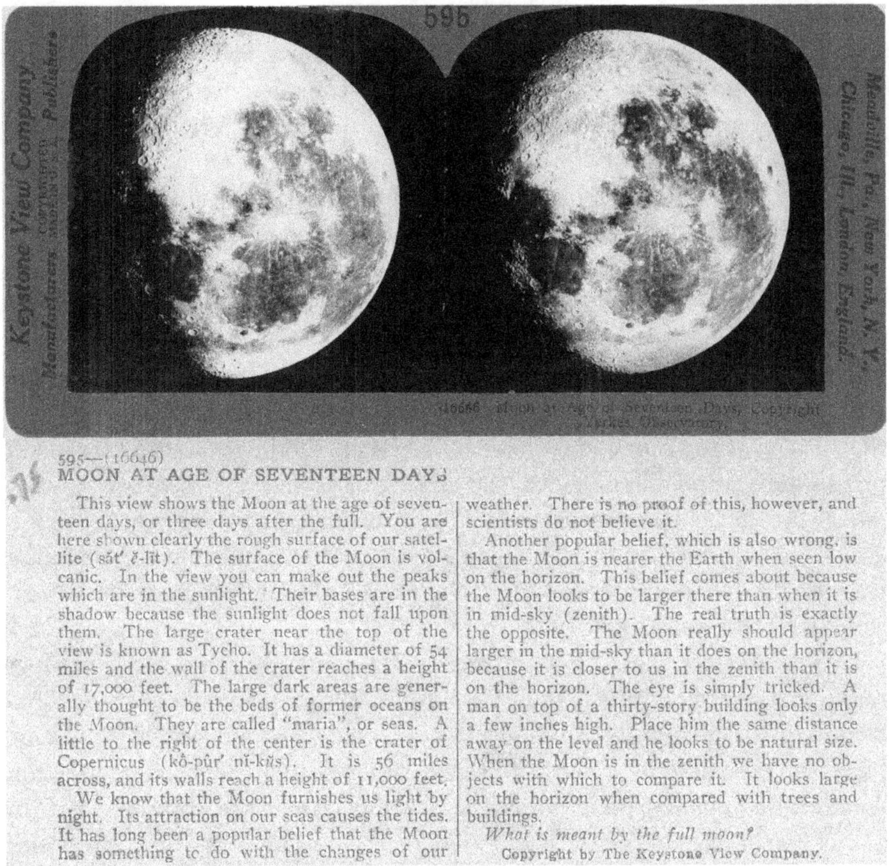

FIGURE 0.10 595 – (16646), *Moon at age of seventeenth days*, Keystone View Company, CARMEN PÉREZ GONZÁLEZ'S COLLECTION

the concept of a human being behind a camera, or more precisely, a human being behind a telescope coupled with a camera: *how* was the camera actually operated from the spacecraft? What was the role of the astronauts in the photographic act? Beyond that, a new kind of lunar photographs and photographers was born: photographs taken directly on lunar soil, by pioneer astronaut-photographers who were selected to step their foot on the lunar surface. The authors focus specially in the central role that the Hasselbald camera, which turned into the space camera, played in this fundamental chapter of the history of lunar exploration.

The first photographs of the far side of the moon were obtained by the Soviet Luna 3 probe in October 1959. The first successful US lunar probes (Ranger 7 in July 1964 followed by Ranger 8 and Ranger 9 in February and March 1965) took a series of photographs before being destroyed on impact with the surface of the moon. A Soviet probe (Luna 9) was able to obtain the first panorama photographs after safely landing on the moon in February 1966. Several US Surveyor probes and Lunar Orbiter photo-reconnaissance probes followed. The latter were able to scan the lunar surface using two different cameras (one for close-ups and another for wide-field views). The first Lunar Orbiters concentrated on finding possible Apollo landing sites. Additional photographs from lunar orbit obtained during the Apollo missions were of considerable interest. Among these, the "Earthrise" image obtained by William A. Anders, Apollo 8 module pilot, is considered one of the most important photographs of the twentieth century.

Launched on 18th June 2009, Lunar Reconnaissance Orbiter (LRO) was the first instalment of NASA's Robotic Lunar Exploration Program designed to map the surface of the moon and characterize future landing sites.[66] LRO's seven instruments were able to collect the as yet most comprehensive set of lunar data,[67] and in December 2010 the most accurate topographic map of the moon surface to date, based on LRO images, was released. A global map of the moon was also produced with a resolution of 100 m per pixel of the Lunar Reconnaissance Orbiter Camera (LROC).[68]

In this chapter, the authors introduce yet another kind of lunar photography in a section entitled "High Resolution Astrophotography: Amateur Astronomers". The authors also remarks on the fact that amateur astronomers have been imaging the moon with video cameras since the 1960s. In 1973, the American company Fairchild Imaging developed their first commercial CCD. Consisting of 100 × 100 pixels, it was used in 1974 to produce the first astronomical photo ever taken by a digital camera: an image of the moon captured with a 20-centimetre telescope. Video offers certain advantages over traditional emulsion-based photography and CCD imaging using refrigerated astronomical cameras. Specialized digital video cameras coupled to a telescope produce excellent high-resolution images of the moon. A high-resolution image can show the finest details of a celestial object independently of the instrument used. The recent advent of CCD cameras that can be operated in video mode, taking 10 or more images per second for periods of up to a few minutes, can be used with excellent results. Atmospheric turbulence (seeing) affects all high-

66 The Lunar Crater Observation and Sensing Satellite (LCROSS) was launched along with the LRO spacecraft. LCROSS was mainly designed to search for water ice on the Moon's surface.

67 More than 192 terabytes of data were collected, the equivalent of all other planetary missions combined.

68 For detailed information on the LROC, see: <http://www.lroc.asu.edu/>.

resolution images. Usually a video file with 1000 or more images is obtained. This will include frames seriously degraded by seeing and others that are less affected. Image-processing software (freeware) can be used to stack and align the best frames in order to produce a low-noise high-resolution image. The best amateur images of the moon have resolutions of 300-500 m/pixel. These images are far better that some professional Earth-based telescopic atlases.

As the authors explain, in 2009 several well-known astrophotographers[69] created an image, which has eclipsed any other so far taken of the moon by ground-based astronomers. This world-record image was part of an ambitious project to create the largest ground-based mosaic image of the moon.[70] The moon was imaged using high-end astronomical video cameras coupled to different large amateur telescopes. Each telescope targeted a small portion of the lunar surface. These high-resolution images were assembled in a giant mosaic of the gibbous (9-day-old) moon. Almost 1000 different areas were imaged, corresponding to 1.2 million captured video frames and 1.1 terabytes of data. The 288 individual image panes were selected on the basis of their overall quality and assembled in a final high-resolution 87.4 megapixel mosaic.[71]

This volume offers readers a visual trip from the very first images drawn in the early 17th century, and the first photographs taken of the moon in the 1840s, through images taken from our planet by contemporary amateur astrophotographers, to those taken from space, and finally from the lunar surface. A visual trip contextualized from the standpoint both of the history of science and technology, and of art history. The latest images selected for the book are the most scientifically useful because of the richness and relevance of the data we can extract from them, but they also remain as strikingly beautiful as ever.

69 Ten different astrophotographers used a range of 9.25-inch to 14-inch Schmidt-Cassegrain telescopes working between 5.5 and 7.8 meter focal length and high frame rate digital astronomical video cameras to create the world record lunar mosaic image.

70 The mosaic moon image was submitted to the 2009 Guinness Book of Records, <http://www.lunarworldrecord.org/>.

71 The resolution of the mosaic image is better than 1 km on the moon surface.

CHAPTER 1

One World is not Enough

Remarks on the History of Selenography

Pedro M.P. Raposo

1 Introduction

Among the first known drawings of the surface of the moon there are three that Leonardo da Vinci left in his notebooks. Leonardo wrote extensively about the moon, commenting on his own observations and discussing old and coeval ideas about our satellite. One of the ancient ideas that attracted his attention, and also his criticism, was that the moon was a mirror that reflected the Earth's oceans and mountain ridges. Leonardo aptly argued that when the moon is in the East it would reflect different spots from those it would show when it is above us or in the west, adding that an object reflected in a convex body takes up only a small portion of that body.[1]

The invention of the telescope[2] and its use in astronomical observation would play a central role in demonstrating that the moon is neither a mirror of the Earth, nor an immaculate, ethereal body as it was conceived in the Aristotelian traditions, but rather a world on its own right, with its own topography; an actual place, where earthlings, in principle, could go. In fact, musings and speculations about the possibility of travelling to the moon accompanied the emergence of selenography.[3] The systematic description and representation of the moon, or in fact of any other celestial body (although the moon holds the primacy in that regard), favored the de-centration necessary to accommodate the idea that the Earth itself is but a body moving in space among others, not a static center in the order of the cosmos.

1 Gibson Reaves, Carlo Pedretti, 'Leonardo da Vinci's Drawings of the Surface Features of the Moon', *Journal for the History of Astronomy* xviii (1987): 55-58.

2 On the early history of the telescope see Albert Van Helden, Sven Dupré, Rob Van Gent, Huib Zuidervaart (eds.), *The origins of the telescope* (Amsterdam: KNAW Press, 2011).

3 See the classic study by Marjorie Hope Nicolson, *Voyages to the Moon* (New York: The Macmillan Company, 1948), and also: David Cressy, 'Early Modern Space Travel and the English Man in the Moon', *The American Historical Review* 111 (2006): 961-982; Allan Chapman, 'A World in the Moon – Wilkins and his Lunar Voyage of 1640', *Quarterly Journal of the Royal Astronomical Society* 32 (1991): 121-132.

 | DOI 10.1163/9789004298873_003

This chapter presents an overview of the history of the observation and mapping of the moon from the invention of the telescope onwards, but mainly prior to the predominance of photography in the study of our satellite. It is not meant to convey an exhaustive listing or comparison of historical efforts to map it, to discuss the relative accuracy of particular representations of the lunar surface, or to scrutinize in depth the establishment of a standard lunar nomenclature; all of the above has been aptly done elsewhere.[4] Nor is the objective here to discuss the relative merits of visual observation and photography, as this topic will be addressed elsewhere in this book. The goal of this chapter is to provide a general backdrop by pinpointing and illustrating some major themes that bear an especial liaison with the history of observing, imaging, and mapping the moon. The latter is indisputably enmeshed with the improvement of the telescope both as an observing and as a measuring device, but it cannot be reduced to a simple history of increasing technical sophistication and accurate representation, or of growing consensus regarding cartographic conventions and nomenclature.

Early telescopic observation and representation of the moon in the seventeenth century was, first of all, a matter of affirming a new way of producing natural knowledge, particularly in astronomy, by empowering the senses with a newly invented optical device and seeking consensus and authority through the deployment of observational evidence.[5] However, the uneven moon revealed by the telescope to Galileo Galilei and others was more than the subject of methodological discussion. It was also a centerpiece in the cosmological debate that opposed the Aristotelian notions of a stark division between the ethereal heavens and a corruptible, central Earth, to the Copernican system, in which the Earth itself was but a planet orbiting the Sun accompanied by its satellite. A satellite that, as the telescope revealed, bore features similar to those of its parent orb.

Observing and representing the lunar surface also provided opportunities for individual observers to flaunt their skills in producing and presenting

---

4 See William P. Sheehan, Thomas A. Dobbin, *Epic Moon – A History of Lunar Exploration in the Age of the Telescope* (Willman-Bell, 2001); Ewen A. Whitaker, *Mapping and naming the Moon: A History of Lunar Cartography and Nomenclature* (Cambridge University Press, 2003). A synopsis of the history of selenography is presented in the first chapter of Znedek Kopal, Robert W. Carder, *Mapping of the Moon: Past and Present* (Dordrecht: Springer Science+Business Media, 1974). See also Scott L. Montgomery, *The Moon and the Western Imagination* (University of Arizona Press, 2001), and the exhibition catalogue by William B. Ashworth, Jr., *The Face of the Moon. Galileo to Apollo* (Kansas City: Linda Hall Library, 1989).

5 See Albert van Helden, 'Telescopes and Authority from Galileo to Casssini', *Osiris* 9 (1994): 8-29, and the other references to Van Helden's work throughout the text.

astronomical knowledge in the way inaugurated by Galileo in *Sidereus Nuncius* (1610). Such was the case of Johannes Hevelius and his *Selenographia* (1647), which can be read as a statement of the author's skills in instrument making, observing, drawing, and engraving.

Of course that rigorous mapping of the moon entailed much more than the affirmation of individual prowess in astronomy. Mapping and naming our satellite was to take possession of it at a symbolical level. And presenting the moon to a sovereign or a patron in the form of an accurate representation was to declare that nothing was beyond their reach. In this regard, it is very telling that the first known moon globe, made by Christopher Wren in the 1660s, was produced with the intent of fostering the support of King Charles II to the emerging Royal Society.

Once its erstwhile image as an immaculate ethereal body was dethroned, mapping the moon was no more abstruse than mapping any terrestrial area, and in fact early selenography had a very close connection with terrestrial mapping. Keen selenographers such as Pierre Gassendi, Michel Van Langren and Tobias Mayer were engaged in surveying works. When they looked at the moon, they saw not only a territory to be thoroughly mapped just like those they surveyed on Earth, but also a potential aid for the accurate measurement of longitudes on land and sea. And just as terrestrial maps are never value-free or neutral depictions of the territories they represent,[6] moon maps also served to extoll sovereigns and certain forms of political order, even though they predominantly became visual testimonies of an almost self-referential pantheon celebrating the heroes of astronomy and selenography (and later, of space exploration). But the issues of who was admitted into that pantheon (and by whom), in what part of the moon each name was inscribed, and to what feature or features in particular it was given, were by no means neutral.

The mapping of the moon was not strange to beliefs and debates on the plurality of inhabited worlds either. Again, once detached from the Aristotelian realm of celestial purity, its suitability for hosting life became a compelling topic for intellectual debate, and a teaser for the public imagination.

2 A Rough Moon: Early Telescopic Observations and Renditions

In May 1609, if not before, Galileo heard about the newly invented telescope and promptly set out to build one himself, using ordinary spectacle lenses. The

6 J.B. Harley, Paul Lexton (ed.), *The New Nature of Maps: Essays in the History of Cartography* (Johns Hopkins University, 2002).

result of this first effort was a very modest device that rendered a magnification of three times. But sometime in November that year he completed a twenty-powered telescope, and began his first astronomical research project: a thorough study of the moon. Between 30 November and 18 December 1609 he observed and drew the moon as it went through its phases, producing eight drawings.[7]

Galileo announced his telescopic findings about the moon in his famous *Sidereus Nuncius,* the short illustrated book published in Venice in 1610 that turned Galileo into an international celebrity almost overnight. According to the then prevailing geocentric cosmology of Aristotle, the heavens were perfect and unchanging, and the heavenly bodies were perfectly smooth and spherical. But it was not an immaculate moon that Galileo saw through the telescope. In *Sidereus Nuncius*, Galileo notes that the brighter regions of the moon are uneven and full of hollows and protuberances, just like the surface of the Earth itself with its lofty mountains and deep valleys.

The section of *Sidereus Nun*cius dedicated to the moon is illustrated with five copperplate engravings, one of which a duplicate. The engravings are not spectacular, and less so are the woodcuts of a pirate edition printed in Frankfurt also in 1610, whose wide dissemination further contributed to confuse scholars and cast doubt over Galileo's drawing and observing abilities (Fig. 1.1).

Nevertheless, the pictures play a very important role, as they illustrate a compelling verbal description of the motion of light and shadow over the moon, which revealed its rugged surface. Above all, the images convey visual evidence obtained through an optical instrument. By adopting this approach, Galileo departed from a prevailing tradition in astronomical illustration centered on diagrams. His images of the moon testify to new possibilities for transmitting pictorial statements that were made possible in the Renaissance by naturalism, perspective and printing.[8]

An influence of cartographic conventions of the late Renaissance for the representation of coastlines, islands, peninsulas, etc. may be found in the way Galileo outlined lunar features,[9] but his images were not meant to be lunar maps. In fact, he did not employ any nomenclature, nor did he include any depiction of the full visible surface of the moon (he actually made only one sketch of the full moon). Galileo's goal was to illustrate the Earth-like nature of

---

7 Galileo Galilei, Albert van Helden (ed./transl.), *Sidereus Nuncius or the Sidereal Messenger* (Chicago, London: The University of Chicago Press, 1989).

8 Mary G. Winker, Albert van Helden, 'Representing the Heavens: Galileo and Visual Astronomy', *Isis* 83 (1992): 195-217.

9 Scott L. Montgomery, *The Moon and the Western Imagination*, p. 125.

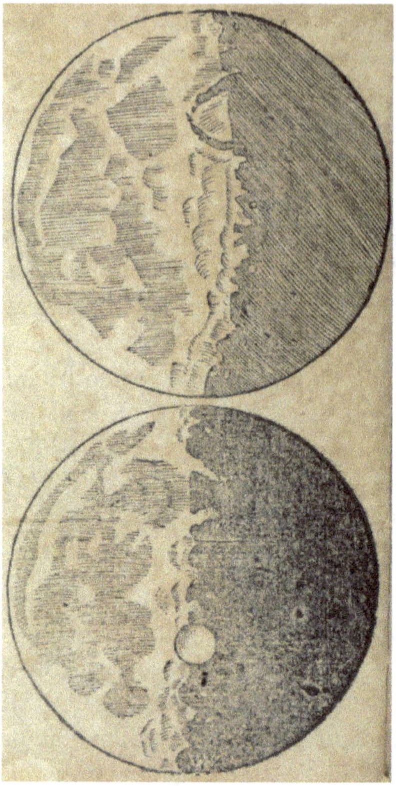

FIGURE 1.1
Woodcuts from the Frankfurt pirate edition of Galileo Galilei's *Sidereus Nuncius* (1610).
COURTESY, ADLER PLANETARIUM.

our satellite. His lunar engravings must be regarded as vehicles of a visual rhetoric that affirmed a new way of producing knowledge: by empowering the senses with optical instruments that rendered hitherto invisible things visible. This was a bold methodological move that collided with the Aristotelian precept of making deductions and inferences from information gathered by the unaided senses, and whose acceptance would be made difficult by the scarcity of good instruments.

There were also issues of status involved. When he published *Sidereus Nuncius*, Galileo was a mathematician, and mathematicians were expected to handle models of planetary orbits and motions qua calculating devices, not to produce statements about the nature or the order of the celestial orbs. But Galileo claimed that his telescope revealed things in the heavens as they really were, albeit at that time no optical theory could demonstrate that the instrument did not deceive the senses.[10]

10 Galileo Galilei, Albert van Helden (ed./transl.), *Sidereus Nuncius or the Sidereal Messenger*, p. 89.

It is well known that Galileo was not the first to produce a telescope rendering of the moon. Thomas Harriot made the first known telescopic sketch of the moon on 5 August 1609 using a six-powered telescope, four months before Galileo's first lunar observation. But this drawing remained virtually unknown until its publication in 1965.[11] In 1611, Harriot produced a sort of telescopic map of the moon, a pen-and-ink drawing with letter and number annotations, which served as position guides, and which may be seen as a rudimentary form of nomenclature. Harriot did not record the existence of the lunar mountains and craters, which at first led some scholars to belittle his renditions. A much more interesting view is that Harriot's drawings, contrary to Galileo's, were akin to maps, conveying approximate delineations and relative locations of visible features. Harriot was a mathematician and surveyor, and had assisted Walter Raleigh in the establishment of the colony of Virginia in America. It is likely that for an observer with such a background the patches of light and dark in the moon would appear as lands and oceans, and be outlined in a way that resembles how coastal lines were represented at the time. Harriot's lunar renderings can thus be seen as an early example of the interplay between surveying and cartography, which must be placed in the context of English colonial exploration.[12]

It has also been suggested that Harriot's moon maps may have been investigation tools to study the orbital motion of the moon and the lunar librations, in the sequence of a similar investigation carried out by William Gilbert, better known for his work on magnetism. Gilbert produced one of the few known pre-telescopic renderings of the moon, and might have done so in order to detect the lunar librations, which would provide evidence favorable to his Copernican ideas. A fellow Copernican who was aware of Gilbert's work, Harriot could have sought to pursue this line of inquiry.[13] If that was the case, his renderings must be also be regarded as an early case of the intersection between cartography, and telescope-based cosmological and astronomical investigation.

Harriot's works did not have an impact though. It was *Sidereus Nuncius* that effectively marked the beginning of telescopic research in general, and of selenography in particular, even if, as remarked above, it was not Galileo's intention to provide an accurate description of the lunar surface. In the two decades

11 *Ibidem*, p. 9.

12 Amir R. Alexander, 'Lunar maps and coastal outlines: Thomas Hariot's [sic] mapping of the moon', *Studies in the History and Philosophy of Science* 29 (1998): 345-368.

13 Stephen Pumfrey, 'Harriot's Maps of the Moon: New Interpretations', *Notes and Records of the Royal Society of London* 63 (2009): 163-168

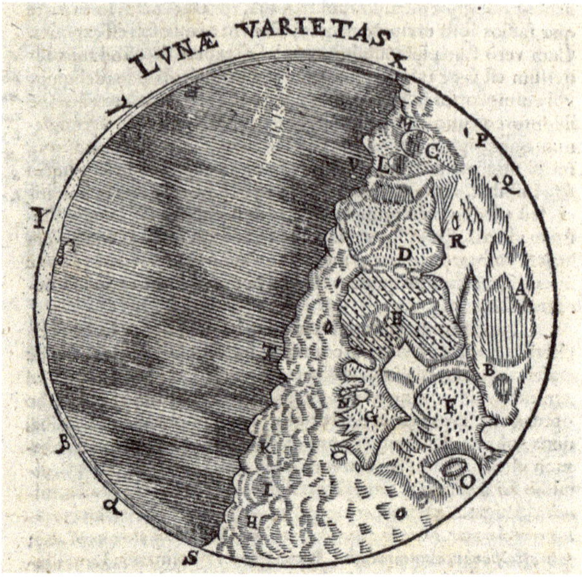

FIGURE 1.2
Rendering of the Moon in Christoph Scheiner, *Disquisitiones Mathematicae* (Ingolstadt, 1614).
COURTESY, ADLER PLANETARIUM.

or so that followed the publication of *Sidereus Nuncius,* several telescopic renderings of the moon were published – for example, by Cristoph Scheiner (see Fig. 1.2.), Charles Malapert, Giuseppe Biancani, Christoforo Borri, and Francesco Fontana.

In general, they consisted of the same kind of schematic rendition that provided not an accurate representation of the visible face of our satellite, but simply a visual notion of its Earth-like appearance.

A first program that explicitly entailed a comprehensive mapping the moon was motivated by the practical value of selenography to terrestrial cartography. In 1628, Pierre Gassendi successfully determined the difference of longitude between Aix-en-Provence and Paris after observing the lunar eclipse of 20 January that year. Gassendi and his collaborator Nicolas de Peiresc realized that an accurate map of the moon could be used to time the passage of the Earth's shadow across particular lunar features from different locations, for the purpose of determining their difference in longitude with greater accuracy. Motivated by this prospect, Gassendi and Peiresc started to work on a lunar atlas. Gassendi even developed a system to name telescopic features on the moon. But the whole project was abandoned after the death of Peiresc, resulting solely in three engravings of different lunar phases, which were commissioned to the artist Claude Mellan, and apparently reproduced in very limited numbers. But other ambitious selenographic projects were soon to be undertaken.

3 The Rise of Selenography: Van Langren, Hevelius, and Riccioli

Michael Van Langren, or Langrenus in the Latinized form, was a member of a prominent Flemish family of cartographers and globe makers.[14] Van Langren served as a court mathematician and cosmographer, first to the Belgian royalty, and later to Philip IV of Spain. In this capacity, he was particularly interested in the accurate measurement of longitude for the production of maps, and well aware of the efforts to find an effective way of determining longitude at sea, for which Philip III had offered a prize of 6,000 ducats.[15]

Van Langren conceived the idea of timing sunrises and sunsets of lunar peaks and crater crests in the waxing and waning moon for the purpose of determining longitude through lunar observations, without having to wait for a lunar eclipse. This idea would prove impractical, because the aforementioned lunar features became illuminated or vanished too slowly.[16] It was, nevertheless, the driving force behind Van Langren's foray into selenography, which resulted in one of the first actual lunar maps. The map, of which different versions survive, was based on thirty drawings covering different lunar phases. It was published as a broadside in 1645. The most elaborate version consists of a map generally conformed to the cartographic conventions of the time, displaying artistic and textual décor, and showing quotations and descriptions of voyages and discoveries over uncharted areas.[17]

Van Langren's moon map constitutes a cartographic celebration of the rule of Philip IV of Spain, presenting a nomenclature system in which the larger and more prominent formations bear the names of Catholic sovereigns. The main highland areas are generally designated with the term Terra, and associated with dignity, honor, justice and other virtues, which were probably meant to represent the virtues of Van Langren's patron Philip IV and other Catholic rulers and dignitaries. Headlands and capes are associated with saints, while

14 Johannes Keunig, 'The Van Langren Family', *Imago Mundi* 13 (1956): 101-109.

15 On the history of the determination of longitude at sea see: Richard Dunn, Rebekah Higgittt, *Finding Longitude: How ships, clocks and stars helped solve the longitude problem* (London: Collins and Royal Museums Greenwich, 2014); William J.H. Andrews (ed.), *The Quest for Longitude* (Harvard University, 1996).

16 Ewen A. Whitaker, 'Selenography in the seventeenth century', in: R. Taton, C. Wilson (ed.), *The General History of Astronomy, Vol. 2: Planetary Astronomy from the Renaissance to the rise of astrophysics, Part A: Tycho Brahe to Newton* (Cambridge University Press, 1989): 119-143, esp. pp. 127-129; Michael Friendly, Pedro Valero-Mora, Joaquín Ibáñez Ulargui, 'The First (Known) Statistical Graph: Michael Florent van Langren and the "Secret" of Longitude', *The American Statistician* 64 (2010): 174-184.

17 Scott L. Montgomery, op. cit., p. 163.

astronomers and mathematicians are represented in minor craters. Terms such as Oceanus, Lacus, Aestuaria, Mare and Portus are used to name watery features, further hinting at the Earth-like nature of the moon mapped by Van Langren – who, to seal his own contribution of selenography, named a prominent crater and a watery feature after himself (respectively, Langrenus and Mare Langrenianum; the crater bears his name to this day).

As a selenographer, Van Langren faced serious competition from Johannes Hevelius, a wealthy brewer, prominent citizen of Danzig, and dedicated astronomer, who set up one of the most remarkable early telescopic observatories in his home. Hevelius considered the renderings of the moon in *Sidereus Nuncius* to be unsatisfactory, which led him to think that Galileo either lacked a good telescope, or had failed to give his observations all the deserved attention, or was not versed in the art of picturing and drawing.[18] Hevelius set out to produce a rigorous telescopic description of the moon, for which he carried out systematic observations of our satellite from November 1643 through April 1645. These observations provided the basis for *Selenographia*, which he published in 1647.

Selenographia contains three major moon maps showing a full lunar disc, and engravings of forty different lunar phases, arranged in order of increasing lunar age. One of the three main maps shows the telescopic appearance of the full moon, whereas another one map adds topographical detail observable in other phases, depicted with the "rows of termite hills" convention of coeval terrestrial maps. For the first time, libration was accounted for in lunar maps, by means of a double border. A third *plena facies* map presents Hevelius's lunar nomenclature (see Fig. 1.3.), which was strikingly different from that of Van Langren (who complained that Hevelius knew about his work without having credited him).

Hevelius considered that names associated with Mediterranean geography used by the ancients were familiar to scholars and historians. He assumed that such a nomenclature would make it easier to remember lunar features, as it was reminiscent of terrestrial cartography. But he also cautioned his readers not to take the association between terrestrial and lunar places too literally.

In 1646, the year before *Selenographia* was published, Francesco Fontana had published *Novae coelestium terrestriumq[ue] rerum observations,* which includes engraved illustrations for nearly every phase of the lunar cycle.[19] But it was *Selenographia* that effectively inaugurated the genre of the lunar atlas. However, *Selenographia* is more than an early lunar atlas. It is also an affirma-

18 Mary G. Winker, Albert van Helden, op. cit.

19 William B. Ashworth, Jr., *The Face of the Moon*, p. 6.

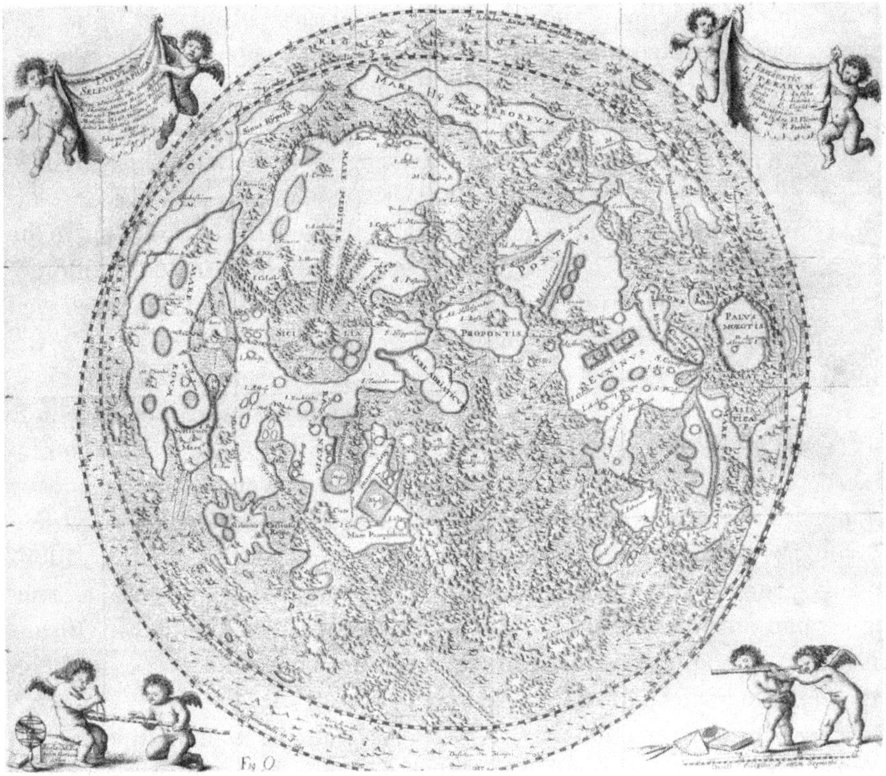

FIGURE 1.3 Plena facies moon map from Johannes Hevelius, *Selenographia* (Gdansk, 1647). COURTESY, ADLER PLANETARIUM.

tion, on the one hand, of meticulous observation with good instruments as the key for the advancement of astronomy, and on the other hand, of Hevelius himself as the embodiment of the ideal astronomer. Adding to the selenographic material proper, *Selenographia* includes not only observations of the satellites of Jupiter, eclipses, sunspots, among others, but also engravings of Hevelius's instruments, which are shown fully assembled and exploded, in order to highlight their fine construction. Hevelius himself is depicted carrying out observations. This was all part of a carefully crafted visual rhetoric that Hevelius would later refine in *Machina Coelestis* (1673, 1679) as a way of reaffirming his credibility in the sequence of disputed positional observations of the comet of 1665, and the ensuing controversy with Robert Hooke on the use of telescopic sights in positional astronomy.[20]

20 Janet Vertesi, 'Instrumental images: the visual rethoric of self-presentation in Hevelius's *Machina Coelestis*', *British Journal for the History of Science* 43 (2010): 209-243.

Hevelius was a fellow of the Royal Society, and used epistolary communication to stay connected with a wide network of informants, with whom he exchanged information. In *Selenographia*, a work that caused much expectative amidst the members of the Royal Society, he extolled his threefold range of skills as an instrument maker, observer, and engraver, which he considered essential to assure a proper transfer of knowledge about the moon.[21] Hevelius thus combined the roles of the natural philosopher and the lowly artisan in the production of what he intended to be an accurate, naturalistic description of the moon. This is why he misunderstood the renderings of the moon presented in *Sidereus Nuncius*, whose purpose, as already remarked, was to show the rough and "earthly" character of the moon, not to represent it rigorously.[22]

Selenographia became an enduring reference for the observation and study of the moon. Hevelius's *plena facies* maps were widely circulated and reproduced, sometimes appropriated, if not plagiarized, in works bearing claims to originality. Such was the case of a map published by the telescope maker Eustachio Divini in 1649, which the author claimed to be based on observations involving large telescopes and a reticule to measure position features, and another map included in *La Dioptrique Oculaire* (1671) by Chérubin d'Orléans (1613-1697), who presented it as the result of drawings made with a newly invented pantograph coupled to a telescope.[23]

Four years after the publication of *Selenographia*, the Jesuit mathematician Giovanni Battista Riccioli published *Almagestum Novum*, which, as the title suggests, was fashioned as a great new compendium of astronomy containing a thorough analysis of the three major cosmological systems under debate in the seventeenth century, that is, the Ptolemaic, the Tychonic, and the Copernican systems. As a member of the Society of Jesus, Riccioli upheld the doctrine of a central and fixed Earth. In *Almagestum Novum*, he advocates for his own modified version of the Tychonic system, with the moon, the Sun, Jupiter and Saturn orbiting a central Earth, while the remaining planets (Mercury, Venus and Mars) orbit the Sun.[24]

Almagestum Novum includes two lunar maps, which were elaborated by Riccioli's fellow Jesuit Francesco Grimaldi. Grimaldi carried out observations

---

21 Janet Vertesi, 'Picturing the moon: Hevelius's and Riccioli's visual debate', *Studies in History and Philosophy of Science* 38 (2007): 401-421.

22 Mary G. Winker, Albert van Helden, op. cit.

23 Ewen A. Whitaker, *Mapping and naming the Moon,* pp. 57-60, 75-78.

24 For a detailed analysis of the arguments presented in *Almagestum Novum* see Christopher M. Graney, *Setting Aside All Authority: Giovanni Battista Riccioli and the Science Against Copernicus in the Age of Galileo* (University of Notre Dame Press, 2015).

of the moon with a telescope referred by Riccioli as an "optimum" instrument, and engraved the plates for the maps. One of them is a topographic, full-moon map with more detail than Hevelius's or Van Langren's. However, Riccioli credits Van Langren, Hevelius, and other astronomers whose work he consulted and which he incorporated in his own selenography. Hevelius's *Selenographia* was fashioned as the product of an individual observing venture, carried out by a self-sufficient astronomer, whose goal was to deliver an accurate representation of the moon as it was seen through his telescope. Riccioli had a different purpose. He intended to capture the universal features of the lunar surface, instead of portraying the moon as observed in particular circumstances[25] – an attitude akin to the truth-to-nature approach in natural history, which sought to evince the underlying type of the species, rather than the individual variations of the specimen.[26]

The second moon map in *Almagestum Novum* presents Riccioli's lunar nomenclature (Fig. 1.4). This nomenclature included a few general terms that had already been employed by Van Langren and Hevelius, such as "terra" and "oceanus", but more parsimoniously, without identifying mountains, peninsulas, or craters. As to the naming of specific features, Riccioli used the names of astronomers, scholars and writers (ancient and modern alike) who had given some contribution to the study of the moon. He distributed the names by octants numbered clockwise from the top of the lunar disc, according to a chronological hierarchy. Names from antiquity are placed on the top of the map, in the first four quadrants, whereas more recent names appear in the lower area. Names are also clustered together according to pedagogical, chronological or cosmological affinities. Riccioli's nomenclature is thus a historical narrative of thought about the moon. Additionally, it includes names referring to the effects the moon on the on human health and temper, and on the weather.

Riccioli considered Hevelius's system of nomenclature inadequate due to the poor correlation between lunar and terrestrial features. He was also more cautious as to assuming similarities between the moon and the Earth. The moon charted in *Almagestum Novum* is an unmistakably uninhabited world. The inscription above Riccioli's and Grimaldi's lunar maps reads "neither do men inhabit the moon nor do souls migrate there".[27]

Riccioli's nomenclature became the basis of the nomenclature system still used today, superseding that of Hevelius. This has been explained in different

25 Janet Vertesi, op. cit., 2007.

26 See Lorraine Daston, Peter Galison, *Objectivity* (MIT Press, 2007).

27 "Nec Homines Lunam incolunt, nec Anime in Lunam migrant".

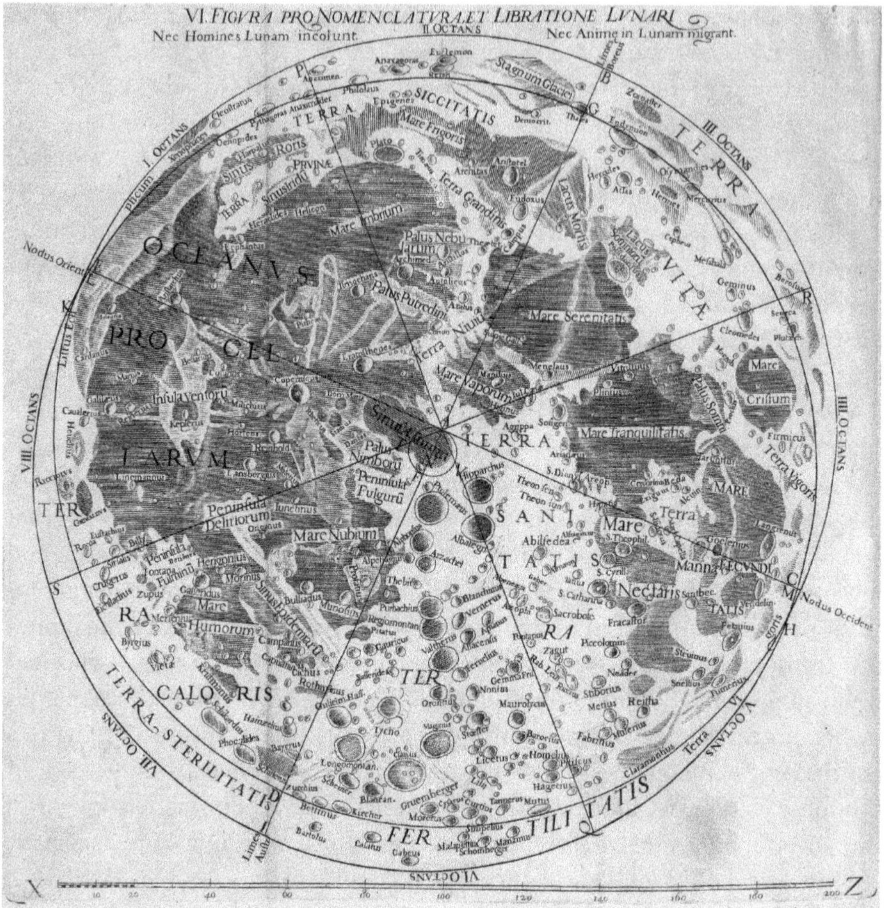

FIGURE 1.4 Nomenclature map in Giovanni Battista Riccioli, *Almagestum Novum* (Bologna, 1651). COURTESY, ADLER PLANETARIUM.

ways. A simple explanation is that Riccioli's system was more practical. In many cases, Hevelius's nomenclature resulted in cumbersome appellations (for example, Celenorum Tumulus, or Montes Caoibacarani), and he often gave a single name to a whole group of craters, which generated confusion among later selenographers.

An alternative explanation underlines a certain ambiguity in Riccioli's iconography, which would have made it appealing to readers of conflicting cosmological allegiances. The frontispiece of *Selenographia*, which displays Galileo as the paragon of observation, would appeal essentially to the Copernican side of the debate. The frontispiece of *Almagestum Novum* shows Riccioli's system outweighing Copernicus's, but also the later getting closer to the light of God, and next to a rendering of the mythological watchman Argus holding a telescope, which might be interpreted as an avowal of telescopic observation as a key to the advancement of astronomy.[28] It has even been suggested that Riccioli could be a cryptic Copernican who used his iconography and lunar nomenclature to hint at his real cosmological allegiance. On the one hand, he relegated the supporters of the heliocentric system to the Sea of Storms (Oceanus Procellarum); on the other hand, some of them have their names associated with prominent features.[29]

As far as the prevalence of Riccioli's nomenclature is concerned, its practical nature seems to have been the decisive element, given that Hevelius's and Riccioli's maps both had a wide diffusion. One can also imagine that the names of erstwhile and coeval authorities in astronomy would have a greater resonance among those engaged in astronomical pursuits, as they certainly would like to see their names inscribed on the moon as well. Juan Caramuel y Lobkowitz once wrote to Gassendi regarding a prospective map of the moon: "all of our friends will be there, you yourself, and Peiresc, and Mersenne, and Naudé".[30] Like Van Langren had done before, Riccioli added his own name and Grimaldi's to their map. After all, what best reward for selenographers than having their names inscribed on the lunar pantheon of astronomy?

4 Tangible Moons: The First Lunar Globes

The idea of making spherical models of the Earth and the heavens originated in Ancient Greece. The earliest known extant globe is the famous Farnese Atlas, a carved marble sculpture depicting Atlas carrying the heavens on his shoulders, dated from around 150 AD.[31] A tradition of engraved, hollow celestial globes developed in the Islamic world from the ninth century onwards.[32] In Western Europe, the production of globes was boosted by the development of the printing press (prior to 1500, globes were unique pieces, drawn, engraved by hand, or painted) and by the emergence of the European seaborne empires, which entailed a dramatic revision of ancient geographical traditions and es-

28 Janet Vertesi, op. cit., 2007, p. 406.

29 Ewen A. Whitaker, *Mapping and Naming the Moon*, p. 65.

30 *Ibidem* (quote), pp. 34-35.

31 See Bradley E. Schaffer, 'The Epoch for the Constellations on the Farnese Atlas and Their Origin in Hipparchus's Lost Catalogue', *Journal for the History of Astronomy* xxxvi (2005): 167-196,

32 See Emily Savage-Smith, *Islamicate Celestial Globes: Their History, Construction, and Use* (Washington, D.C.: Smithsonian Institution Press, 1985),

tablished the backdrop for the cosmological debates and astronomical inquiries that marked the beginnings of modernity.

From the sixteenth century onwards, globes were generally printed and produced in a pair comprising a terrestrial and a celestial globe. They played an important role in the circulation and development of the new cosmographical knowledge that was central to the European maritime expansion. Globes became symbols not only of learning, but also of wealth and might. Terrestrial globes were often included in portraits of royalty and state officials to signify authority and possession of foreign lands.[33] Having both the Earth and the heavens inside a room within a hand's reach conveyed the sense that no place on Earth or star above was beyond the scope of human ambition for knowledge and power. No less with the moon, if it could be similarly represented.

When Michel Van Langren began his selenography investigations, his intention was to construct a lunar globe, but he eventually opted to produce a flat map, which was easier and cheaper to produce.[34] The first recorded moon globe was made by the English mathematician, astronomer and architect Christopher Wren, who engaged in lunar observations while in the post of Savilian Professor of Astronomy in Oxford. Wren regarded his lunar survey as an attempt to advance the telescopic astronomy opened up by Galileo,[35] particularly by fashioning it as a quantitative undertaking based on accurate measurement. Wren was an active participant in a milieu of British observers that cultivated micrometric measurements with the telescope.[36] Another prominent figure in this milieu was Robert Hooke, whose famous *Micrographia* (1665) includes a drawing of the lunar formation nowadays known as Hipparchus, accompanied by small drawings of two craters and the multitude of stars in the Pleiades.[37] This may be considered the first detailed telescopic drawing of a particular lunar feature. Such precise astronomical observations, together with the microscopic minutia lavishly portrayed in *Micrographia*, testify to Hooke's belief in instruments as redeemers of fallen mankind through the empowerment of the senses, and to the empiricism championed by the then emerging Royal Society.[38]

33 Sylvia Sumira, *Globes: 400 Years of Exploration, Navigation and Power* (Chicago, London: The University of Chicago Press, 2014)

34 Ewen A. Whitaker, 'Selenography in the seventeenth century', p. 127.

35 James A. Bennett, *The Mathematical Science of Sir Christopher Wren* (Cambridge University Press, 2002), pp. 38-41.

36 *Ibidem.*

37 On Hooke's telescopic observations see Allan Chapman, *England's Leonardo: Robert Hooke and the Seventeenth Century Scientific Revolution* (CRC Press, 2004), esp. pp. 77-79.

38 Jim Bennett, 'Hooke's instruments', in Jim Bennett, Michael Cooper, Michael Hunter, Lisa Jardine, *London's Leonardo: The Life and Work of Robert Hooke* (Oxford University Press,

The project for the lunar globe served to foster the patronage of King Charles II to the Royal Society. Wren was also requested to make drawings of insects and other small living creatures with the microscope, a project that was left for Hooke and which resulted in *Micrographia*. Wren was enticed by Hevelius's *Selenographia*, which he wanted to supersede through a program of accurate micrometric measurements of the lunar surface. He was likely influenced by John Wilkins too. Wilkins was a leading figure in the establishment of the Royal Society,[39] and the author of *Discovery of a world in the Moone* (1638), a short book in which he stressed similarities between the Earth and the moon and conjectured about travelling to our satellite.[40] Wilkins collaborated with Wren in his lunar observations, and the resulting moon globe was probably produced under his direction.

The globe was completed in August 1661. It was 10-inch in diameter, made of pasteboard, molded in relief, and painted. An inscription read: "To Charles the Second, King of Britain, France and Scotland, for whom Dr. Christopher Wren has created the new world of this Selenosphere, because, for one of His magnitude, "one [world] is not enough"".[41] The globe was added to the King's cabinet. Christiaan Huygens, who also observed and drawn some lunar features, saw it there during a visit to London in 1663, recording that the globe was very pleasant to look at with all its marks and small round valleys.[42] Wren promised to make a larger and more accurate version of the globe for the Royal Society, but there's no record of a second globe having been made, and the original one is lost. At least a visual testimony of it has reached us: the globe is depicted in the portrait of Wren hung at the Sheldonian Theatre in Oxford.[43]

The German astronomer Tobias Mayer was similarly committed to anchor selenography on accurate micrometric measurements. Mayer undertook surveying work during his youth and nurtured a particular interest in longitude

---

2003), pp. 63-104, cf. p. 65; Michael Hunter, 'Hooke the Natural Philosopher', in Jim Bennett, Michael Cooper, Michael Hunter, Lisa Jardine, *London's Leonardo: The Life and Work of Robert Hooke* (Oxford University Press, 2003), pp. 105-162, p. 126.

39 Lisa Jardine, *On a Grander Scale: the Outstanding career of Sir Christopher Wren* (London: Harper Collins Publishers, 2002), p. 183.

40 Allan Chapman, 'A World in the Moon – Wilkins and his Lunar Voyage of 1640', *Quarterly Journal of the Royal Astronomical Society* 32 (1991): 121-132.

41 Cited by Lisa Jardine, *On a Grander Scale*, p. 180.

42 *Ibidem*, pp. 180-181.

43 J[ames] A. B[ennett], 'A Double Anniversary for John Russell, R.A.', *Sphaera* 2, Autumn 1995, <http://www.mhs.ox.ac.uk/about/sphaera/sphaera-issue-no-2/a-double-anniversary-for-john-russell-r-a/> (16 April 2016).

measurements.[44] In 1747, Mayer began to make an ortographic projection of the lunar surface in order to ascertain what lunar markings would be covered during a lunar eclipse the following year, with the goal of reckoning the longitudes of different observing places.

Mayer considered Hevelius's and Riccioli's maps unsuitable for this kind of work. In 1748, he began a series of observations of lunar features, using nine- to twelve-foot telescopes and a glass micrometer designed by himself. By the end of August 1749 Mayer had completed a comprehensive series of micrometric measurements. He also developed a system of selenographic coordinates to define the positions of lunar markings, and set out to produce a lunar globe on the basis of these works. The Cosmographical Society of Nuremberg issued a prospect for the globe in 1750, but Mayer's departure from Nuremberg, together with his publisher's bankruptcy, led to the demise of the project. A similarly ill-fated undertaking was the elaboration of a map based on Mayer's observations, which was interrupted by his death.

The best known outcome of Mayer's lunar investigations is the set of tables of lunar motion to determine longitude at sea that he developed as a response to the famous Longitude Act of 1714, and which rendered him a prize of 3,000 British pounds. Mayer's lunar tables played a decisive role in the development of the so-called method of lunar distances, which rivaled (but was also used in parallel) with the method of the mariner's chronometer originally designed by John Harrison.[45] Nevertheless, a small map of the moon based on Mayer's observations and some of his selenographic works were published in 1775 in *Opera Inedita Tobiae Mayeri*, under the direction of George Christoph Lichtenberg.

Accurate measurement with an eyepiece micrometer was a central requirement in the production of lunar globes, but an artistic insight was equally important in the making of the earliest surviving printed globes of the moon, which are due to John Russell. Russell is known mainly for his pastel portraits of royalty and other leading figures of the late eighteenth century (including the astronomers William Herschel and Nevil Maskelyne) and for having been painter to King George III.[46] Russell was also a keen amateur astronomer. He owned a six-foot reflector made by William Herschel, and a refractor by John Dollond that he complemented with an eyepiece micrometer. The Dollond refractor proved particularly well suited to Russell's system of lunar measure-

44 Eric Gray Forbes, 'Tobias Mayer (1723-62): A Case of Forgotten Genius', *British Journal for the History of Science* 5 (1970): 1-20.

45 See references in note 15.

46 Sylvia Sumira, *Globes*, p. 174.

ment, which was similar to terrestrial triangulations.[47] Russell's selenographic project was further empowered by his individual idiosyncrasies as a visual artist.[48] Russell began to observe and draw the moon in 1764, an activity that he maintained for roughly four decades. In 1795 he made a four-by-five foot pastel drawing; then he started to work on a globe, which was ready two years later. Russell intended the globe to be sold on a subscription basis, but apparently only a few were made.

Russell's lunar globe (Fig. 1.5) is a stipple-engraved piece, 12-inch (30.5 cm) in diameter. In most of the surviving examples, it is mounted on a complex frame designed to demonstrate the lunar librations, the tilt of the lunar axis, the sunrise and sunset lines in the moon, and also the parallax of the moon as seen from the Earth, for which a small terrestrial globe is attached to an extension of the frame. The full apparatus, generally titled 'Selenographia', constituted a performance piece. Russell was as confident in its suitability to recreate the experience of observing the moon as to suggest that users removed the globe from the frame and pointed a telescope at it.[49]

5 Lunar Atlases, Selenites, and an Itinerant Moon

The works of Tobias Mayer were followed by other selenographers in the German lands, among them the Lilienthal-based magistrate and amateur astronomer Johann Hieronymus Schröter. Schröter was initially led into selenographic investigations by William Herschel's announcement of what he believed were signs of active volcanoes on the far side of the moon. Herschel carried out numerous observations of the Moon, expecting to find observational evidence of our satellite being inhabited.[50] Schröter would be equally propelled by the belief that the moon and the planets were suited to host life similarly to the Earth.

Notwithstanding its quality, Mayer's posthumously published map of the moon was just 7.5 inches to the moon diameter, which motivated Schröter to

---

47 J[ames] A. B[ennett], op. cit.

48 Frances Terpak, 'Objects and Contexts', in Barbara Maria Stafford, Frances Terpak (eds.), *Devices of Wonder: From the World in a Box to Images on a Screen* (Los Angeles: Getty Research Institute, 2001), pp. 143- 364, cf. p. 198.

49 See also Omar Nasim's contribution to this volume and his references to the moon globe made in 1838-9 by Hofrättin Wilhelmine von Witte.

50 Michael J. Crowe (ed.), *The Extraterrestrial Life Debate: Antiquity to 1915. A Source Book* (University of Notre Dame Press, 2008), pp. 169-181; Michael J. Crowe, *The Extraterrestrial Life Debate, 1750-1900: The Idea of a Plurality of Worlds from Kant to Lowell* (Cambridge University Press, 1988), pp. 59-70.

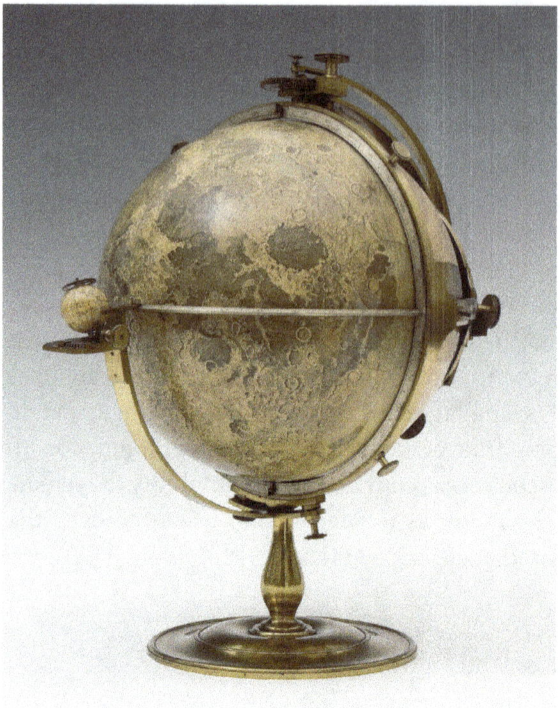

FIGURE 1.5
Selenographia moon Globe by John Russell, Inv. 52085
© MUSEUM OF THE HISTORY OF SCIENCE, UNIVERSITY OF OXFORD.

produce a new map. Soon Schröter decided to sketch as many individual lunar formations as possible.[51] It was an ambitious program that entailed drawing lunar features under all conditions of illumination, measuring the heights and depths of the most important elevations and craters, investigating the existence of a lunar atmosphere, and tracking down possible changes on the lunar surface.

Schröter used a 2.5-inch Dollond refractor and several reflecting telescopes. He also designed a special device to observe the moon through the telescope using a grid of small squares, while using the other eye to draw on a board. The results of Schröter's observations were presented in a two-volume work titled *Selenotopographische Fragmente zur genauern Kenntniss der Mondfläche* (*Fragments of selenography for a precise description of the lunar surface*) (Fig. 1.6).

Schröter garnered recognition as a selenographer and also as a planetary observer, with Lilienthal becaming an important node in the European networks and circuits of astronomy. It was the place chosen for the meeting of

51 William Sheehan, Richard Baum, 'Observation and inference: Johann Hieronymous Schröter, 1745-1816', *Journal of the British Astronomical Association* 105 (1995): 171-175.

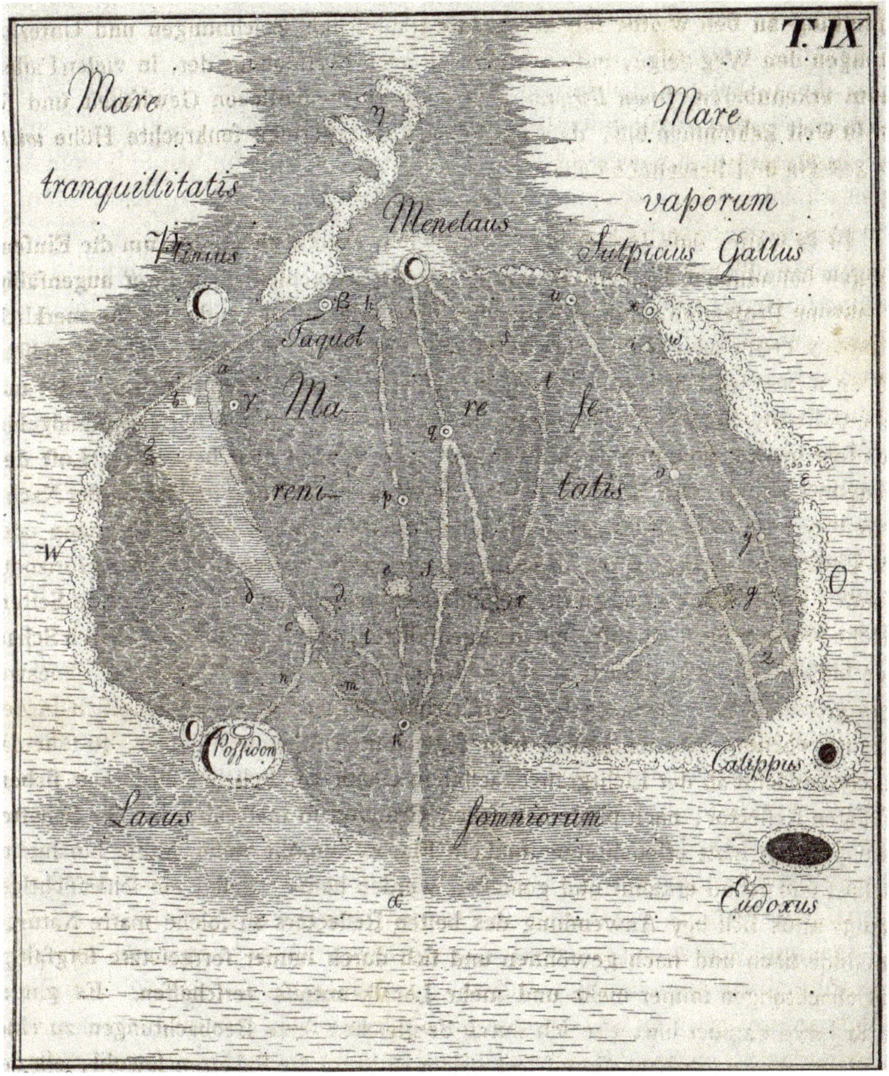

FIGURE 1.6 Plate from Johann Hieronymus Schröter, *Selenotopographische Fragmente ...*, vol. 1 (Lilienthal, 1791). COURTESY, ADLER PLANETARIUM.

astronomers held in 1800 under the leadership of Franz Xavier von Zach that established a group of astronomers from all over Europe, later known as the "celestial police", and which was expected to conduct a coordinated search for a planet between Mars and Jupiter.[52]

52 See Clifford Cunningham, *Discovery of the First Asteroid, Ceres: Historical Studies in Asteroid Research* (Springer, 2016), esp. pp. 1-24.

Another German selenographer inspired by the idea of life on the moon was Franz von Paula Gruithuisen, a physician turned astronomer. Gruithuisen displayed a great ability in observing and drawing the moon, but was greeted with ridicule by other astronomers, who mercilessly rebuked his sensational claims about an alleged lunar city.[53] The idea of life on the moon would become a sensational topic in popular culture in the 1830s, due to the so-called "Moon hoax", a series of satirical articles published in the *New York Sun* by a writer named John Locke. The articles described a fictional moon populated by winged "selenites", purportedly on the basis of observations made by John Herschel, who at the time was carrying out astronomical work in South Africa. The articles were intended to satirize the idea of a plurality of worlds and its advocates, especially Thomas Dick, who went as far as to present a table with the population of the various bodies in the solar system (with the moon faring at 4,2 billion inhabitants). But more than making his readers laugh at such follies, Locke essentially contributed to spread the belief in selenites.[54]

In general, selenographers of the remainder of the nineteenth-century eschewed such flights of fancy, embracing the idea of a barren moon with a dormant landscape, and seeking to represent it realistically and with great precision, in accordance with the ideals of objectivity and precision that prevailed in the period.[55] Such was the case of Johann Heinrich Mädler and his patron Wilhelm Beer. They fostered a particular interest in planetary cartography, producing elaborate maps of Mars.[56] Mädler also observed the moon for nine years, an undertaking that resulted in a map in four parts, each representing a quadrant of the lunar disc in great detail. Beer and Mädler also published a book titled *Der Mond*, covering topics such as the moon's orbital motion, librations, eclipses, tides, and the possible existence of a lunar atmosphere.

Beer and Mädler were revered as authorities in lunar science for decades, but the acme of nineteenth-century selenography in the German lands, and of selenography based on visual telescopic observations in general, is arguably to be found in the work of J.F. Julius Schmidt. Initially inspired by Schröter's *Fragmente,* Schmidt maintained a systematic program of lunar observing for more than three decades. He generally used telescopes of greater aperture than

53 William P. Sheehan, Thomas A. Dobbin, *Epic Moon*, pp. 75-85; Michael J. Crowe, *The Extraterrestrial Life Debate*, pp. 202-204.

54 Michael J. Crowe, *The Extraterrestrial Life Debate*, pp. 202-215.

55 Lorraine Daston, Peter Galison, op. cit., and also Norton M. Wise (ed.), *The Values of Precision* (Princeton University Press, 1997).

56 On the mapping of Mars see K. Maria D. Lane, *Geographies of Mars; Seeing and Knowing the Red Planet* (University of Chicago Press, 2011).

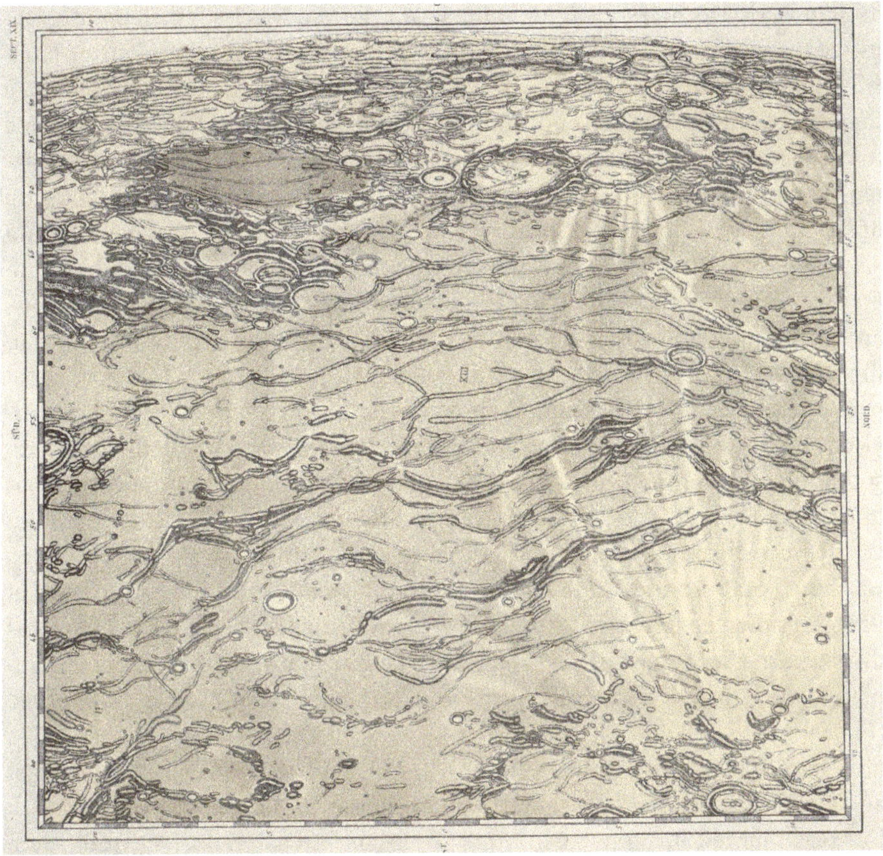

FIGURE 1.7 Plate from J.F. Julius Schmidt, *Charte der Gebirge des Mondes* (Berlin, 1878). COURTESY, ADLER PLANETARIUM.

those used by his predecessors, which allowed him to observe smaller details. His observations rendered more than 1,000 sketches of the lunar surface, as well as height and depth measurements. Schmidt's selenographic investigations resulted in *Charte der Gebirge des Mondes,* published in 1878. *Charte...* presents a map of the moon in twenty-five square sections with no overlaps, showing south at the top (Fig. 1.7).

Schmidt also set out to publish a manuscript atlas left by Wilhelm Lohrmann, a surveyor from Dresden who had already proposed the idea of constructing a lunar map in twenty-five sections. Lohrmann was the first selenographer to use hachures to indicate the length and steepness of slopes, by then a common technique in terrestrial cartography. Only four sections of Lohrmann's map had been published during his lifetime, but Lohrmann left the complete map in manuscript form. In 1878, the same year *Charte*… was issued, Schmidt published the whole twenty-five sections of Lohrmann's map under the title *Mondkarte in 25 Sektionen.* This work was now overall obsolete, but Schmidt's *Charte*… would prove an enduring and respected source for selenographic matters, together with the works of Beer and Mädler. In 1876, the British amateur astronomer Edmund Neison published *The Moon*, an English adaptation of *Der Mond* with some modifications based on Neison's own observations and also some of Schmidt's work. Neison's book became an influential selenographic source on its own right, although a smaller book also titled *The Moon* and containing an artistic map in four quadrants, published by T.G. Elger in 1895, proved more accessible to a wider readership.

While engaged in his selenographic investigations, Julius Schmidt supervised the construction of a large plaster model of the moon's nearside corresponding to a diameter of 19 feet.[57] The globe was produced in the early 1850s by Thomas Dickert, a curator at the National History Museum in Bonn. Schmidt verified the accuracy of all measurements and positions in the globe, which was based on Beer and Mädler's map, with some features added on the basis of Schmidt's own observations. The globe was eventually taken to the U.S.. It was displayed at the Steinway Hall in New York before being installed at the Columbian Field Museum in Chicago. It is now in the Science Museum in London. (Fig. 1.8).

According to a description left by Oliver Cummings Farrington, a geologist and curator at the Field Musem, the prevailing color was a "dull yellow", with the seas marked in "gray-green" and the bright streaks radiating from some craters in "light yellow".[58] The globe was displayed with a regular orientation – that is, with north at the top – and provided a realistic, albeit exaggerated, representation of the lunar topography, constituting, according to Farrington, a very useful tool for the study of selenography. It was also a showpiece that, although lacking the dynamical functionality and the artistic flair of Russell's 'Selenographia', was better suited to produce an impression that the moon had been brought into the show room, within a hand's reach – even if it did not draw crowds to it, as is suggested by a note in *Scientific American* regarding the public display of the globe in New York in 1881.[59] In any case, in a time when museums sought to enhance their role in public education, popular astronomy publications multiplied, spectacular lectures about astronomy illustrated

57 On this globe see also Michael Geffert's contribution to this volume.

58 Oliver C. Farrington, 'The Schimdt-Dickert Moon model', *Science* 9 (1899): 35-36.

59 "The colossal representation of the moon, which has been on exhibition at Steinway Hall in this city, during the past week, does not appear to have attracted anything like the attention it deserves", in 'A Gigantic Artificial Moon', *Scientific American* 44 (1881): 225.

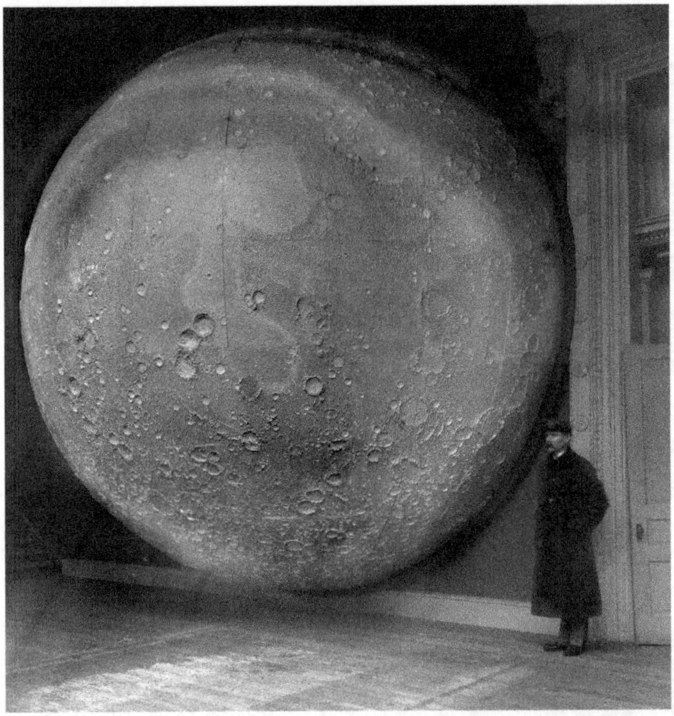

FIGURE 1.8
Lunar globe in the Columbian Field Museum, c. 1898.
COURTESY, THE FIELD MUSEUM, CSGEO12536.

with projected images conquered new audiences, and societies of amateurs burgeoned,[60] ersatz moons had necessarily to grow bigger and more conspicuous, and similar gigantic models would find a place in museums and planetariums over the ensuing decades.

6 Conclusion

In 1865, upon the initiative of William R. Birt, a selenographer with a particular interest in transient lunar phenomena, the British Association for the

60 On the context for popular and amateur astronomy in the late nineteenth century and the early twentieth century, see, for example: Agustí Nieto-Galan, ""... not fundamental in a state of full 'civilization". The Sociedad Astronómica de Barcelona (1910-1921) and its popularization program," *Annals of Science* 66 (2009): 1-32; Pedro Ruiz-Castell, *Astronomy and Astrophysics in Spain (1850-1914)* (Cambridge Scholars Publishing, 2008), esp. pp 110-114; Klaus B. Staubermann, "Making stars: projection culture in nineteenth-century German astronomy," *British Journal for the History of Science* 34 (2001): 439-451. See also Charlotte Bigg, Kurt Vanhoutte (eds.), 'Spectacular Astronomy', thematic issue of *Early Popular Visual Culture* (Vol, 15, Issue 2, May 2017).

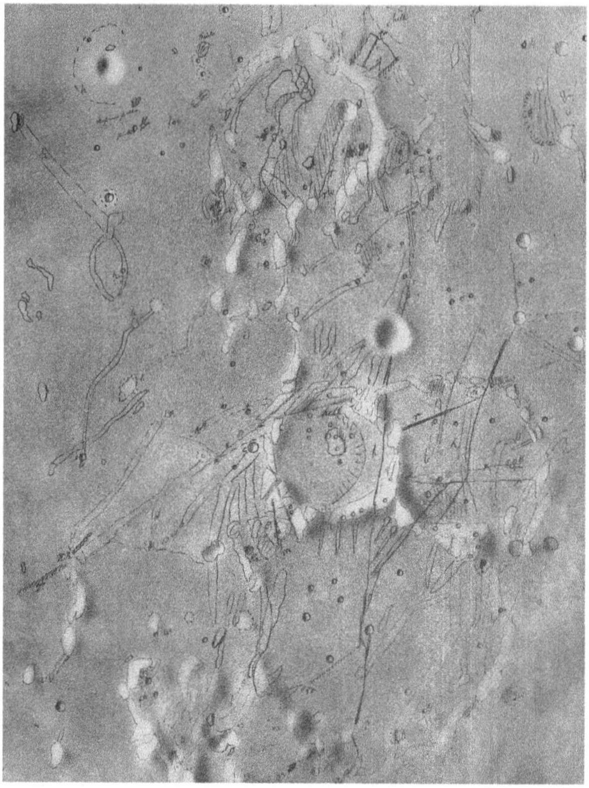

FIGURE 1.9
Plate from Johann Nepomuk Krieger, *Mond-Atlas* (Vienna, 1912). COURTESY, ADLER PLANETARIUM.

Advancement of Science approved a resolution to produce a very detailed map of the moon, for a 200-inch lunar diameter, with five-degree square divisions. Only a few divisions were completed, and the project was eventually abandoned. A few other selenographic projects based on visual telescopic observations were undertaken in the first decades of the twentieth century, but, to a great extent, Schmidt's *Charte* had sealed the age of pre-photographic lunar cartography.[61]

Visual observation and photography were then used as complementary approaches until the first decades of the twentieth century, with transcripts of photographs being used as an aid.[62] Johann Nepomuk Krieger combined photography and hand-drawn features in the production of his *Mond-Atlas,* the

61 Ewen A. Whitaker, *Mapping and Naming the Moon*, p. 134.

62 Znedek Kopal, Robert W. Carder, *Mapping of the Moon*, p. 33.

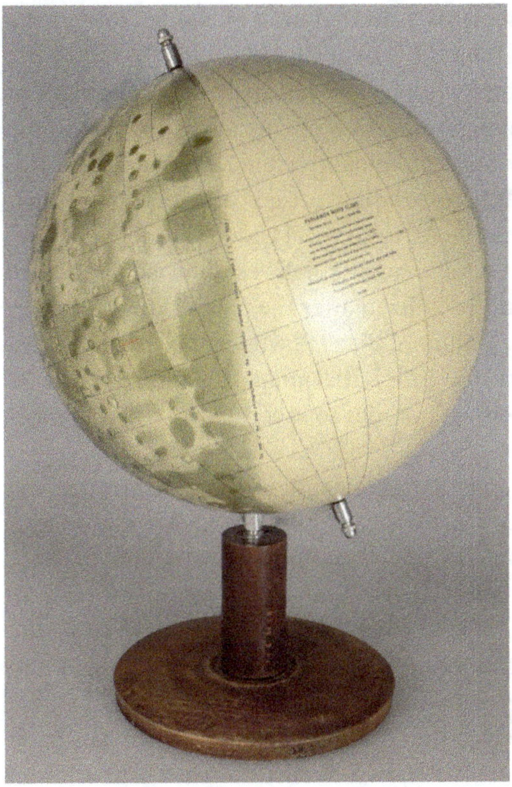

FIGURE 1.10
Globe showing topographical features on the far side of the moon (Pergamon Press, Oxford, 1963, Adler Planetarium collections FTZ-4). COURTESY, ADLER PLANETARIUM.

first volume of which was published in 1898 (Fig. 1.9).[63] Another example is the map included in *The Moon,* a work published in 1931 by the director of the Lunar Section of the BAA, Walter Goodacre, who combined fine detail obtained in good observing conditions with positions derived from photographs taken at the Mount Wilson Observatory.[64]

Photography would eventually prevail, albeit there was a limitation that neither visual observation, nor photography, nor a combination of both, could overcome per se: the fact that only roughly 60% of the lunar surface is accessible to observers on Earth. The Space Age would reveal a full-body moon, making it possible to map the Earth's satellite in its entirety. In October 1959, the Soviet spacecraft Luna 3 returned the first photographs of the far side of the moon. In East Germany, the globe maker Paul Räth produced a lunar globe

63 Jodi Lacy, Marvin Bolt, *Mapping the Universe* (Chicago: Adler Planetarium & Astronomy Museum, 2007), p. 65.

64 David M. Harland, *Paving the Way for Apollo 11* (Springer/Praxis Publishing, 2009), p. 15.

using the new data, which covered just a limited area of the far side of the moon around the eastern limit of the visible near side (Fig. 1.10).[65]

With a few gores still blank, Räth's globe conveyed, at once, a statement of the Eastern Block's primacy in taking possession of the far side of the Moon, and the promise that the endeavor was to be completed. Newly observed features had been named after Soviet heroes, which raised concerns about lunar nomenclature among the International Astronomical Union (founded in 1919).

A nomenclature grounded on the scheme originally established by Riccioli and Grimaldi had been significantly stabilized after the painstaking work carried out by the amateur astronomer Mary A. Blagg,[66] who worked through discrepancies between other historically influential maps (including those of Beer and Mädler, Schmidt, and Neison) in order to ascertain correspondences between different appellations.

But now there were newly observed lunar formations to name. Around the time of the first lunar landings, between 1967 and 1971, extensive discussions on the nomenclature of the far side of the moon took place within the IAU. A particular debate on whether names of astronauts and cosmonauts should be given to features on the far side of the moon led to the decision of using the names of nine astronauts and nine cosmonauts (in each case, six dead and three alive).

Riccioli and Grimaldi had turned the near side of the moon into a pantheon of astronomers and selenographers; the now unveiled far side of the moon served as an extension of that pantheon, accommodating the feted heroes of space exploration through a careful negotiation. Barren and dormant as Mädler, Schmidt and others had aptly imagined and portrayed it, the moon was never a neutral territory. After all, as it has been noted elsewhere,[67] the ancient idea of the moon as a mirror that reflects the Earth was not necessarily wrong; it was simply not to be taken literally.

---

65 Jodi Lacy, Marvin Bolt, op. cit., p. 69.

66 On Mary Blagg and other women astronomers see Marilyn Bailey Ogilvie, 'Obligatory Amateurs: Annie Maunder (1868-1947) and other British Women Astronomers at the Dawn of Professional Astronomy', *The British Journal for the History of Science* 33 (2000): 67-84.

67 Scott L. Montgomery, *The Moon and the Western Imagination*, 117.

CHAPTER 2

Mapping Time Rather than Mapping Space

The Moon in Persian Astronomy during the Naseri Period (1848-1896)

Carmen Pérez González

For Laura Morala Forte

Iran has a long history of astronomy with renowned astronomers such as Abd al-Rahman al-Sufi, Al-Biruni, Nasir al-Din al-Tusi.* Three lunar craters are named after these three great figures.[1] Persia also had well-known observatories, such as the Maragha Observatory, established in 1259 with Naser al-Din al-Tusi as its first director, and the Samarqand Observatory.[2] According to George Saliba, the first astronomical trend in Persia seems to have begun in pre-Islamic times with the production of the *Zij-ash-shahranyar* written in Pahlavi, later translated into Arabic, sometime during the 8th century; the second trend was concerned directly with astrology, the third with Ptolomy's *Almagest* and its

* Special thanks are due to Tofigh Heidarzadeh, for sharing with me literature, contacts to leading historians of astronomy in Iran, and for his support, and also special thanks to Fariba Payerband Sabet. Thanks are due as well to Volker Remmert; Erhard Scholz; Evan Siegel; Pouria Nazemi; Fahrhad Atai; Maryam Ehktiar; Negar Navabi; Ghazaal Bozorghmehr; Debora Velleuer, Rana Javadi; Mohammad Reza Tahmasbpour; Keihan Darvishi, Houman Sarshar; Elmar Seibel; Kazem Kokakam; Guillaume Blanchard, Henk Zoomers, Peter Louwman, Huib J. Zuidervaart, Kamram Arjomand, and Francis Richard.

1 Al-Sufi: *Azophi* (22.1 S; 12.7 E; D: 47.05; 1935); Al-Biruni: *Al-Biruni* (17.9 N; 92.5 E; D: 77.05; 1970); Naser al-Din Tusi: *Nasireddin* (41 S; 0.2 E; D: 52.05; 1935). In all, 24 craters of the moon have names of Arabic and Islamic origin, bearing witness to famous Arabian or Persian scholars. For a detailed list of these craters with full names of the scholars they refer to and their coordinates see: "Illustrious Names in the Heavens: Arabic and Islamic Names of the Moon Craters", in *Muslim Heritage. Discover the Golden Age of Muslim Civilization* (<http://muslimheritage.com/article/illustrious-names-heavens-arabic-and-islamic-names-moon-craters>. Accessed March 9, 2016).

2 For an insightful overview of the history of Iranian astronomy from the Maragha to the Samarqand School, see: George Saliba, "Persian scientists in the Islamic world: astronomy from Maragha to Samarqand," in *The Persian Presence in the Islamic World,* ed. Richard G. Hovannisian et all (Cambridge: Cambridge University Press, 1998), 126-146.

 | DOI 10.1163/9789004298873_004

derivatives, and the fourth with criticism critique of the theoretical foundations of Greek astronomy.[3]

In his knowledge of the fixed stars, Abd al-Rahman al-Sufi was one of the leading astronomers of his time. He left a number of works, the most important one being his *ketab sowar al-kawakeb al-tabeta* (*Book on the Constellations of the Fixed Stars*).[4] In this treatise he gives a full description of the classical system of constellations, both according to the "scientific" Greek classification, as laid down in Ptolomy's *Almagest*, and to the popular Arabian tradition, adding his own observations and providing corrections to both traditions. Drawings of all the constellations and tables of the individual stars of each constellation with coordinates (for the epoch 10 October 964) are included. This book became a classic, with many copies being made through the centuries.[5] Al-Sufi's book is arranged according to the 48 constellations, in the order defined in the *Almagest*. For each constellation he adds two drawings, one showing the constellation as seen in the sky and the other as seen on the globe. As noted by Paul Kunitzsch,

> this representation of the *globe view* drawings, however, is not correct; instead of here showing the constellations in the true convex, outer, view, he simply repeats the *sky view* pictured reversed, as seen in a mirror.[6]

Al-Sufi's drawings of the constellations were continually adapted by the copyists to their contemporary local contexts. Nevertheless, it is important to note

3 Saliba, "Persian scientists in the Islamic world," 139-140.

4 Ishan Hafez, "Abdul-Rahman al-Sufi and his Book of the Fixed Stars: A Journey of Rediscovery" (PhD Thesis. Centre of Astronomy, James Cook University, Townsville, Australia, 2010). Paul Kunitzsch, "Abd-al-Rahman B. 'Omar Sufi," *Encyclopedia Iranica,* I/2 (1982): 148-149; Paul Kunitzsch, "The Astronomer Abu 'l Husayn al-Sufi and his Book on the Constellations," *Zeitschrift für Geschichte der arabische islamischen Wissenschaften: ZGAW* 3 (1986): 56-81; Paul Kunitzsch, "Sufi Latinus," *Zeitschrift der Deutschen Morgenlandischen Gesellschaft: ZDMG* 115 (1965): 66-74.

5 Some of these copies are: 13th century: one copy at the National Library, Paris, MS the Arabe 2489; another copy at Aya Sofya Library, Istanbul, MS 2595 (in Persian); 14th century: British Library, Or. 5323; 15th century: one copy at MET, 13.160.10; another copy at National Library, Paris, MS Arabe 5036; and one whose date is not sure, but probably 15th century, Institute of Oriental Studies, St. Petersburg, MS C724; late 16th century: Malik Library, Tehran, MS 6037; 17th century: one copy at Majlis Library, Tehran; another copy kept at the National Library of Russia, St. Petersburg, MS Ar. n.s. 191.

6 Paul Kunitzsch, "The Constellations", in *Images of Islamic Science. Illustrated manuscripts from the Iranian world,* 1, ed. Ziva Vesel et al. (Paris: Institut Français de Recherches en Iran, 2009), 104-105.

that the iconography illustrating his work was largely inspired by models inherited from antiquity. As stated by Anna Caiozzo,

> Human, animal, or figurative representations served since ancient times as instructive models for the astronomer, thus assisting him to locate and memorise the positions of the stars composing the 48 constellations.[7]

Three hundred years after al-Sufi's death, the Maragha Observatory, founded in 1259 just south of Tabriz, represented a new high point in the science of astronomy in the Islamic world. The Maragha Observatory's main purpose was to make exact astronomical observations aiming at correcting deficiencies in existing *zij* tables. *Zij* tables are handbooks of astronomical observations containing listings of the periodic positions of the planets as seen from various points on the earth.[8] As well as a huge library of around four hundred thousand volumes,[9] the Maragha Observatory possessed many unique instruments, including terrestrial and celestial spheres, a large armillary sphere, and climatic maps of the earth. The observatory was also a pioneer in giving instruction in the natural sciences. It seems that almost a hundred of al-Tusi's students were instructed in astronomy and the natural sciences, and that the local ruler's funds were used for that purpose.[10] This fact is of extreme importance, for it suggests, as Sayili argues,[11] that even if the local religious scholars (the *ulama*) did not support the teaching of natural sciences, the local ruler did.

In Islamic astronomy, lunar crescent visibility was an important observational subject of study. The Muslim calendar is lunar, and each month begins with the first sighting of the lunar crescent above the horizon.[12]

The Samarqand Observatory was built 160 years after Maragha. As at Maragha, the main activity was to produce a *Zij* like the *Ilkhani Zij*. Several leading astronomers of the time, the most distinguished of them being Giath al-Din Jamshid Kashini, worked there alongside its director Ulugh Beg. Kashini had

7 Anna Caiozzo, "Iconography of the Constellations", in *Images of Islamic Science. Illustrated manuscripts from the Iranian world*, 1, ed. Ziva Vesel et al. (Paris: Institut Français de Recherches en Iran, 2009), 107.

8 The *Zij-I Ilkhani* included data derived from observations done during 12 years, and was completed in 1272. See Saliba, "Persian scientists in the Islamic world".

9 Sayili, *The Observatory in Islam*, 194.

10 Sayili, *The Observatory in Islam*, 219.

11 Sayili, *The Observatory in Islam*, 219.

12 See: David A. King, *Astronomy in the Service of Islam* (Aldershot: Variorum, 1993); and Hamid-Reza Giahi Yazdi, "Nasir al-Din Tusi on lunar visibility and an Analysis with Modern Altitude-Azimuth Criteria," *Suhayl* 3 (2002-3): 231-243.

some knowledge of the Maragha Observatory, and therefore, as stated by Tofigh Heidarzadeh, represented a kind of connecting link between these two main centres of medieval astronomy.[13] The main achievement of the Samarqand Observatory was the *Zij* of Ulugh Beg, probably the most precise and complete *Zij* compiled during the Islamic Period.

The decline of the Samarqand Observatory was followed by a long period of relatively uninteresting astronomical activity until an early (unsuccessful) attempt to introduce modern astronomy into Iran took place through the agency of Christian – first Jesuit and later Capuchin – missionaries to Persia during the Safavid Dynasty (1501-1722). By 1624 the Italian traveller Pietro della Valle[14] had already prepared a Persian translation of the Jesuit astronomer Christoforo Borri's treatise on Tycho Brahe's system.[15]

Simultaneously with this western incursion many Iranians – due to the centuries-old political, social and cultural relations between medieval India and Iran – moved during the Safavid Era to India, particularly to Bombay (Mumbai), Calcutta (Kolkata), and the cities of Awadh province. Abu al-Fath Muhammad Sultan, the last Safavid king, fled to Sind and settled in the mid-1790s at Lucknow – to him Mirza Abu Talib dedicated his tract, *Miraj al-Tawhid*.[16]

---

13 See: Tofigh Heidarzadeh, *From the Maragha School to the Dar al-Funun. A Historical Review of Astronomy in Iran from the 13th to the 19th Century*, Paper presented at the Symposium on Science and Technology in the Turkish and the Islamic World, Istanbul, 3-5 June, 1994. (unpublished).

14 For Pietro Della Valle and his contact to Iranian mathematicians and astronomers, see: John D. Gurney, "Pietro della Valle: The Limits of Perception," *Bulletin of the School of Oriental and African Studies* 49 (1986): 103-11; Della Valle Pietro, *Viaggi di Pietro Della Valle il pellegrino, con minuto ragguaglio di tutte le cose notabili osservate in essi: descritti da lui medesimo in 54 lettere familiari all'erudito suo amico Mario Schipano, divisi in tre parti cioè: la Turchia, la Persia e l'India* (Torino: 1843), Volume 1.

15 See: Aydin Sayili, "An early seventeenth century Persian manuscript on the Tychonic system," *Anatolia* 3 (1958): 84-87; and Kamram Arjomand, "Transfer of Knowledge to Iran in the Safavid Era: A Letter on Modern Astronomy Explaining the Tychonic Model," *Majjalah-i tarikh-i Ulum* 10 (2012): 1-26 (in Persian).

16 For insightful research on this topic, see: Razaulah S.M. Ansari, "The First Comprehensive Book in Indo-Persian on Modern Astronomy," in *Proceedings of the Fifth International Conference on Oriental Astronomy*, held in Chiang Mai, Thailand, 4-8 Oct. 2004, edited by K.Y. Chen, et al. (Chiang Mai: Chiang Mai University, 2006), 81-91; Razaulah S.M. Ansari, *Introduction of Modern Astronomy in India during 18th-19th centuries* (New Delhi: Institute for the History of Medicine and Medical Research, 1985); Razaulah S.M. Ansari (ed.), "Modern Astronomy in Indo-Persian Writings," in *History of Oriental Astronomy, Proceedings of the Joint Discussion-17 at the 23rd General Assembly of the International Astronomical Union,* organised by the Commission 41 (History of Astronomy), held in Kyoto,

Consequently, as Ansari explains, writings of Indian scholars in Indo-Persian were collected and transported to Iran quite easily. A copy of the tract *Risalah dar Athbat-i hay'at-i jadid* (Tract on the Proofs of Modern Astronomy) is extant in the Library of Gharb in Hamadan (Iran), and another tract of modern astronomy and science – according to Ansari, the *Risalah dar Hikmat-i jadid wa afrinish-i sitaragan* (A Tract of Modern Natural Philosophy and the Existence of the Stars), written in 1851 – is held at Tehran University Library.[17]

Another key event of the early 17th century was the invention of the telescope in Holland in 1608, and its prompt diffusion through Dutch expeditions and Christian missions – especially Jesuit missions to Asia – which often carried telescopes as diplomatic gift to the kings of their host countries. In fact the Vereenigde Ostindische Compaignie (V.O.C., Dutch East Indies Company) played a fundamental role in the spread of the telescope throughout Asia (see the introductory chapter in this volume, page 21, footnote 48). The historical reconstruction of the arrival of the telescope in the different countries of Asia is of fundamental importance, as it marks the moment at which local astronomers were technologically enabled to practise observational astronomy – more precisely to pursue telescopic observations of the Moon – and to draw through the telescope.

Nevertheless, so far as Iran is concerned, the story of the telescope remains a mystery – although some clues have been found as to when and how the instrument may have arrived (or been invented) on Iranian soil.

1 The Introduction of the Telescope to Iran

Records point to the fact that already in the 1660s at least one telescope had arrived in Iran or that one had even been built there. An article published by the Iranian journal *Yadegar* in 1945, and reprinted years later in *Nojum,* the leading Iranian astronomical journal,[18] informs us that a telescope may have been constructed in Iran by the French Capuchin Raphaël du Mans.

August 25-26, 1997, ed. S.M. Razaulah Ansari (Dortrecht: Kluwer Academic Publishers, 2002), 133-144.

17 Ansari, "Modern Astronomy in Indo-Persian Writings".

18 I am indebted to Dr. Tofigh Hakemzadeh for referring me to this article and for facilitating contacts with several Iranian scholars who have helped me shed light on this important and *mysterious* chapter of the history of observational astronomy in Iran.

The *Yadegar* article tells how, in his *Riaz-ol-Olama* (1695), Mirza Abdollah Esfahani (known as Afandi) wrote as follows of a renowned engineer known as Tanasobi (full name Thābit ibn Qurra al-Ḥarrānī):

> Mullah Mohammad Saleh Ghazvini in his book titled *Navader-ol-Oloom-val-Adab* (*Experts in Science and Literature*) mentions that in his time in Isfahan, there was a Western engineer, named Raphael, expert in technology who had no rival especially in mathematics. He constructed an instrument and learned about that instrument from the western masters; if you look through it at night, you will observe many stars that were not seen before, and especially around some of the planets some stars appear, and the Moon appears in a very strange shape and they claim that there are lands and jungles and towns on the Moon and the Pleiades appears so impossibly huge and myriad of stars in large distance from each other can be seen through it. Briefly, one can see all sorts of wonders in the sky using this instrument. It looks like an Indian reed which is used for making spears, approximately two cubits long; it is denser than the reed and is made of cardboard and on its two sides there are two circular glasses similar to eye-glasses; the secret of that industry lies in those two glasses and their circular and concave shape. When you put your eye to one of the glasses and look through the reed, this effect influences the two confronting glasses and it is strange.[19]

This is the first known description of a telescope in Iranian sources. The writer of *Riaz-ol-Olama* then continues:

> But what Mullah Mohammad Saleh writes about this instrument is nothing new, as right now it is very popular among the people in the west and we have seen it many times in Constantinople and Isfahan and it is, in fact, a type of instrument known as Remote Viewer[20] and we have seen other features in this instrument than what he has written about.

The narrator of the above, Mullah Mohammad Saleh ibn Bagher Ghazvini (known as Roghani), even though the date of his birth and death are not clear, must have been alive at least until 1669, because his translation of *Al-Sahifa*

---

19 Abbas Iqbal Ashtiyani, "Awwalin durbin-i nujum-i jadid dar Iran," ("The First Modern Telescope in Iran") *Yadegar* 10 (1945): 33-36.

20 This would be the "direct translation", but if we want to use the terminology used at the time in Europe, then it would be "spyglass".

al-Sajjadiyya appeared in 1662 and another translation from his hand in 1669. Around that time he must also have written *Navader-ol-Oloom-val-Adab*. He was one of the scholars of the Shah Abbas II (1642-1666) and Shah Soleiman (1667-1695) period, some 50 or 60 years after the telescope was used for the first time to observe celestial bodies in Europe.[21]

However, as stated in the *Yadegar* article, the Raphael mentioned in the writings of Mullah Mohammad Saleh Ghazvini is most probably the French Capuchin Raphaël Du Mans, who led the Capuchin mission to Isfahan from 1646 to 1696[22], and who in 1660 wrote *L'Estat de la Perse en 1660*, a book he dedicated to his minister, Jean-Baptiste Colbert.[23] While in Iran, Raphaël Du Mans worked as a translator at the court of Shah Abbas II and Shah Soleiman, being responsible during his sojourn in Isfahan for most of the correspondence between these monarchs and foreign countries.[24] Since this period corresponds fully with the years Mullah Mohammad Saleh Ghazvini spent in Isfahan, and the French Capuchin was familiar with astronomy and optics, it is not unlikely that the Raphael referred to in Mullah Mohammad Saleh's *Navader-ol-Oloom-val-Adab* was actually Raphael Du Mans.

Raphaël du Mans may well have known his fellow Capuchin Chérubin d'Orleans, a leading astronomer of the time, whose book on the fundamentals of optics, *La dioptrique oculaire*, was published in 1671. In this work, Chérubin d'Orleans printed extraordinarily detailed drawings of a telescope he had himself invented and with whose aid he had apparently drawn an exquisite map of the Moon. This map became the focus of a major controversy, as another leading astronomer of the time, Johannes Hevelius, accused d'Orleans of plagiarizing it.[25] On this issue Guillaume Blanchard writes: "It is interesting that between 1676 and 1681 Chérubin d'Orleans was accused by l'Abbé Claude Comier of plagiarism. Nevertheless, it has yet to be demonstrated that Chérubin

21 Ashtiyani, "The First Modern Telescope in Iran".

22 For an overview of the Capuchin mission to Persia, see: Richard Francis, "Capuchins in Persia", *Encyclopaedia Iranica* IV/7 (2007): 786-788, online: <http://www.iranicaonline.org/articles/capuchins-in-persia> (Accessed February 18, 2016).

23 Raphaël du Mans, *Estat de la Perse en 1660*, per Le P. Raphaël du Mans, Supérieur de la Mission des Capuchins d'Isfahan. Publié avec notes et appendice per Ch. Schefer, membre de l'Institut. Paris, Ernest Leroux Editeur, 28, Rue Bonaparte, 28, 1890.

24 "Raphaël du Mans", *Encyclopaedia Iranica* VII/6 (2004): 571-572.: online: <http://www.iranicaonline.org/articles/du-mans> (Accessed March 9,2016).

25 Ewen A. Whitaker, *Mapping and Naming the Moon. A History of Lunar Cartography and Nomenclature* (Cambridge: Cambridge University Press, 2003), 76-78.

plagiarized Hevelius".[26] Whatever the case, since du Mans and d'Orleans were contemporaries and both were Capuchins, it is not unlikely that they knew each other, or at least that du Mans knew d'Orleans' book.[27] Neither *Le Estat de Perse* nor the two books on du Mans written by Francis Richard,[28] make any reference to either d'Orleans or the telescopes. However, there are several references in du Mans' work to astrolabes, their functioning, and the role of astrologers at court:

> Each year the king spends more than twenty thousand *tomans* to maintain his astrologers, *monagem*, who are always by his side with their astrolabe to take the good time, having influence, to dominate, and to say when it is good to sit, to get up, to leave, to eat, to sleep, to wear this or this color, with the result that he is in their absolute disposal.[29]

We can also read several paragraphs devoted to books kept in the court library, including the following list of the most important books related to Western astronomy, optics and mathematics:

> Here they have Almagestum by Ptolemy in Arabic, the *Sphérique* by Ménélaus and Théodose, all kinds of theories about the movement of the planets such as *Coagé Nesseir* by Mirza Ouloukbec; Euclide in all its works, fragments of Arquimedes and Apollonious and other ancient authors, also the perspective of Ebn Heissen, the books of aritmetics, *elme has-*

26 E-mail communication dated 16.02.2016 with Guillaume Blanchard, author of Chérubin d'Orleans' biography: "Chérubin d'Orleans and his binocular microscope," in *Treasury of Optics. The Collections of the Optisches Museum Jena* (Jena: Optisches Museum of Ernst Abbe Foundation, 2013), 127-144. For a detailed account of Comier's accusations of plagiarism with respect to d'Orleans see: Blanchard, "Chérubin d'Orleans," 132-133.

27 According to Blanchard, it is unlikely that d'Orleans travelled to Iran. He argues that if d'Orleans travelled at all outside France, it would have been possible only between 1642 and 1665. E-mail communication: 16.02.2016.

28 Francis Richard, *Raphael du Mans, missionnaire en Perse au XVIIe s.*, Vol. I: *Biographie. Correspondance*, Vol. II: *Estats et Mémoire;* Moyen Orient & Océan Indien XVe-XIXe s; Société d'Histoire de l'Orient (Paris: Editions L'Harmattan, 1995).

29 Du Mans, *Estat de la Perse*, 166. Original in French: *Le roy chaque année dépense plus de vingt mille tomans pour entretenir ses astrologues, monagem, qui sont tousiours auprès de luy avec leur astrolabe pour prendre la bonne heure, l'ascendant, pour dominifier, et pour dire quand il est bon de s'asseoir, de se lever, de partir, de manger, de se coucher, de vestir telle on telle couleur, de sorte qu'il est dans leur disposition absolue.* Translated by Debora Velleuer.

> *abe,* the algebra, *elegebre,* the optics, *minaser,* and the movements, *gerre sakril.*[30]

Here du Mans also compares and comments on the state of development of mathematics and science in general in Persia and in Europe:

> Mathematics is grown here more general but not in sublime degree as in the Occident. Here, all the parts of mathematics are located, riasi, but they subordinate all these sciences to the judiciary, ehkoum, saying that methods without effect are useless.[31]

Séraphin d'Orleans, who lived and worked with Raphaël du Mans in the Capuchin hospice in Isfahan for several years, came like Chérubin and Raphaël from the North African province of Temian. This makes it even more likely that the three would have known each other. Another book that du Mans may have known was Anton Maria Schyrleus of Rheita's treatise *Occulus enoch & Elliae* (1645), which provided rich and detailed information on how to make telescopes.[32] As Blanchard points out, only a few Capuchins were involved in optics and astronomy during the 17th century, so they must all have known each other.[33] Regrettably, the Capuchin library in Isfahan (near Qa'la-i-tabarruk) was dispersed in the course of the Afghan invasion in the 18th century,[34] so it is impossible to know which books on astronomy and optics du Mans would have been reading.

---

30 Du Mans, *Estat de la Perse*, 164-65. Original quote in French: Ils ont icy l'*Almageste* de Ptolémée en arabe, la *Sphérique* de Ménélaus et Théodose, plusieurs sortes de theories et moiens mouvements de planettes comme de *Coagé Nesseir,* de Mirza Ouloukbec; Euclide en toutes ses oeuvres, quelques fragments d'Archiméde et d'Apollonius et d'autheurs anciens, aussi la perspective de Ebn Heissen, des livres d'arithmétique, *elme hasabe,* d'algèbre, *elegebre,* d'optique, *minaser,* de forces mouvantes, *gerre sakril.* Translated by Debora Velleuer.

31 Du Mans, *Estat de la Perse*, 165. Original quote in French: *Les mathématiques se cultivent icy plus generalment, mais non pas en supresme degree comme en Occident. Icy se trouveront toutes les parties des mathématiques, riasi; mais toutes ces sciences, ils les subordonnent à la judiciaire, ehkoum, disant que les moiens sans l'effet son inutiles.* Translated by Debora Velleuer.

32 Anton Maria Schyteus Rheita, *Occulus enoch & Elliae* (Antwerp: ex officina typographica Hieronymi Verdussij,1645).

33 Another important name for the history of optics in the Capuchin order is the Father Anian , as Blanchard pointed out. E-mail communication: 16.02.2016.

34 Richard, *Raphaël du Mans*, 52, note 120. For general information on the library: Richard, *Raphaël du Mans* I: 50-52.

Another important issue mentioned in the *Yadegar* article is that the Persian astronomers from Isfahan had at least seen a telescope in Constantinople. Blanchard comments:

> I am aware that Chérubin provided lots of telescopes for important people connected to Louis XIV. In the list there is one for the French Ambassador in Constantinople around 1679.[35]

A second possible way for the telescope to have been introduced to Persia would be the Jesuit mission, which was established in Isfahan before the Capuchins arrived. To date, however, no written records have been found to support this hypothesis. The third possible route would have been through the Dutch V.O.C., a subject that needs to be still explored in-depth.[36]

For the moment, we can only conclude that it is feasible that Raphaël du Mans either constructed his own telescope or was given one by a foreign visitor to Isfahan as early as the 1650s. So the technology needed to make lunar telescopic observations would have been available in Persia a mere four decades after it was invented in Holland. Despite this, however, nothing was produced for the next two centuries by using the telescope – or at least, no documentary evidence to the contrary has yet been found. It is important to note at this stage that there was considerable religious objection to the introduction of modern astronomy in Iran, with its central role in establishing the motion of the Earth around our sun. This was exacerbated by the opposition of modern astronomers to the astrology that was considered by the *ulama* to be the essential element of Ptolemy's astronomy as taught in the traditional *madrasas*. Kamram Arjomand explains in detail the roots and results of this controversy in his article "The Emergence of Scientific Modernity in Iran: Controversies Surrounding Astrology and Modern Astronomy in the Mid-Nineteenth Century".[37]

35 E-mail communication, 13.02.2016. For the relationship of patronage by Louis XIV of Chérubin d'Orleans, see: Blanchard 2013, 130-133.

36 For insightful studies of travellers and missionaries in Safavid Iran, see: Rudi Matthee, "Christians in Safavid Iran: Hospitality and Harassment," *Studies on Persianate Societies* 3 (2005): 3-43; Rudi Matthee, "The Safavid Under Western Eyes: Seventeenth-Century European Travellers in Iran," *Journal of Modern History* 13 (2009): 137-171; Rudi Matthee, "Safavid Iran Through the Eyes of European Travellers," *Harvard Library Bulletin* 23/1-2 (Spring-Summer 2012): 10-24; Colin Mitchell, "Les voyageurs français en Perse de 1600 à 1730," *Eurasian Studies* 4/1 (2005): 41-7.

37 Kamram Arjomand, "The Emergence of Scientific Modernity in Iran: Controversies Surrounding Astrology and Modern Astronomy in the Mid-Nineteenth Century," *Iranian Studies* 30/1-2 (1997): 5-24.

The first Persian translation of a treatise on modern astronomy, Mirza Mas'ud's *Tarjome-ye Hei'at* (*Translation of the Science of Configuration*), was published in Iran in 1818.[38] Mas'ud, a governmental official who is best considered a linguist rather than a specialist in astronomy[39], completed the translation as part of his official duties at the court of Fath 'Ali Shah (1797-1834). James Lyman Merrick, an American Presbyterian missionary in Iran from 1834 to 1835, also translated a book on modern astronomy.[40]

2 Dar al-Funun and the Introduction of Modern Astronomy in Qajar Iran

Mighty is he who has knowledge.
By knowledge the old hearts grow young again.[41]

Modern astronomy was not introduced in Persia until the mid 19th century, a decade after the introduction of photography, during the rule of Naser al-Din Shah (1848-1896) of the Qajar Dynasty (1785-1925), a period critically important for the introduction of modern science in general in Iran. Founded by Amir Kabir, Naser al-Din Shah Qajar's first prime minister, the Dar al-Funun (*Polytechnic University* 1851- ca. 1892) was Iran's first secular institution of higher learning.[42] The fields of study represented there were military and medical sci-

---

38 Yoichi, Isahaya, "Gravity of Modernity: reactions to the "New Astronomy" in Iran and Japan," in *Translation, History and Arts: New Horizons in Asian Interdisciplinary Humanities Research,* ed. Meng Ji et al. (Newcastle upon Tyne: Cambridge Scholars Publishing, 2013), 27-47.

39 The authors of the chapter introduced in footnote 40 make an interesting comparative analysis of the first translations of a treatise on modern astronomy done in Iran and in Japan. They arrive at the conclusion that in both cases: the translators should be considered linguists, as they were not specialists in astronomy. Rather than advanced astronomical works, the books could be categorized as being for a general readership; later intellectuals familiar with astronomy declared their dissatisfaction with the first translations. Finally, both translators regarded the works of modern astronomy as objects lying outside their own astronomical traditions: Yoichi 2013, 33.

40 Thanks are due to Tofigh Heidarzadeh for referring me to this manuscript. See, Hadi Tabatabaie's Master Thesis on this topic.

41 Ferdowsi, *Shahnameh.* This poem is inscribed on the upper façade of the main entrance of Dar al-Funun.

42 Maryam Ehktiar, "The Dar al-Funun: Educational Reform and Cultural Development in Qajar Iran" (PhD. diss., New York University, 1994); Maryam Ehktiar, "Nasir al-Din Shah and the Dar al-Funun: The Evolution of an Institution," *Iranian Studies* 34/1-4 (2001):

ences, natural sciences, history and geography, foreign languages and translation, technology (printing, photography and telegraphy), and the arts (painting, music and drama). The institute was planned and built under supervision of the Qajar prince Bahram Mirza. The best students at Dar al-Funun travelled to Europe to complete their education after graduating at the polytechnic institute. Shaped by a similar academic, intellectual and professional curriculum, most of them returned home after a number of years abroad, while hundreds of university-educated westerners moved to Iran to become teachers at Dar al-Funun.

But even if great efforts were put into introducing modern science in Qajar, its initial adoption in Iran was extremely slow. Among the many reasons for this was the resistance of the local scientific community to "modern" astronomy, because traditional Persian astronomy was focused on calendar making and divination. The Qajar prince Ali Quli Mirza Etezad al-Saltane, director of Dar al-Funun and first minister of sciences, played a fundamental role in combating such religious objections. His book, *Falak al-Sa'ada* (Auspicious Heavens), published in 1861, was an open attack on the traditional sciences.[43]

Several European travellers refer to Dar al-Funun in their journals, mentioning the royal astrologers and the importance of their task at court and beyond. The Austrian physician, Dr. Jakob Eduard Polak, personal doctor to Naser al-Din Shah and teacher of medicine at Dar al-Funun, wrote some pages about astrology at court in his travelogue *Persien. Das land und Ihre Bewohner*:[44]

> Gleichen Credits erfreut sich die Astrologie; der Glaube daran steht noch, wie in Europa im Mittelalter, in voller Blüte. Der *munaedschim* (Astrolog) stellt das Horoskop (*táleh*), bestimmt die glücklichen und unglücklichen Stunden einer Unternehmung (*Saat*), sowie die Stunde und Minute, in welcher der König die Stadt verlassen oder in dieselbe einziehen soll. [...] nun bestimmen die Astrologen, nach untereinander getroffener Verständigung, verschiedene Tage, von denen natürlich einer der rechte sein

153-164. For detailed information on preparation for the school and the method and impact of the institution, see: Negin Nabavi, "Dar al-Funun: The First Modern College in Iran" (Master Thesis, St. Anthony's College, 1990).

43 Arjomand, "The Emergence of Scientific Modernity in Iran," 17-18 and 22-23. Arjomand discusses "the firm order that was issued by Naser al-Din Shah, probably on Etezad al-Saltane's recommendation, prohibiting the writing of astrological prophecies in calendars so as to combat superstition and break the predominance of astrology in everyday Iran".

44 Dr. Jakob Eduard Polak, *Persien. Das land und Ihre Bewohner* (Leipzig: J.M. Brodhaus, 1865).

> mußte. Die Sternkundigen bedienen sich der tafeln (*sitsch*) des berühmten Astronomer *shardsche* Nässir aus Märageh, der sogennanten Tafeln des Schdari, und der italienischen Astronomer Cassini zur Bestimmung des Jahresanfangs (*nauruz*), des Eintritts der Jahrzeiten (*fass'l*), der Sonnen- und Mondfisternisse (*kaeschuf-e-aftab u mah*) [...][45]

Another important matter to take into consideration here is the travels not only of the shah[46], but also of Reza Akkasbashi (the first court photographer) and Dar al-Funun's brightest students, to Europe, and the simultaneous impact of the European scholars who took up teaching posts at Dar-al Funun. Fahrhad Atai comments:

> The students' young age was to their disadvantage in a different way as well. In a culture where a firm tradition of ruling according to the family affiliations of senior statesmen was in place, the young graduates of European schools would not be welcome to share power with the older elite. The ruling elite, including the Qajar kings, were more at ease employing foreign advisors and technicians than the European-educated young Iranians. The underutilization of the skills of the returning students may have been due to other factors, which were inevitable in nineteenth century Iranian society. One such obstacle may have been the fact that the mere acquisition of new techniques and sciences would not guarantee their application. For Mirza Mahmud to be able to practice astronomy, for instance, there had to be an observatory available in the country.[47]

The inevitable result of such limitations was that the students were often employed in areas not related to their special expertise. Another obstacle in this particular area was cultural. Most young graduates came from the families of notables, dignitaries, and the Qajar tribe, who had been engaged for decades in military and administrative professions. While technical professions were considered respectable in Europe, such occupations were, however, seen as unfit

45 Polak, *Persien*, 287-288.

46 See: Chapter 4 of Naghmeh Sohrabi, *Taken for Wonder. Nineteenth-Century Travel Accounts from Iran to Europe*: "The Traveling King. Nasir al-Din Shah and His Books of Travel" (Oxford: Oxford University Press, 2012), 73-103; and Pérez González, Carmen; and Sheikh, Reza (2015), "From the Inner Sanctum - Men who were Trusted by the Kings", in *The Eye of the Shah: Qajar Photography and the Persian Past,* ed. Jennifer Y. Chi (New York: ISAW and Princeton University Press, 2016), 132-157.

47 Fahrhad Atai, "The Sending of Iranian Students to Europe, 1811-1906" (PhD diss., University of California, Berkeley, 1992).

for the elite class in Iran. Thus, many students, despite their technical skills, chose civil and military careers over professions related to their training and expertise.[48]

The camera entered Iran as early as 1842, during the Qajar Dynasty. Naser-al-din Shah Qajar, one of the last of the dynasty, was fascinated by the new medium. Only three years after its invention in Paris, his father, Muhammad Shah Qajar, had received, upon his request, two daguerreotype cameras, one each from Queen Victoria of England and Emperor Nicolas I of Russia. Naser-al-din eventually became both an amateur exponent and a patron of photography, establishing the Royal Photography Atelier in the Golestan Palace, and ensuring that Dar-al-Funun would offer classes in the science and art of photography.[49]

With the opening of Dar al-Funun, modern science was suddenly introduced to a country whose traditional schools, completely under the sponsorship of the clergy, were acquainted only with traditional science. Without the necessary background for debating such a development, this presented an abnormal phenomenon, especially with regard to astronomy. The result was a war between old and new, with obstinacy ruling on both sides.[50] Traditional Persian astronomy was mainly used for the calculation of the lunar calendar. Hence, regarding the Moon, traditional astronomers were concerned with mapping time rather than space: more precisely, mapping the lunar surface was an alien interest. How this changed, if at all, with the opening of Dar al-Funun is one of the main research questions of this chapter.

The introduction of astronomy in Iran, when contrasted with the introduction of photography, nicely illustrates Maryam Ekhtiar's statement that

> Some factors that determined the measure of success or failure were the extent to which the court patronized and fostered the development in a certain field, financial resources allocated to the advancement of the discipline, i.e. instructors' salaries, textbooks, and teaching aids, the availability of qualified instructors [...]. Fields that were most needed, or

48 Atai, "The Sending of Iranian Students to Europe", 192-92.

49 Reza Sheikh and Carmen Pérez González, eds., "The First Hundred Years of Iranian Photography" (special issue), *History of Photography* 37/1 (2013); Carmen Pérez González, *Through the Lens of the 19th Century Iranian Photographers* (Leiden: Leiden University Press, 2012).

50 For insightful arguments on this topic, see: Arjomand, *The Emergence of Scientific Modernity in Iran*); and Heidarzadeh, *From the Maragha School to the Dar al-Funun.*

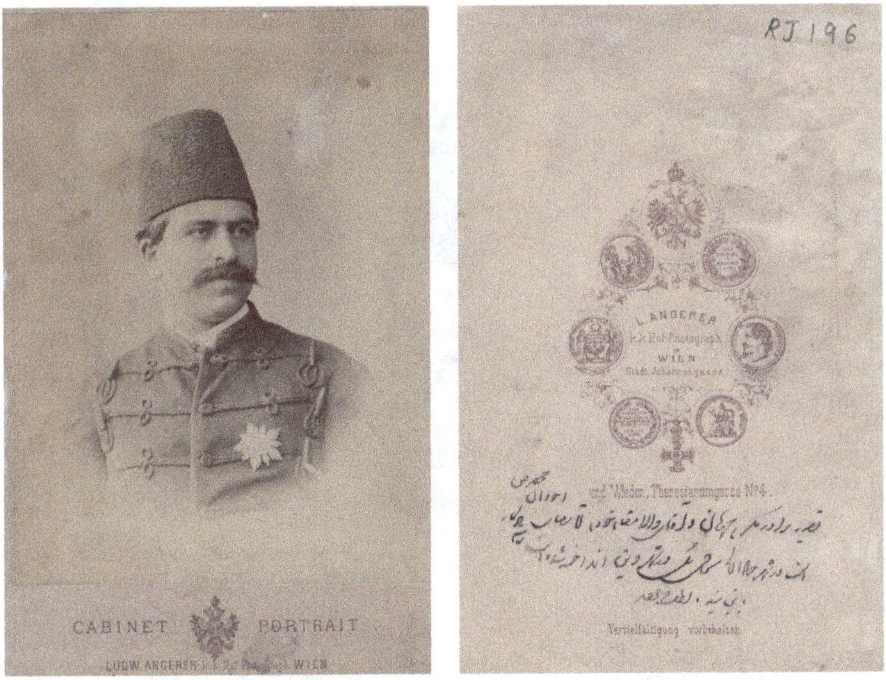

FIGURE 2.1 L. Langerer, portrait of Reza Akkasbashi, first court photographer to Naser al-Din Shah in Vienna, c. 1873, KIMIA FOUNDATION COLLECTION

> favored by the court, developed more rapidly than those that did not enjoy this support.[51]

In order to explore in parallel the introduction of photography and modern astronomy in Iran, their main protagonists must be introduced: the first court photographer, Reza Akkasbashi (active 1858-1878) (Figure 2.1), and the chief court astronomer or *munajjimbashi*, Abdul Ghaffar Najm al-Dawleh (Figure 2.2).

Reza Akkasbashi, the First Court Photographer

Aqa Reza Iqbal al-Saltaneh was Naser al-Din Shah's first court photographer, working most actively from 1858 to 1878.[52] He had taken photography courses

51 Ekhtiar, "The Dar al-Funun," 302.

52 For a detailed chronology on the life and work of Reza Akkasbashi, see: Mohammad Reza Tahmasbpour, *Aqa Reza, the Court Photographer* (Tehran: Akshane-ye Shar Museum and Tehran Municipality Department of Art and Culture, 2006) (in Persian). For general

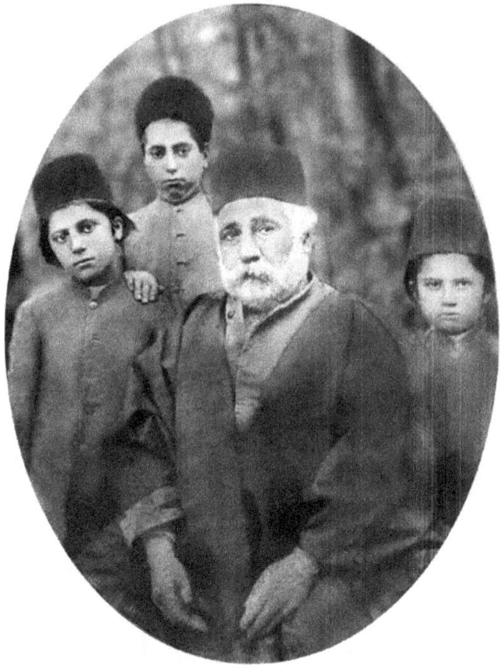

FIGURE 2.2
Unknown court photographer, Abdul Ghaffar Najm al-Dawleh, royal astronomer to Naser al-Din Shah, c.1890. PALACE GOLESTAN LIBRARY.

with the shah, and both started photographing at the same time. In 1862 he was appointed the first court photographer, with the title *Akkasbashi*[53], and from that moment onwards he was known as Reza Akkasbashi. He accompanied the shah on his travels during the 1860s around the country and later on to Europe. Both, king and devoted court photographer enthusiastically experimented with the different photographic techniques at court, under the supervision of hired Western photographers, such as François Carlhian.

Reza Akkasbashi's fascinating legacy included a rare collection of 100 stereographs with various themes now kept at the Kimia Foundation in New York[54]. Mostly taken in the early and mid 1860s, many of them depict the royal

information on this photographer, see: Yayar Zoka, *The History of Photography and Pioneer Photographers in Iran* (Tehran: Sherkate Entesharat Elemi wa Fahrhangui, 1997), 47-57 (in Persian).

53 This title is parallel to the older title of court painter, *Naqasbashi,* and also to court astronomer, *Munajjimbashi.*

54 For more information on this collection, see: Pérez González and Sheikh, "From the Inner Sanctum".

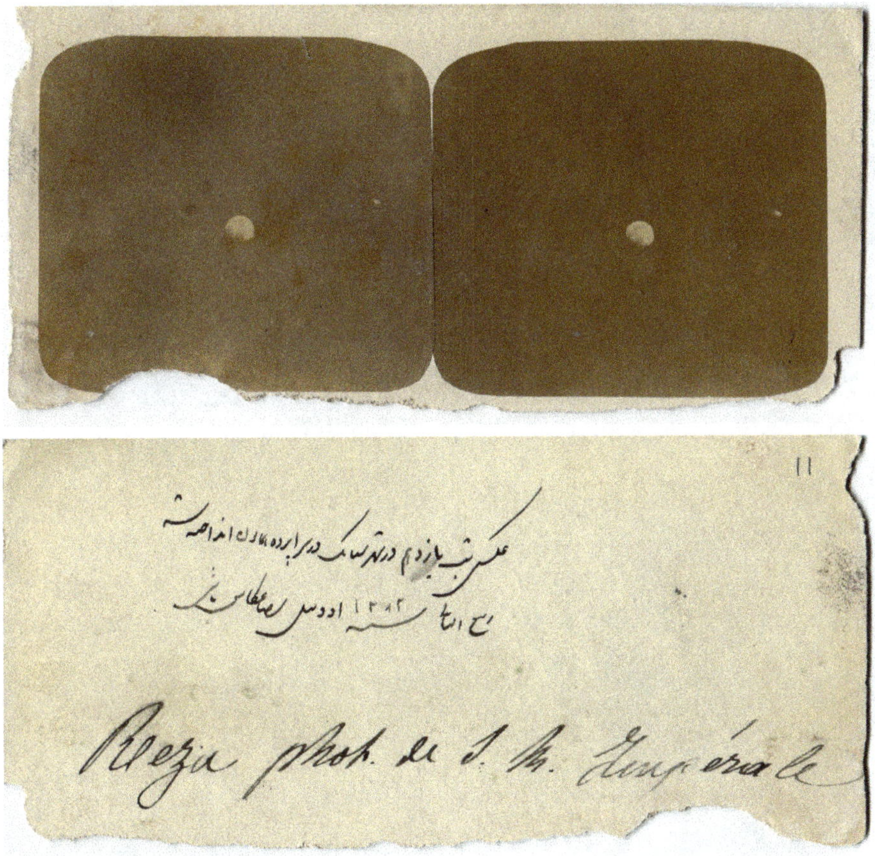

FIGURE 2.3 Reza Akkasbasi, First known photograph of the Moon taken in Iran, c. 1864, KIMIA FOUNDATION COLLECTION

entourage encamped in diverse locations in northern Iran. All these stereographs bear inscriptions on the back, often written by the shah himself or by Reza Akkasbashi. The majority have been dated and are signed: *khanezat Reza* ("from the Inner Sanctum – Reza"). Among them there are two stereographs of the moon (dated c. 1863, figure 2.3), which suggests that Reza Akkasbashi took the earliest photographs of the moon – albeit clearly not through a telescope – from Iranian soil.

The purpose of these images was ostensibly aesthetic and documentary rather than scientific. On the back of the stereographs the photographer himself has written factual information about the date and place of the exposure, and the following image shows the photographic tent with which he was travelling (Figure 2.4). This stereograph is extraordinary from both an historical

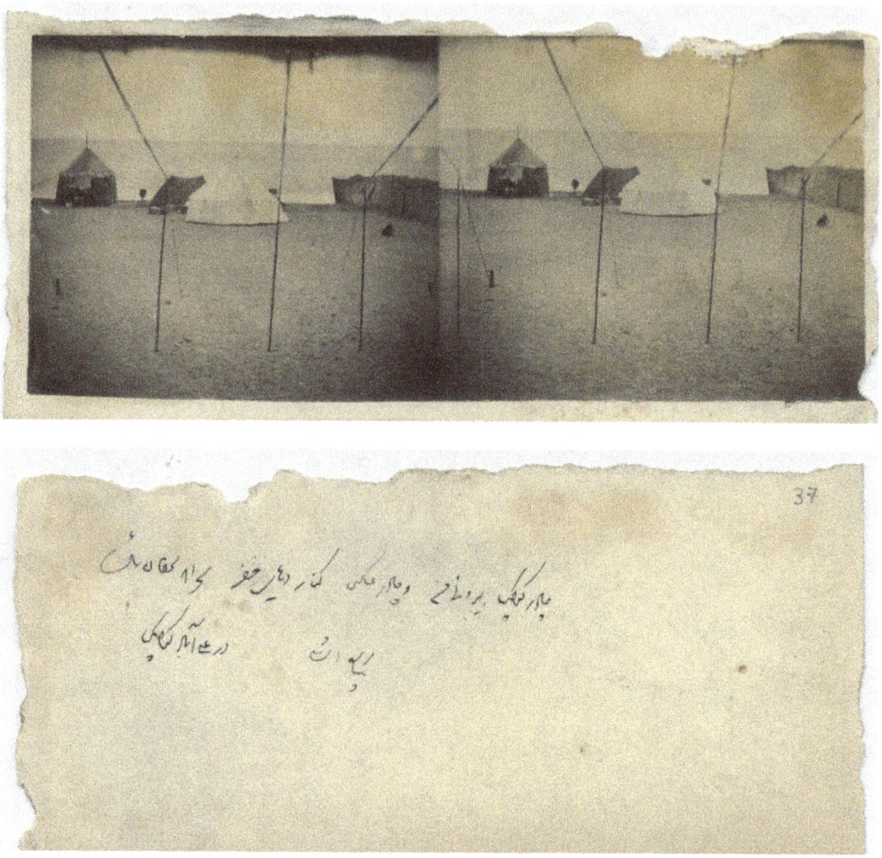

FIGURE 2.4 Reza Akkasbashi, photographic tent in royal camp, northern Iran, c. 1864, KIMIA FOUNDATION COLLECTION

and aesthetic standpoint. The picture was taken from inside a large tent, framing the composition to comprise three tents of various shapes and sizes scattered on the beach, with a view of the sea in the background.[55]

Reza Akkasbashi may well have known and worked together with Abdul Ghaffar Najm al Dawleh – this is a question I am currently researching from primary sources. What is clear is that both men shared space within the walls of the Golestan Palace and Dar al-Funun. In the long extant plan of these two sites by Dr. Jean-Baptiste Feuvrier[56] (Naser al-Din Shah's physician from 1889 to 1892) it can clearly be seen how close Reza Akkasbashi's photographic space at

55 For detailed description and analysis of this image, see: Pérez González and Sheikh, "From the Inner Sanctum", 140.

56 See: Jean-Baptiste Feuvrier, *Trois ans à la cour de Perse* (Paris: F. Juven, 1899).

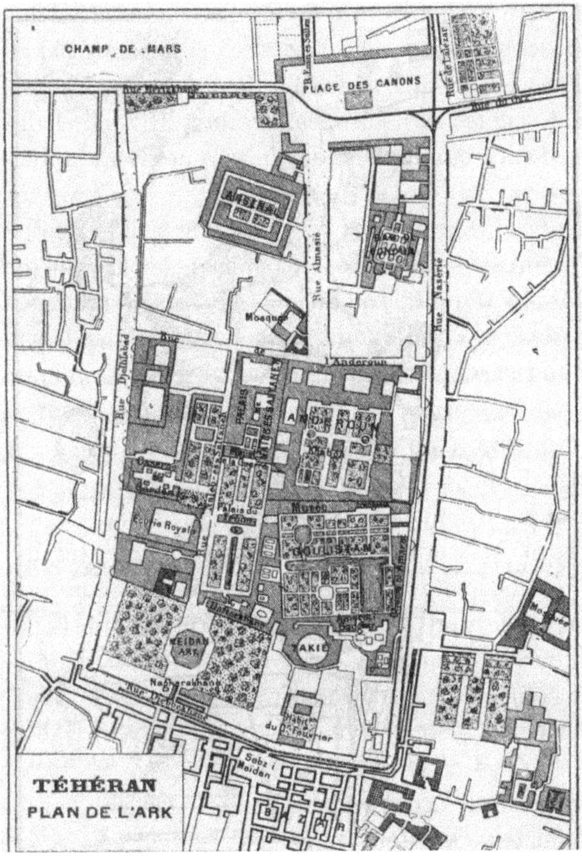

FIGURE 2.5
Dr. Fevrier, Map of Tehran (with the Royal Golestan Palace and Dar al-Funun), c.1890

court was to the Dar al-Funun building where Abdul Ghaffar Najm al-Dawleh worked and taught (Figure 2.5).

Abdul Ghaffar Najm al Dawleh, Court Astronomer and Dar al-Funun Teacher

Haj Mirza Abdul Ghaffar Najm al Dawleh was born in 1839 (some sources give 1843) in Isfahan.[57] His father, the clergyman Mullah Ali Mohammad Isfahani, was a famous mathematician of his time. As Fariba Payerband Sabet explains, in his father's house the young Najm al Dawleh was brought up in surroundings replete with science and learning, especially in mathematics and astronomy. At the invitation of Etezad al-Saltaneh, who had realized his scientific

57 Fariba Sabet Payerband, "Abdull Ghaffar Najm al-Dawleh," in *Nojum* 1 (2009): 54-55 (in Persian).

potential, the mullah moved with his family to Tehran, where he received due respect for his scientific knowledge and sound experience in managing Dar al-Funun. Eventually, Najm al Dawleh became a student there, learning English, French and modern European science with European teachers. His knowledge of mathematics was improved and complemented with astronomy and spherical astronomy.

In 1859, at the age of twenty, al Dawleh took up a teaching post in mathematics at Dar al-Funun. He gradually noticed the lack of course books in all the scientific fields taught at Dar al-Funun. Etezad al-Saltaneh insisted that such books should be written in Persian[58], and upon his request, al Dawleh began writing textbooks not only on the subjects he himself taught – mathematics, spherical astronomy etc. – but also on allied fields such as geography, topography, mechanics, painting, industrial map designing, population census, and technology. At the age of 29 he had written some 20 books.[59]

One of al Dawleh's innovations in compiling calendars was to make a logical connection between the date of the Prophet's Migration with the months and days of the solar calendar. He assigned the date of Prophet's Migration from Mecca to Medina as the basis of the modern calendar, known at the time as the Jalali Calendar and now as the Solar Hijri calendar. Writing on the 39th anniversary of Naser al-Din Shah's reign, in the year 808 of the Jalali Calendar (1885-1886), he mentions the Solar Hijri Calendar in a marginal note: "year 1256 of Solar Hijri Calendar".

Working on calendars was a family occupation in Najm al Dawleh's family, as he noted in the Persian Lunar calendar for 1874:

> These days, since seventy years ago, in Tehran or Isfahan, whoever has studied mathematics and Iranian and western astronomy and metaphysics, the beginning and the starting points have been in my father's family and they have learned their knowledge there, either directly or indirectly.[60]

Naser al-Din Shah, who was interested in astronomy, became himself a pupil of al Dawleh's. On his return to Iran from one of his trips to Europe, he brought

58 Atafallah Atami, "Introduction of modern astronomy in Qajar Era," *Nojum* 1 (2000): 28-31 (in Persian).

59 See Table A2. "A Selection of Textbooks and Other Books by Iranian Instructors and Students at Dar-al Funun" in Ehktiar, "The Dar al-Funun," 314-16. In this table we can read the titles of textbooks authored by Najm al-Dawleh and other Dar al-Funun instructors.

60 Sabet Payerband, "Abdull Ghaffar Najm al-Dawleh," 55.

him a modern French book on astronomy, which he requested his master to translate. Al Dawleh regularly taught the shah from this book, titled *Aseman* (the Sky). Unfortunately the translation was never published, and only a few copies of the book were sent to his brother, Fakhr al-Dawleh (1883-1955), and to the governor of Isfahan Zel-i al-Sultan (1850-1918).

In addition to working on calendars, and translating and writing books on astronomy, al Dawleh was active in observational astronomy. He was one of the few people who observed the 1874 Venus Transit. About this event he writes in his *Badayet al-Nojum* (*Novelties in Astronomy*):

> On 9th December 1874[61], Venus moved across the Sun once more and people were sent to all corners where this event could be observed. In Tehran, too, some people came from Berlin and on top of the Sepah Salar Garden pavilion they installed certain instruments and I, along with my father and the late Etezad al-Saltaneh, witnessed the transit of Venus across the face of the Sun. Based on this transit they measured the exact distance between the viewpoint and the Sun and thus specified the approximate record they had registered before.[62]

The first major government-funded German scientific enterprise sent expeditions to observe the transit of Venus from Zhifu in China, the Kerguelen Islands, Auckland, Mauritius, Isfahan (Persia) and Luxor (Egypt). The expedition to Isfahan was purely photographic, and included four persons: the photographer Gustav Theodor Fritsch, the astronomer Ernst Becker and two other photographers, the German geographer Franz Stolze and Hugo Buchwald. The party went by train through the Baltic provinces to Tsaritsyn (now Volgograd), took a steamer down the Volga to Astrakhan, across the Caspian Sea to the Persian port of Rasht, and then a caravan of fifty-eight animals to Tehran. After an audience with the shah, the scientists continued their trip to Isfahan, where a "Garden palace" outside of the city was used as a lodge, and the instruments were set up in the vicinity (Figures 2.6, 2.7 and 2.8).[63] The caption to a photograph of this expedition printed in the *Illustrierte Zeitung* identifies the four

61 The 1874 transit of Venus, which took place on 9 December 1874 (01:49 to 06:26 UTC) was the first of the pair of transits of Venus that took place in the 19th century, with the second transit occurring eight years later in 1882.

62 Sabet Payerband, "Abdull Ghaffar Najm al-Dawleh," 54-55.

63 This photograph from Azita & Elmar's collection was one of 17 of the Venus Expedition printed as lithographs in the *Illustrierte Zeitung,* No. 1658, 10 April 1875, 269.

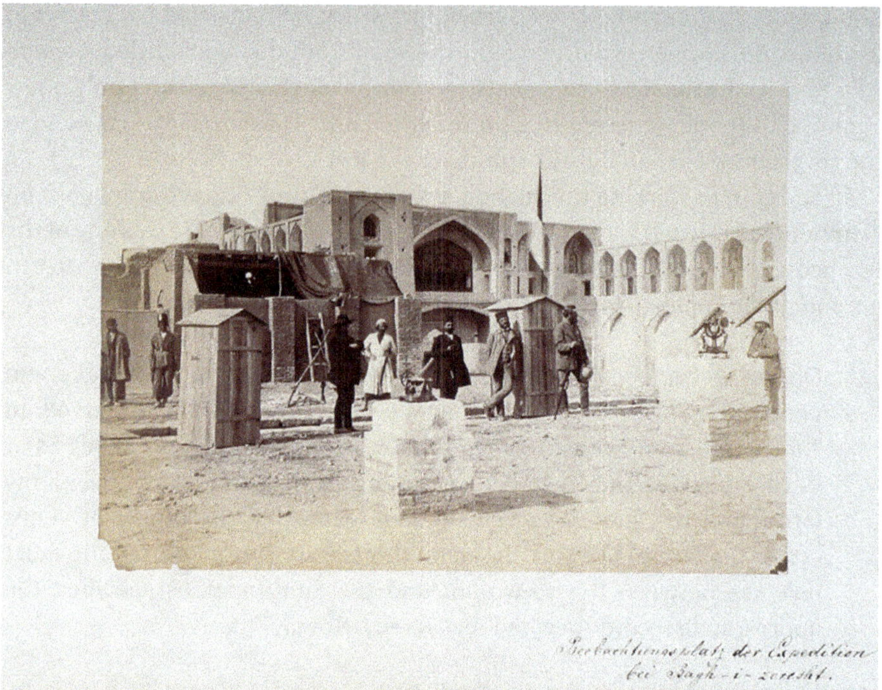

FIGURE 2.6 Unknown photographer, Observatory place in Isfahan, Venus Transit, 1874, AZITA AND ELMAR SEIBEL COLLECTION

expedition members mentioned above, and also "Hoeltzer" – i.e. Ernst Hoeltzer, a foreign photographer active in Iran, who had settled in Isfahan.

Gustav Fritsch's comments on local architecture are not without sarcasm:

> Soon a building was raised on the western part of the platform, which excelled through its simplicity, when compared with the nearby castles of Shah Abbas, of a weird architectural style, resembling most closely a shielded battery.[64]

Pouria Nazemi notes that Naser al-Din Shah's physician, Dr. Joseph Désiré Tholozan, wrote a report on the German Transit of Venus Expedition in a gov-

64 Hilmar W. Duerbeck, "The German Transit of Venus expeditions of 1874 and 1882: organization, methods, stations, results," *Journal of Astronomical History and Heritage* 7/1 (2004): 8-17, 12.

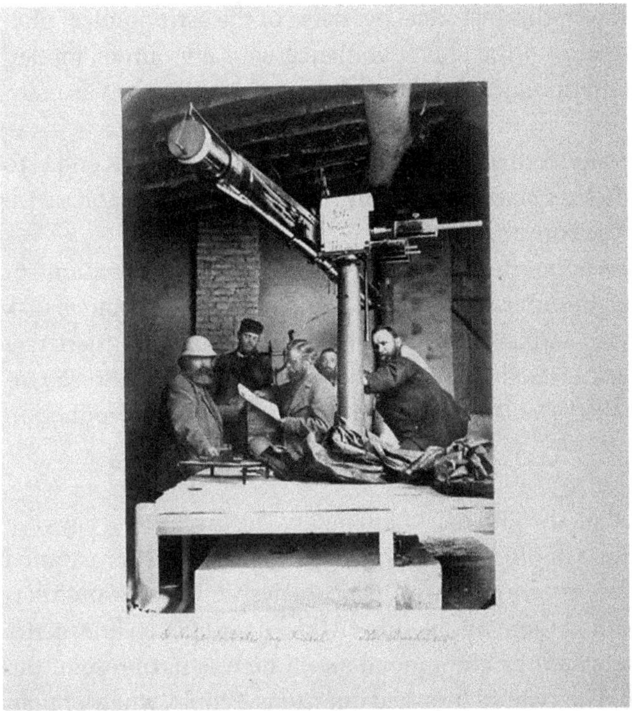

FIGURE 2.7 Unknown photographer, Observatory place in Isfahan, Venus Transit, 1874, AZITA AND ELMAR SEIBEL COLLECTION

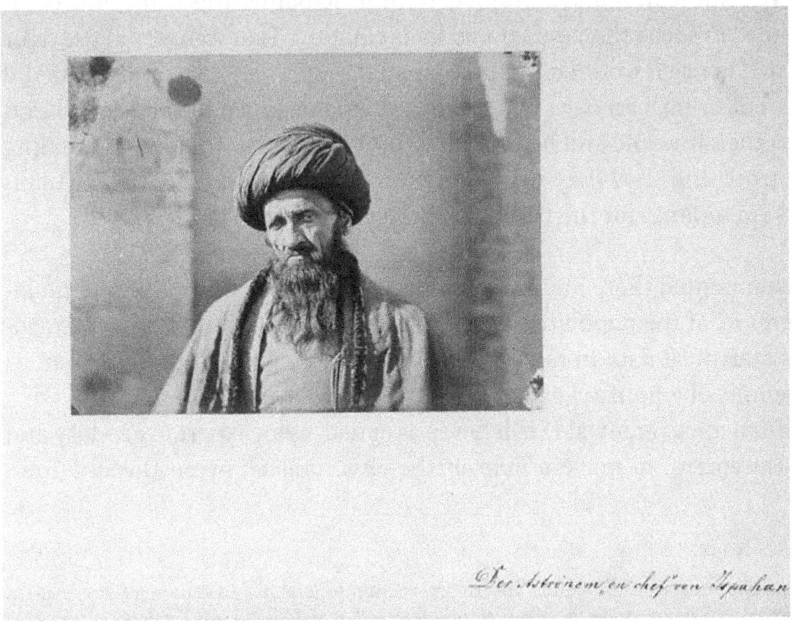

FIGURE 2.8 Unknown photographer, Chief astronomer from Isfahan, German Venus Transit expedition to Isfahan, 1874, AZITA AND ELMAR SEIBEL COLLECTION

ernment journal.[65] However, this historical overview of the astronomical phenomenon only refers briefly to the shah's audience with a group of foreign scientists and does not mention the Iranians who accompanied the Germans to Isfahan.

Among al Dawleh's other contributions was adjusting the Hijri calendar to the Christian calendar. In his *Comparative Essay* (1903) he designed tables with simple instructions for the conversion.

In 1867, al Dawleh was appointed by the shah to register the inhabitants of Tehran. Working with eight of his students, he completed this enormous task with the utmost diligence, precision and sensitivity within less than two months, in spite of a lack of facilities. He presented the results to Etezad al-Saltaneh in a booklet called *The Population of the Capital*, in the introduction to which he noted:

> […] the task started on the 16th day of Ramadan. The first several days the group members were mobilized all together so that they accustomed themselves with the task and to check inaccuracies, and this devoted servant also accompanied them all the time. At the door of each house, the elder of the house would be summoned and if he was not present, the elder wise woman. They would be asked questions. People, whether man or woman, feared and refrained from articulating the truth and commenced to speak their delusional hallucination. That was why at the door of each house it would take approximately a quarter of an hour to ask the resident to take an oath in a thousand languages [to tell the truth] and even then it would still be probable that the person would not be telling the truth and thus they were asked various questions in different manners [in order to for the person in charge to infer the truth] […][66]

He also commented there on the distribution, and the quantitative and qualitative increase, of the population and even discussed Malthus' theory. Categorizing his statistical data in tables, he calculated the population of Tehran as 155,736 people, of whom 147,256 were peasants and 8,480 military.

After this achievement, al Dawleh was assigned, as an expert in geodesy and land measurement, to make a map of the city. Assisted by 20 Dar al-Funun

65 Pouria Nezami, "German Venus Transit Expedition to Isfahan," in *Nojum* 3 (2012): 42-45 (in Persian). I thank Pouria Nazemi for sending me digital copies of Dr. Tholozan's report.

66 Payerband, "Abdull Ghaffar Najm al-Dawleh," 54-55. Original: Abdull Ghaffar Najm al-Dawleh, *The Population of the Capital.*

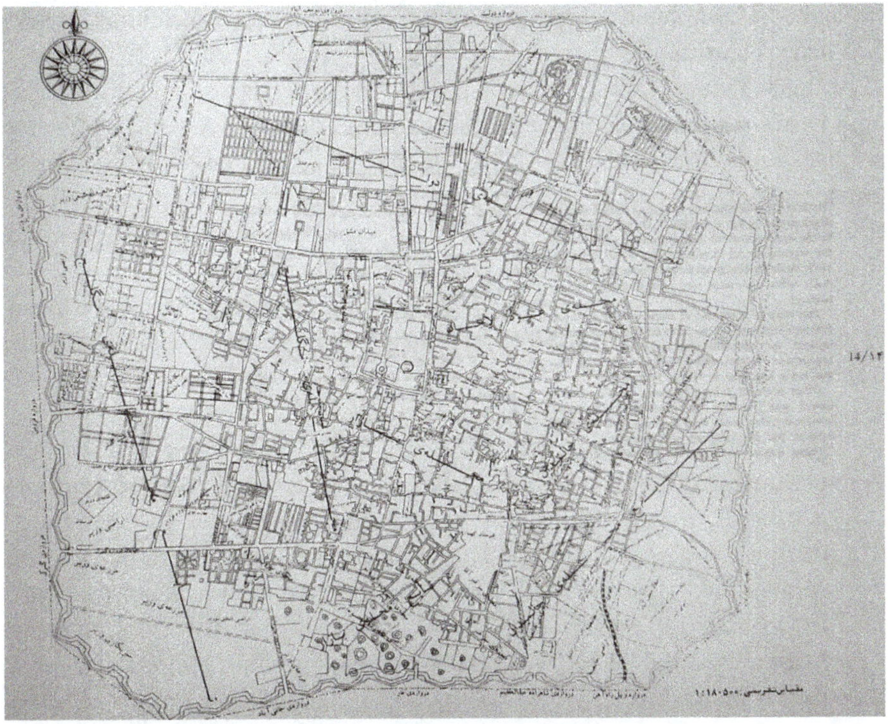

FIGURE 2.9 Map of Tehran by Abdull Ghaffar Najm al-Dawleh, 1887

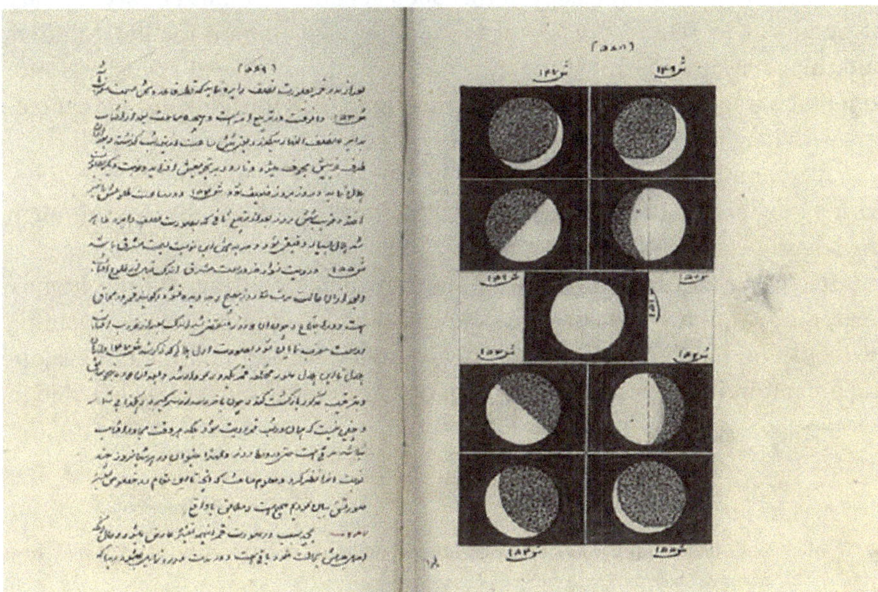

FIGURE 2.10 Page of the book *Qanun-i-Nasiri* by Abdul Ghaffar Najm al-Dawleh, with the phases of the Moon, 1865

students, he completed this project in 1888, after having worked on it for several years (Figure 2.9).

Al Dawleh's most important book is a treatise in Persian on modern astronomy, written at the insistence of the Minister of Education Ali Quli Mirza and named *Qanun-i Nasiri* after the name of the shah. Chapter IV of the treatise provides a detailed description of the motion of the Moon, its apparent diameter, crescents, different phases (Figure 2.10), orbits, craters, lunar eclipses and solar eclipses, as well as the lunar occultation of the planets.

The author was clearly well versed in the work of contemporary English and French astronomers. As M.N. Mathur observes, an important conclusion to the treatise is that

> [it] particularly shows that the Eastern mind was not averse to Western learning and that it responded favourably to its acquisition without forgetting its own traditions, as is clear from the works produced in those days: (1) *Risalah dar Ahwal-I Mulk-I Farang wa Industan* (in Persian) by Muhammed Hussayn ibn Abdul Azim al-Husayni al-Isfahani (d. 1790), (2) *Risalah dar hay at-i Jadid* by Abu Talib al-Hussayni, (3) *Majmu a-i Shamsi* by Abul Khayr ibn Mawlwi Ghiyathuddin, (4) *Miftah al-Aflak* (anonymous).[67]

I have been able to study a copy of this book kept at the Majlis Library (Library of Congress) in Tehran.[68] Chapter VI contains a section on the lunar craters and their physical resemblance to earthly volcanoes, as well as insights into lunar topography[69], together with a table (Figure 2.11) introducing 12 lunar craters with heights and names, presumably observed by al Dawleh himself.

A further interesting item is a drawing of a generic lunar crater (Figure 2.12), with a long explanation of the form of the lunar craters as observed through the telescope.

After 50 years of research, teaching, translating, writing, editing literary texts, discussing technological, astronomical and economic issues, preparing textbooks on mathematics, geometry and spherical astronomy, and developing the natural sciences, especially that of astronomy, Najm al Dawleh died in 1908.

67 M.N. Mathur, "Qanun-i Nasiri. A Persian treatise on modern astronomy," *Studies in History of Medicine & Science* IX/3-4 (1985): 153-159, 156-7.

68 Thanks are due to Rana Javadi for providing me with a digital copy of the manuscript.

69 Abdull Ghaffar Najm al-Dawleh, *Qanun Nasiri* (Tehran: 1865), 608-611.

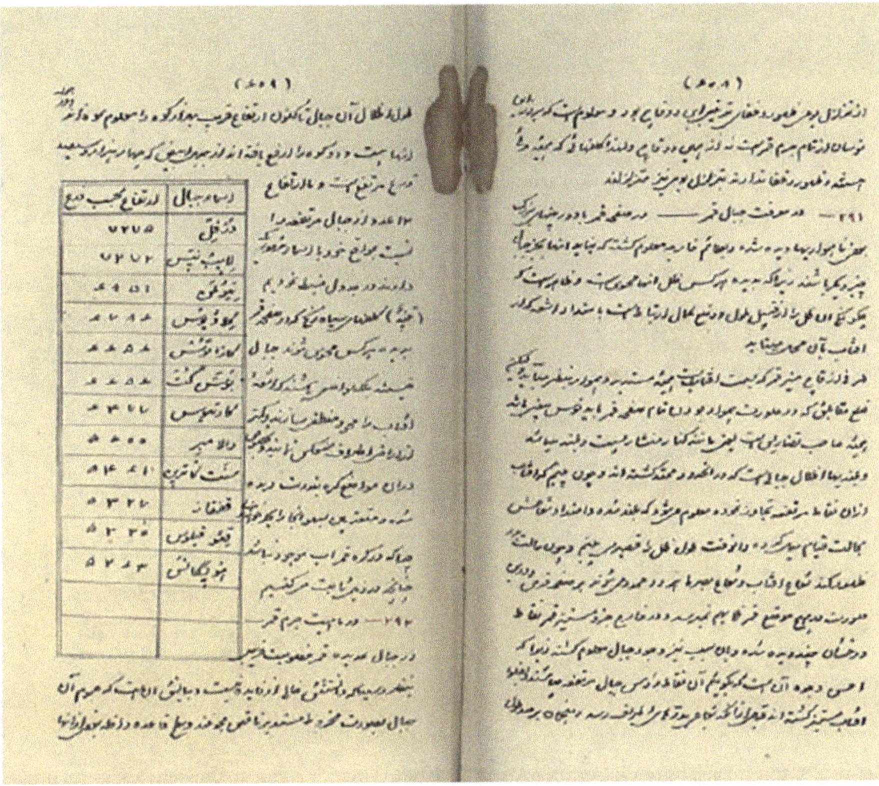

FIGURE 2.11 Page of book *Qanun-i-Nasiri* by Abdul Ghaffar Najm al-Dawleh: a section on lunar craters, 1865

Mahmud Khan Qomi, an Astronomy Student in Paris and Brussels

The second most important figure in the introduction of modern astronomy in Iran is Mirza Mahmud Khan Qomi (Figure 2.13), who became a student at Dar al-Funun.

After two years studying there, he completed his *Naser Calendar,*[70] which attracted special attention. In 1858 he was sent, together with 39 other Iranian students, to Europe. The group was under the supervision of the Iranian Ambassador to England and France, Amir Nezam Garusi. Qomi was assigned to study astronomy in Paris, where he attended courses at the École Polytechnique and the Sorbonne, studying under the head of the Paris Observatory, Urbain Jean Joseph Le Verrier. He was the only astronomy student in the Iranian group, and can be considered one of the first Iranians to be educated in

70 Thanks are due to Fariba Payerband Sabet for sending me a digital copy of the book.

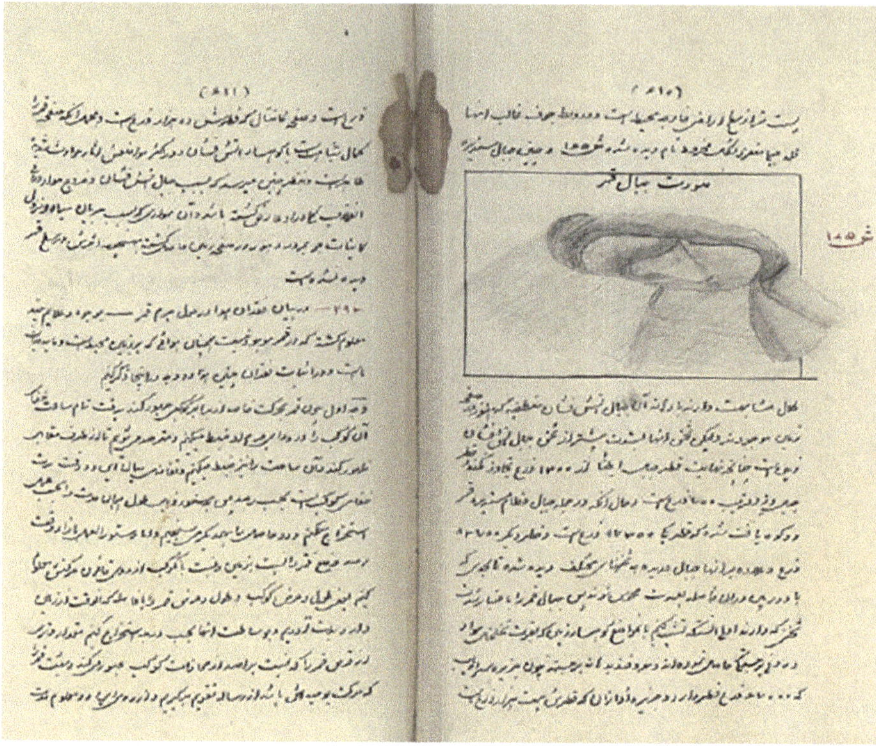

FIGURE 2.12 Page of book *Qanun-i-Nasiri* by Abdul Ghaffar Najm al-Dawleh: a section on lunar craters, 1865

the modern science. Five years later, after his graduation in Paris, he went to Belgium for further studies.

The focus of research at the Paris Observatory was on the quest for asteroids in the Asteroid Belt. In 1860, Fariba Payerband Sabet notes, a group of astronomers there, including Qomi, discovered an asteroid which they named Danae 61.[71] To date, however, no document has been found at the Paris Observatory regarding this matter. According to James Lequeaux of the Paris Observatory, "We know of only one student of Le Verrier during all his lifetime: Aimable Gaillot."[72] Sabet explains further that the feedback on the Iranian students' sojourn in Europe was published in *Dawlat Aliyeh Iran*. Most of the reports there

71 Fariba Payervand Sabet, "Mahmod Khan Ghomi," *Nojum* 10 (2009): 53-55 (in Persian).

72 E-mail communication: 14.3.2016. James Lequeaux has written extensively on Le Verrier's biography. See: James Lequeaux, *Le Verrier – Magnificent and Detestable Astronomer*, Springer, 2013.

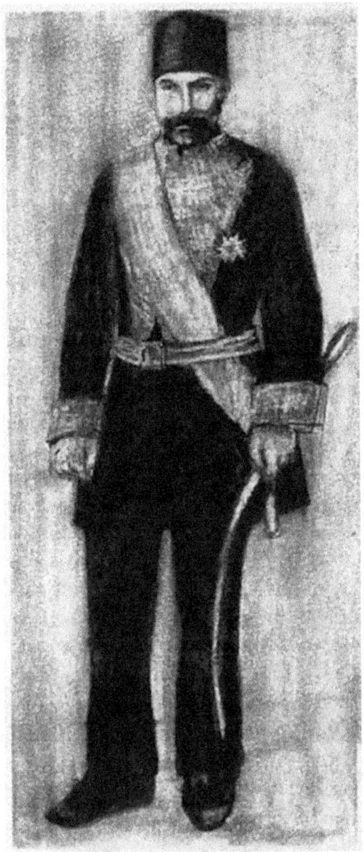

FIGURE 2.13
Portrait of Mahmood Khan Ghomi, royal astronomer to Naser al-Din Shah, c. 1870. PRIVATE COLLECTION.

were written by the leader of the group, Courvoisier. Of these, several are devoted to Mahmud Khan Qomi, among them the following (Figure 2.14)[73]:

> [...] The astronomer Mirza Mahmood is one of the students dispatched by the state taking lessons in the listed sciences and is attempting to conclude his studies in this field. He is a talented, enthusiastic young man and his educators are content with his efforts, and he has made considerable progress in the science of astronomy. To finish his mission these days, he is in need of especial instruments for observation, which were not provided to him, as expected, in the French Observatory.[74]

73 Thanks are due to Fariba Paveryand Sabet for sending me several pages of the *Dowlat Alayhe Iran* where information about Mahmod Khan Qomi's travel and studies was printed.

74 Payerband, "Mahmod Khan Qomi," 53-55. Translated by Ghazaal Bozorghmehr.

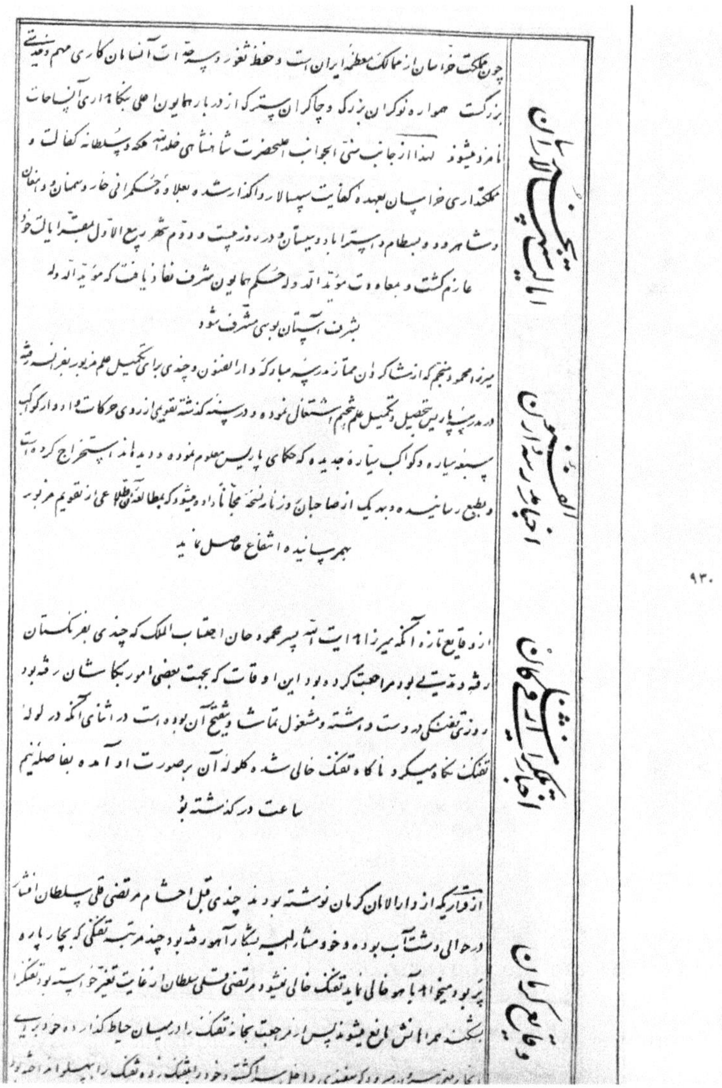

FIGURE 2.14 Newspaper's page reporting on Mahmud Khan Qomi's trip to Europe.

A significant detail is the advice given to Qomi by Mahmud Khan Naser al-Molk, the Persian minister in London. He admonished him

> to cultivate detachment, because you have been in Europe for several years, and if you do not have the proper attitude you will soon die of sorrow because of the state of affairs in Iran.[75]

75 Atai, "The Sending of Iranian Students to Europe," note 427.

Later in this section it will become evident how wise and premonitory the Persian minister's words were.

It may not be surprising that in spite of his training as an astronomer, like many other leading Iranian students of the period, Qomi became a diplomat.[76] Atai continues that, upon his return, he was granted the title of Khan, and the position of *Sarhang*, and was made deputy director of the Telegraph Bureau.[77] He is said to have been dismissed from this post by Naser al-Din Shah, but was later given the title of Mushir al-Wizarah and dispatched by Mirza Husayn Khan Sipahsalar as chargé d'affaires and consul to Baghdad. He also served as consul general in Trabzan (Trebizond) for a while. During the tenure of Sa'd al-Dawlah he served as Minister of Trade, receiving the title of Mushawir al-Mulk, and headed the courts dealing with commercial disputes. After serving in this post, he remained in retirement until his death.[78]

While still holding the position of Minister of Telegraphs, Qomi tried in vain to convince the shah to build a Royal Observatory. He would take his observational equipment and telescopes to the top of the Shams al-Emareh tower in Tehran (Figure 2.15), where Naser al-Din Shah would often briefly accompany him.

Qomi would tell the shah about the wonders lying in the night sky, always attempting to convince him to build an observatory. On one of those nights Qomi told the shah:

> May his highness stay in perfect health for long and may his just reign stay forever. His majesty is well aware that that under the protection of his Royal care and the provisions of Iran supreme state, I spent years of study in the West and in the school of polytechnics I was educated in astronomy. In the countries of France and Belgium I worked and researched using modern instruments. This day I see it as a responsibility to compensate for this generosity. His highness is well aware that these days in foreign countries there are constructions and buildings built for research on heavenly spheres and planets and these constructions are the pride of nations and countries and carry with them the name of kings. This humble servant would like to ask for a budget from the royal treasury to construct an observatory near the capital city of Tehran and collect the most recent tools and instruments in Iran and to hire and educate a team, and to adorn it with the honorable name of your highness' kingship. So we can do research and be proud to work in the Royal Nasseri

76 Atai, "The Sending of Iranian Students to Europe," 167-168, footnote 474.

77 Atai, "The Sending of Iranian Students to Europe," 135.

78 Atai, "The Sending of Iranian Students to Europe," 135-36.

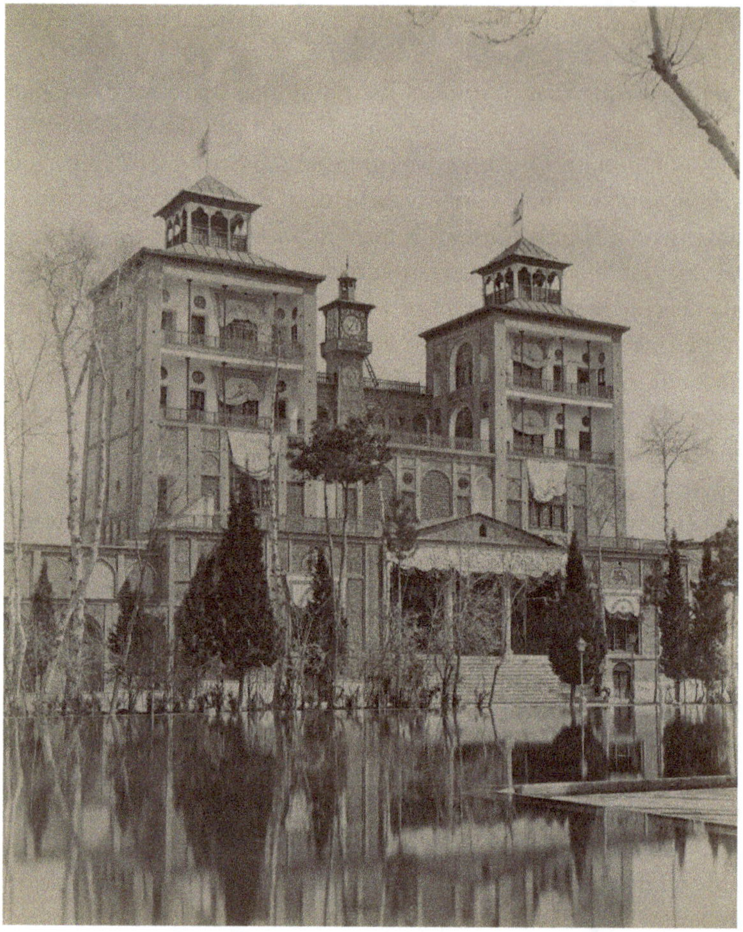

FIGURE 2.15 Antoin Sevruguin, Shams al-Emarol Tower, Palace Golestan, Tehran, "Observatory" at court, c. 1880s, AZITA AND ELMAR SEIBEL COLLECTION

> Observatory and your peasants and servants worship the Shah for his highness' glory and grace and care for modern science. It is not becoming of our power and prestige to possess of the modern tools and instruments any less than other nations in the world.[79]

Naser al-Din Shah replied:

79 Payerband, "Mahmod Khan Qomi," 53-55. Translated by Ghazaal Bozorghmehr.

> Moshaverol-Molk, you are still young and naïve and full of yourself because you have recently returned from foreign lands. At this moment it suffices to tell you that we cannot dissipate a fortune in ruling the country young man. We are spending endlessly on lands and court servants and peasants and the embassies of foreign states and every day we have heard nothing but curses. Imagine we spend our wealth on heavens and skies! No! We do not need the Western instruments of entertainments in the heavens. You must understand that money cannot be spent on air and skies. At the moment you'd better prepare to travel to Baghdad again. We will appoint you once again as the chief ambassador. Strive to be successful in this job and not in the business of the air.[80]

Naser al-Din's second travelogue tells of his visit to London and of his interest in the Royal Observatory there – not necessarily, however, for its astronomical use:

> We now descended, and the Heirs-Apparent, with their wives, took leave and departed, as I wished to go to the Observatory. I proceeded to the open ground of the Naval College, in the middle of which there was a large ship of war fully rigged, for the exercises of the naval boys, who were practicing their manual drills there. About five hundred naval pupils, too, were drawn up in line. We stopped a while to see them exercise; and then, mounting a carriage, drove off to the Observatory. The tower of the Observatory is built on a high hill, and is reached by stone steps. Large telescopes are mounted in a species of turrets, which are made to revolve by machinery, so that the telescopes point in any required direction. It has a celebrated Chief Astronomer, who has so often ascended into the air in a balloon. Its view over the city of London and the environs of the river Thames is magnificent.[81]

Naser al-Din Shah is probably referring here to the Greenwich Observatory and to Sir Sir George Biddel Airy, the Chief Astronomer, who held that title from 1835 to 1881. Later on, the travelogue relates how the shah used a telescope several times in both Greece and Russia, not to observe celestial bodies, but as an aid to viewing distant ruins and landscapes. Naser al-Din's son, Mozaffar

80 Payerband, "Mahmod Khan Qomi," 53-55. Translated by Ghazaal Bozorghmehr.

81 Shah of Persia, A *Diary Kept by His Majesty the Shah of Persia, During his Journey to Europe in 1878* (Translated from the Persian by special permission of His Majesty, by Albert Houtum Schindler and Baron Louis de Norman) (London: Richard Bentley and Son, 1879), 174.

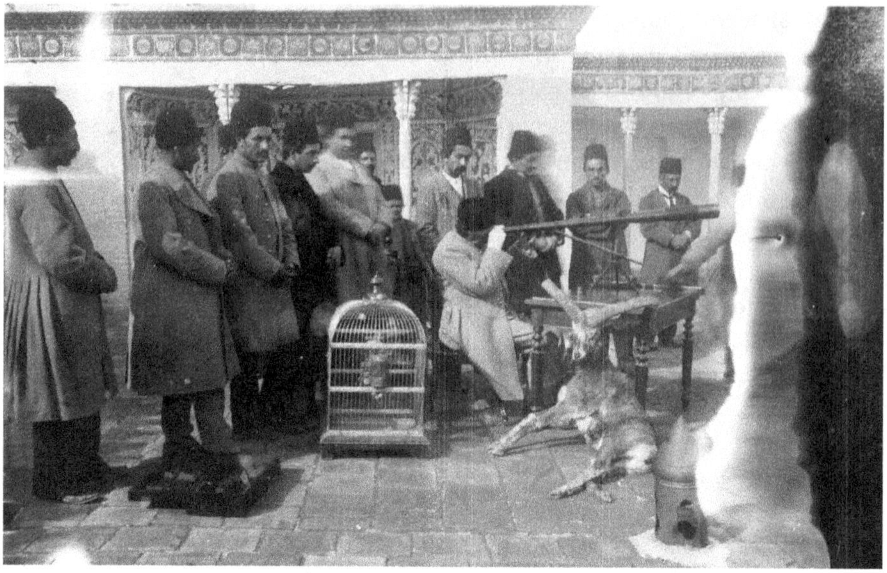

FIGURE 2.16 Unknown photographer, Mozaffar al-Din Mirza with a telescope in Tabriz, c. 1890,
PRIVATE COLLECTION

al-Din Mirza, when he was still a prince and living in Tabriz, owned telescopes as well (Figure 2.16).[82]

In spite of all his attempts at persuasion, Mahmud Khan Qomi never managed to build a Royal Observatory, and his life was largely spent on tasks far removed from astronomy.

Ironically, then, it seems that the telescope, originally conceived as a *spyglass* and diplomatic toy, failed to develop on Iranian soil into anything more scientifically oriented. The most highly trained of all the Iranian astronomers, Mahmud Khan Qomi, became a diplomat rather than a professional scientist, and spent his life in the service of a monarch who was more interested in the perpetuation of his royal image on earth than in celestial issues. The shah's dismissive attitude, especially to the building of a modern observatory in Tehran, jeopardized the introduction of modern astronomy in Iran throughout his long reign (1848-1896).

82 I thank Mohammad Reza Tahmasbpour for finding this photograph for me.

CHAPTER 3

Japanese Lunar Drawings, Maps and Photographs before the 1870s

Tsuko Nakamura

1 Introduction

Since the seventh century, the Japanese astronomy was under the exclusive influence from China until the first encounter with Europeans in 1543. Traditional Chinese astronomy was characterized by the Yin-yang natural philosophy, astrology and the calendar making and improvement, whereas it took little interest in the scientific structure of the cosmos and physical nature of celestial bodies.[1]

In 1613 the captain John Saris (c.1580-1643) of the British East India Company sailed to Japan and visited the first Shogun Iyeyasu Tokugawa (1543-1616) to request establishing a bilateral trade relation. As one of numerous tributes, Saris offered to Iyeyasu a telescope with a gold-coated tube on a silver mount[2], which was the earliest telescope ever imported into Japan. Considering that a few years were necessary at that time for a sailing ship to reach Japan from Europe via the Cape of Good Hope, this is a typical example of quick cultural transfer.[3] Although historical documents record that a considerable number of telescopes were brought to Japan during the first half of the 17th century by Spanish, Portuguese, Dutch and British traders, none of them seem to have survived, including the Saris' one.

Even after the first introduction of a telescope of 1613 (the telescope was then called *to-o megane* in Japanese, meaning a far-seeing glass), Japanese astronomers had long shown no interest in directing telescopes to celestial bodies to know their true character. This was caused by a long-standing tradition inherited from Chinese astronomy. The primary duties of professional astronomers of this time, who served at the Emperor's court, were routine

1 Shigeru Nakayama, *A History of Japanese Astronomy* (Cambridge: Harvard University Press, 1969).

2 M. Earnest Satow, *The Voyage of Captain John Saris to Japan* (London: The Hakluyt Society, 1900).

3 Engel Sluiter, "The first known telescopes carried to America, Asia, and the Arctic, 1614-39." *Journal for the History of Astronomy* 28 (1997): 141-145.

 | DOI 10.1163/9789004298873_005

astronomical calculations for the yearly issue of the calendar, and the visual sky watching for astrological divination and its reporting to the court. Their indifference to the astronomical use of telescopes was probably also due to that most then telescopes had neither high magnifying powers nor good glass quality enough to scrutinize the surfaces of the Moon and planets.[4] Moreover, telescopes were so expensive that the ordinary people of that period could not afford to enjoy astronomical observations as a hobby.

2 Early Use of Telescopes for Astronomical Observations

From the seventh century through the middle of the 17th century, the Japanese had continued to issue calendars simply based on the Chinese luni-solar calendrical theory. However, in the 1680s Harumi Shibukawa, previously a master of the Shogun's *go*-game championship, proposed an original calendrical theory, and in 1684 it was eventually adopted by the Shogunal government as the official national calendar, called the Jokyo calendar. Thanks to this achievement, Shibukawa was nominated to be the first Shogunal astronomer.

He was also eager to improve old Chinese and Korean star maps and catalogues by re-observing star positions with an armillary sphere. As a result he added 61 new constellations comprising 308 stars to the traditional Chinese constellation system. In this respect, Shibukawa can be regarded as the first Japanese who practiced astronomy as science. To identify barely visible stars in his established constellations, he used a telescope.[5] Although according to his writing, he must have pointed his telescope to the Moon as well, but neither such a record nor a drawing is left.

In 1720 a historic decree was issued by the eighth Shogun Yoshimune Tokugawa, announcing that the Shogunate lifted the century-long import ban of the Western books relating to the Christianity.[6] This was because Yoshimune, the owner of rational and scientific spirit, had recognized the importance of European science and technology through the books translated into Chinese by the Jesuit missionaries in China. Yoshimune ordered an optician based at Nagasaki

4 Tsuko Nakamura, "The earliest telescope preserved in Japan," *Journal of Astronomical History and Heritage*, 11/3 (2008), 203-212.

5 Tateki Tani, *Tani Jinzan's Jinki Roku* II (Tani Jinzan's Dialogues with Shibukawa Harumi, II) (Tokyo: publisher unknown, 1910). In Japanese.

6 e.g., Richard Mason and John Caiger, *A History of Japan*, revised edition (Tokyo: Tuttle Publications, 1997).

to produce a special large telescope for astronomical use and he himself observed the Moon, the Sun, and planets with it.

Due to the Yoshimune's epoch-making political decision, non-authoritative Japanese citizens could also soon have chances to learn the knowledge of Western astronomy, via Chinese-translated books (the Japanese then had not yet acquired the ability of reading European languages). Outcomes of Western astronomy that most impressed the Japanese were such as the Copernican heliocentricism, advanced European observational instrumentation, and telescopic drawings of celestial bodies. As far as we know, it was Goryu Asada that first sketched the Moon through a telescope.

Asada used to be a medical doctor who served a daimyo governing a small local fief as his official physician. Asada was so enthusiastic in studying Western astronomy that he implored resignation from his position, but could not obtain permission. So one night he finally fled to the largest commercial city of Japan, Osaka, changed his old name into Goryu Asada, and started a private school for teaching Western astronomy and medical science. This school attracted many talented students, where with their master they eagerly studied advanced Western astronomy, as represented by the Kepler's elliptic motion theory of planets.[7]

In parallel with theoretical astronomy, Asada and his disciples also payed special attention to the astronomical instruments invented by Tycho Brahe whose drawings had been included in Chinese-translated books, and telescopes for astronomical observations. In the Asada's letter draft of 1779 directed to an unknown recipient,[8] he explains his sighting the Moon in detail, with a primitive sketch of the half moon (Figure 3.1).

Asada used an Italian reflector telescope. He describes as follows:

> The terminator of the bright and dark sides for the half moon is much more complicatedly rugged than in the visual observation; There exist a lot of circular features (craters) that look like flat-bottomed kettles or ponds whose western sides are rimmed by the dark shadow; Some circular features have hills or shadowed peaks in the center; Although the lunar surface resembles the Earth's mountains and seas, there seems to

7 Shigeru Nakayama, "Asada Goryu", *Dictionary of Scientific Biography* 1 (New York: Charles Scribner's Sons, 1970). Thomas Hockey, *Biographical Encyclopedia of Astronomers* (New York: Springer Publishing, 2009).

8 Toshio Kage, *Collected Works of Asada Goryu*, Letter No.10 (Ed. T. Kage, in Japanese) (Oita: Historical Documents Office, 1999).

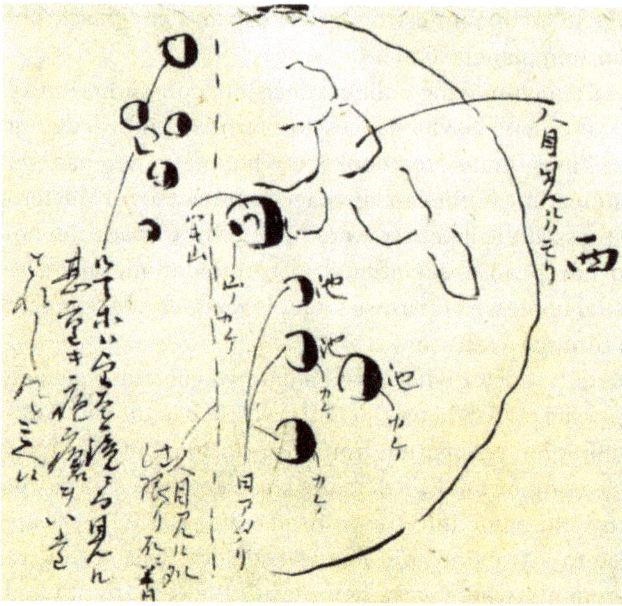

FIGURE 3.1
A primitive sketch of the Moon with comments, drawn by Asada in 1770.
Note: Kage 1999.

> be no soil, no water, no vapor; Rather it looks like the surface of a crushed copper craft or the human skin suffered from small pox; etc.

It is interesting to note that Asada attempted to estimate the depth of a crater utilizing the width of the rim shadow. But after all, Asada's main concern was to apply the knowledge of Western astronomy to the improvement of the luni-solar calendar, but not to reveal the physical nature of celestial bodies. In honor of Asada's astronomical achievements and contribution, the nomenclature committee of the International Astronomical Union named a small lunar crater Asada. The image on the next page shows a photograph of that crater taken by the Japanese Lunar Orbiter Kaguya of JAXA in 2008 (Figure 3.2).

3 Lunar Observations by Telescope Makers

The early telescopes invented in Europe consisted of a convex and a concave lenses (the Dutch-type or Galilean telescope) or two convex lenses (the Keplerian telescope). In the latter half of the 17th century, however, the Schyrlean telescope became much more popular, devised by Anton M. Schyrle of Reita in

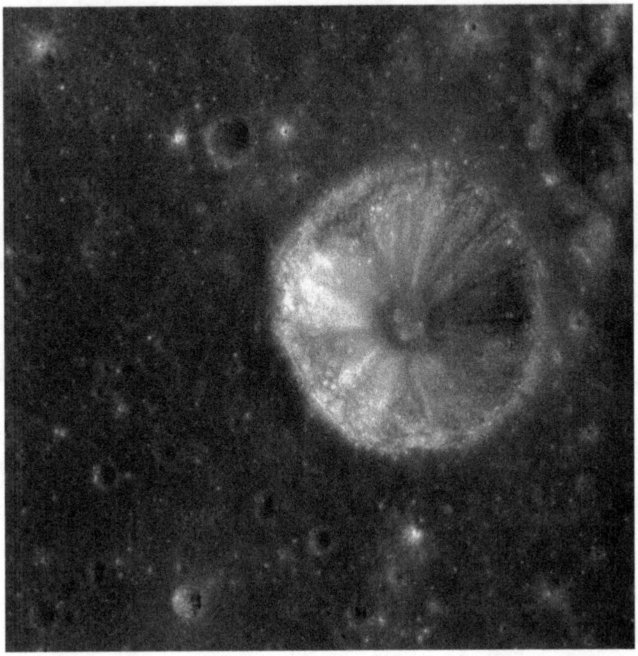

FIGURE 3.2
Asada crater on the Moon, photographed by the Kaguya Lunar Orbiter in 2008. The diameter measures 12 km, located at the position of 7.30N and 49.90E.
COURTESY OF JAXA/SELENE, JAPAN.

1645.[9] A typical Schyrlean telescope comprises four convex lenses (the eye-piece having three lenses) and is characterized by a wider field of view, giving erect images. From the middle of the 17th century, non-European telescopes with four convex lenses, similar to the Schyrlean type, had already been fabricated probably in the southern coastal part of the continental China, though it is unclear whether the Jesuit missionaries taught the production technique to the Chinese people or they invented it by themselves. The oldest of such telescopes, made in 1650 or before, is preserved at the Tokugawa Fine Art Museum, Nagoya.[10] It is certain that production of four-convex-lens telescopes had already started at Nagasaki around the end of the 17th century. Since then, this lens configuration became the standard of the Japanese telescopes, but without particular technical innovation until the 1870s.

9 Henry C. King, *The History of the Telescope* (New York: Dover Publications, 1979); Inge Keil, *Augustanus Opticus: Johann Wiesel (1583-1662) und 200 Jahre optisches Handwerk in Augsburg* (Berlin: Akademie Verlag, 2000).

10 Nakamura, "The earliest telescope preserved in Japan," 203-212.

3.1 *The Iwahashi Family*

Zenbei Iwahashi of Osaka was the only optical artisan who succeeded in launching the telescope making on a commercial basis from around 1790. His good customers were the Shogunal astronomers, land-surveyors for fiefs of daimyos, rich and curious merchants in large cities, and captains of trading and whale catcher boats. Accordingly, one can now find out several tens of Iwahashi telescopes preserved at museums and private collectors, whose typical magnifying power was 10-15. As the drawing tube of Iwahashi's telescopes is characterized by eye-catching colorful painting, they are still popular at antique auctions worldwide. The Iwahashi family continued their telescope business over five generations, till the end of the 19th century.

To promote his telescopic sales, Zenbei held Star Party twice in 1793 by inviting well-known scholars and intellectuals who lived in the then capital of Japan, Kyoto, to show them the Sun, the Moon, planets, and the Milky Way through his telescopes. With the same purpose in mind, he also published in 1802 a rotating planisphere and a booklet explaining how to use it, along with the introductory knowledge of astronomy. In the book Zenbei included his telescopic observations of celestial bodies.[11] Figure 3.3 shows his sketches of the crescent moon. Because of the drawing tradition at that time in Japan and the wood-cut printing publication, both figures are far from realistic but depicted in a pattern-like style.

But from the left of the figure, one can understand that Zenbei was most impressed by conspicuous ray craters like Tycho Brahe and Copernicus, and the shadowed darkness of the lunar mare regions.

3.2 *Reflector Maker Kunitomo*

Tobei Kunitomo was born in 1778 at a local village called Kunitomo near Kyoto, where the residents had all been engaged in gun works. In his 40s, Kunitomo had established his status as a skillful and talented master of the Shogunal gunsmith. In addition to musketry and artillery weapons, he devised various mechanical gadgets over his career, such as air guns, air-pumped lanterns, automatic archeries, prototypes of the bicycle and the flying glider, and so forth.

All Japanese telescopes produced before Kunitomo's times were refractors with lacquered paper or wooden tubes. When Kunitomo stayed at Edo, the shogun's capital (old Tokyo), during 1817-1822, he happened to encounter a portable Gregorian telescope imported from Britain. The owner of the telescope, one of the leading daimyos, permitted Kunitomo to take a look at the telescope

11 Zenbei Iwahashi, *Heitenngi Zukai* (Illustrated Explanation of Planisphere and Introduction to Astronomy, in Japanese) (Osaka: publisher unknown, 1801).

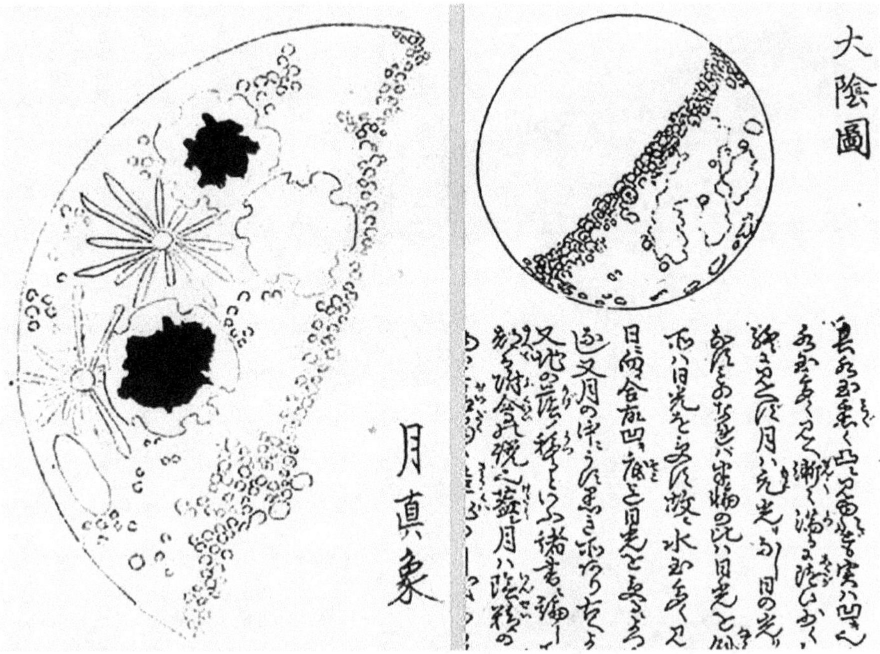

FIGURE 3.3 Lunar sketches of the Moon by Zenbei Iwahashi in 1793 (*Heitengi Zukai* 1802).

and measure sizes of its main parts, as he knew well Kunitomo's talent and curiosity to the Western instruments. It is very likely that this experience gave a strong impact to Kunitmo as a specialist of metal work, because he had never seen before such a telescope all parts of which were made of brass and bronze.

After Kunitomo's return to his home village, he was inundated with administrative duties of the gunsmith union for more than ten years. Around 1832, however, he suddenly began to cope with developing a Gregorian telescope on his own. Following a few years' painstaking trials and errors, Kunitomo finally succeeded in completing his first product in 1834, now preserved at the Ueda City Museum of Nagano prefecture (Figure 3.4).

It still provides fairly clear images of the Moon and planets at the magnifying power of 50-60, and a Foucault's test of the primary bronze mirror revealed a reasonably good parabolic surface.[12] So far, four telescopes including the first one of Ueda are known to survive, though the documents and letters relating to Kunitomo indicate that he produced at least several similar telescopes for

12 Yoshio Tomita, Tsuko Nakamura, et al., "Investigation of the Gregorian telescopes made by the gunsmith Tobei Kunitomo in the 1830s", *Kokuritsu Tenmondaihou* (National Astronomical Observatory Report) 4 (1998): 9-41, in Japanese (abstract in English).

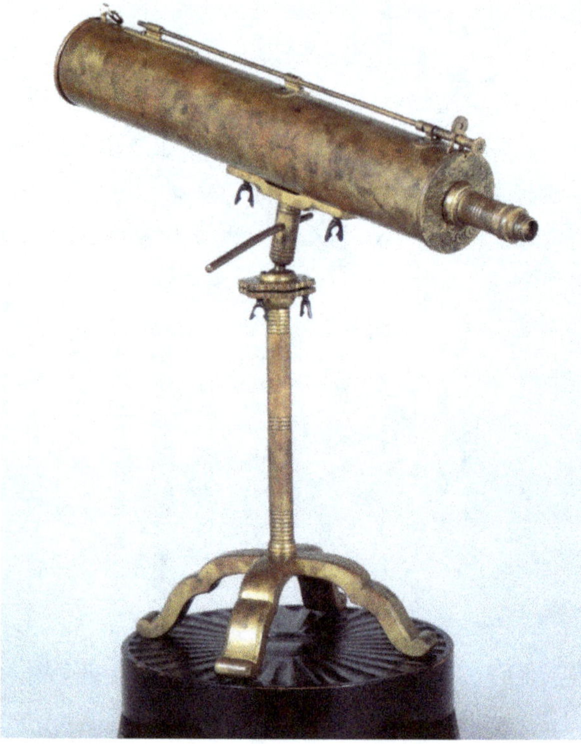

FIGURE 3.4
The first Gregorian reflector telescope produced by Tobei Kunitomo.
COURTESY OF UEDA CITY MUSEUM.

selling them. An interesting thing is that, Foucault tests of the existing four telescopes showed that all the primary mirrors had the parabolic surface close to that of modern telescopes.[13] This fact indicates that, contrary to the past skeptical opinions, Kunitomo had surely established the knowhow in paraboloid polishing of the bronze mirror.

If that is the case, this finding brings about a puzzling question about how Kunitomo could master the means of the parabolic mirror polishing. There was no such technical background at all in Japan of the Kunitomo's era, other than producing traditional refractor telescopes by combining eye-glass-like lenses. Since he was basically an uneducated artisan from the countryside, he could of course read neither Dutch books imported at Nagasaki, nor had he intellectual tutors who were familiar with the knowledge of European science and technology.[14] Because of the reasons, his accomplishment in producing Gregorian reflectors has often been attributed to his ingenuity and incessant

13 Tomita, "Investigation of the Gregorian telescopes", 9-41 and his private communication.

14 Tsuko Nakamura, "Telescope making and its Western influence in the modern history of Japan: Kunitomo's reflector telescopes," *Proceedings of the Third International Conference*

invention so far, but this is really one of the unanswered enigmas relating to Kunitomo.

Kunitomo's workshop notebooks tell us that during the course of developing his reflector telescopes, he frequently pointed them to the Sun and the Moon. He repeated such tests in order to examine the completeness of the parabolic surface of the primary mirror by checking the image quality of celestial bodies. However, he made such observations not only for the testing purpose of his products. His documents obviously show that he bore also a strong motive for investigating the true nature of the Sun, the Moon and planets. Such a typical example is his solar observations.[15] His sketches of the solar disk and attached memoranda show that he distinguished the penumbra of the sunspot from the umbra, indicating his carefulness in observations and superiority of his telescopes. His most conspicuous achievements were that he left consecutive sketches of the sunspots (158 days), ranging over 13 months from the beginning of 1835; just nine years after S.H. Schwabe of Germany started continuous monitoring observations of sunspots in an attempt to discover the then hypothesized near-Sun planet Vulcan. Needless to say, Kunitomo did not know the Schwabe's project at all.

Analysis of the Kunitmo's words cited from his book collection suggests that he probably wanted to confirm the possible seasonal variation of the sunspot number, which was written in an introductory book of astronomy he read.

The Moon was another target of Kunitomo's frequent observations. The next image shows Kunitomo's sketches of the Moon drawn in September of 1836 (Figure 3.5).

Explanations of the drawings state that the observations were made on the ninth age of the lunar phase. However, we must call attention to numerous small craters painted far away from the terminator in Figure 3.5, where it is generally difficult to recognize such craters near the half moon because of the uncontrasted visibility. Therefore, it will be reasonable to suppose that both the figures, the right one in particular, were not the sketches for one night but composed from a few nights observations. Nevertheless, detailed features of Figure 3.5 remind us that Kunitomo's sketches are much more objective compared with those by other Japanese like Asada and Iwahashi.

---

on Oriental Astronomy (2000): 135-142 (Fukuoka: Fukuoka University of Education, Ed. M. Hirai).

15 Issei Yamamoto, "Kunitomo and his astronomical activities in the pre-Meizi era," *Isis*, 26/2 (1937) 330-335.

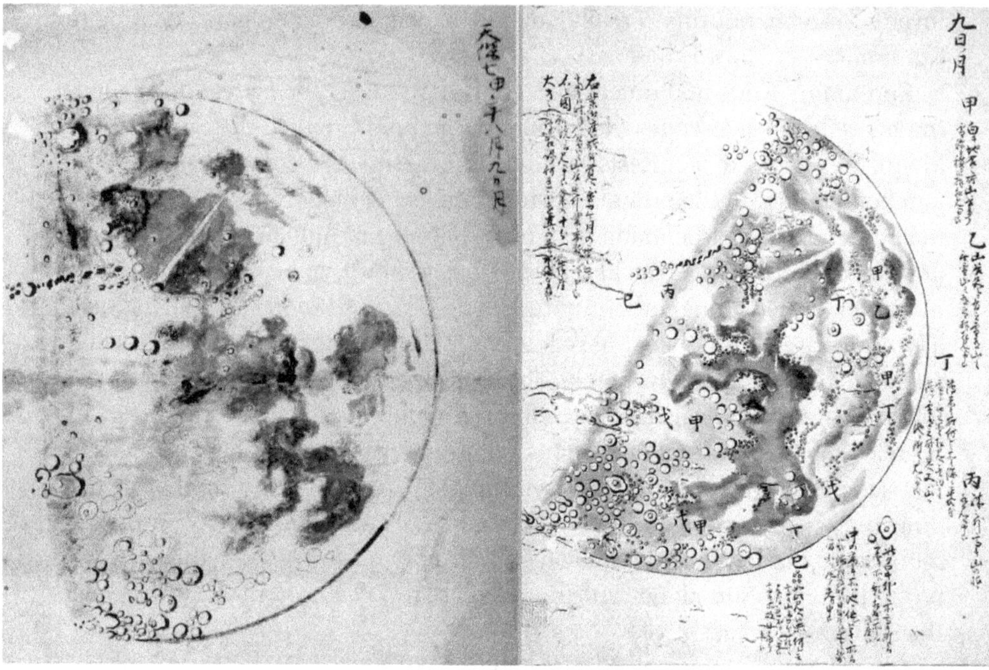

FIGURE 3.5 Kunitomo's sketches of the lunar disk observed in September of 1836, NAGAHAMA-JO CITY MUSEUM

4 Lunar Maps Made by Feudal Samurai Lords

Although in Japan there exist some other records of lunar observations made during the 1790s-1840s, all of them including Kunitomo's are basically no more than sketches made on a single night. On the other hand, what is introduced in this Section is not a sketch of the Moon but a genuine lunar map.[16]

4.1 *The Lunar Map* Taiiin no Zu *Drawn by Masatami Hotta*

Figure 3.6 shows the lunar map now owned by a local collector in Kumamoto, Kyushu. This historical object, entitled "*Taiin no Zu*" (Figure of the Moon) was painted in the traditional style of a hanging screen: an upper part with a figure and a section below containing an explanation in Chinese. The explanation here says that this map was drawn using a telescope in September 1813, proba-

16 Tsuko Nakamura, "A lunar map made in 1813 by a Japanese feudal warlord," in *Mapping the Oriental Sky*: Proceedings of the 7th International Conference on Oriental Astronomy, ed. Tsuko Nakamura et al. (Tokyo: National Astronomical Observatory of Japan, 2011), 82-87.

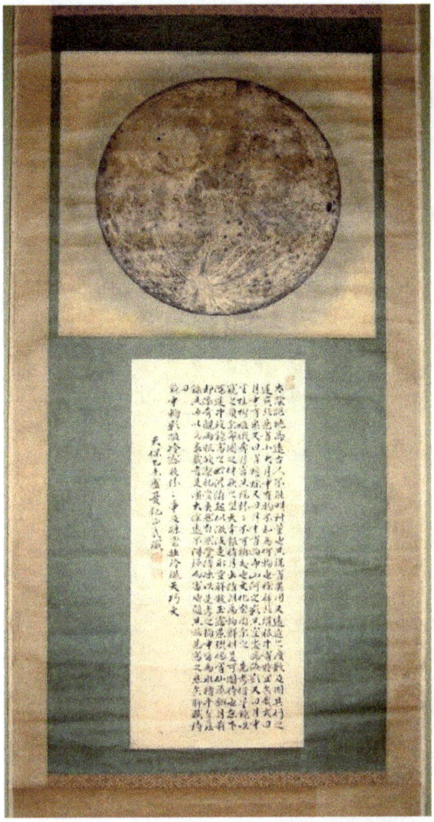

FIGURE 3.6
The lunar map painted in the Japanese traditional style of a hanging screen. This was observed and composed by Masatami Hotta in September 1813. Below the map is an explanation in Chinese with Masatami's signature, whose sentences were written in 1835; the reason of such a delay is unknown.
COURTESY OF S. SHIMADA.

bly in Edo (Tokyo), by Masatami Hotta , a feudal lord who governed a small district close to Kunitomo's village.

The map was made in response to a request from Masazane, who was Masatami's father. In spite of the lengthy sentences, most of them consist of decorative and literary expressions and there is little scientific description. The diameter of the lunar globe measures 38.5cm, and the surface features, the craters, mountain ranges and the high-land and low-land (mare) regions, are very delicately and elaborately drawn in blue, silver, and green, although those colors now look fairly dull as a result of aging.

In Figure 3.7 one can notice the famous crater Plato as a distinct ellipse on the upper right, and very numerous tiny craters in the high-land regions. Very conspicuous are two bright-ray craters, namely Kepler on the lower left and Copernicus on the lower right.

Figure 3.8 represents the lower left part of the lunar disk. Here we can see the famous crater Tycho, with its magnificent ray system extending up to the region of the maria.

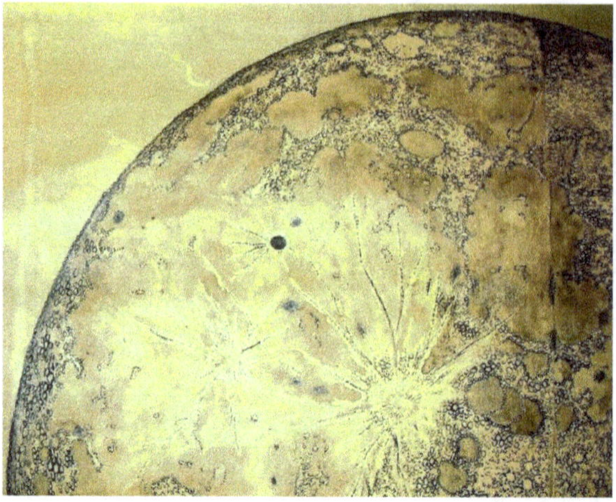

FIGURE 3.7
The upper left quarter of Taiin no Zu.
COURTESY OF S. SHIMADA.

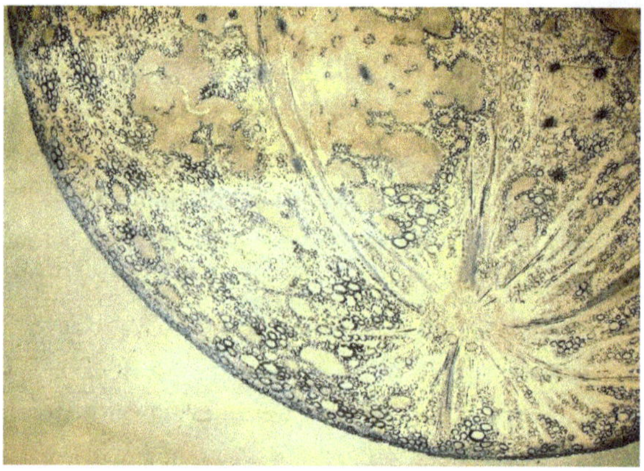

FIGURE 3.8
The lower left quarter of Taiin no Zu.
COURTESY OF S. SHIMADA.

From figures 3.7 and 3.8, it is highly likely that Masatami used an imported telescope with a high magnification and one that gave an erect image. Since it is well known that in the latter half of the eighteenth century the British optician James Short manufactured many portable Gregorian reflecting telescopes[17] similar to the one produced by Kunitomo and some of these were exported worldwide (including to Japan), we speculate that Masatami used such a telescope, or a comparable one made by Dutch artisans.[18]

17 Harriet Wynter and A.J. Turner, *Scientific Instruments* (London: Studio Vista, 1975).

18 Huib J. Zuidervaart, "The reflecting telescope in the Netherlands," *Annals of Science* 61 (2004): 407-452.

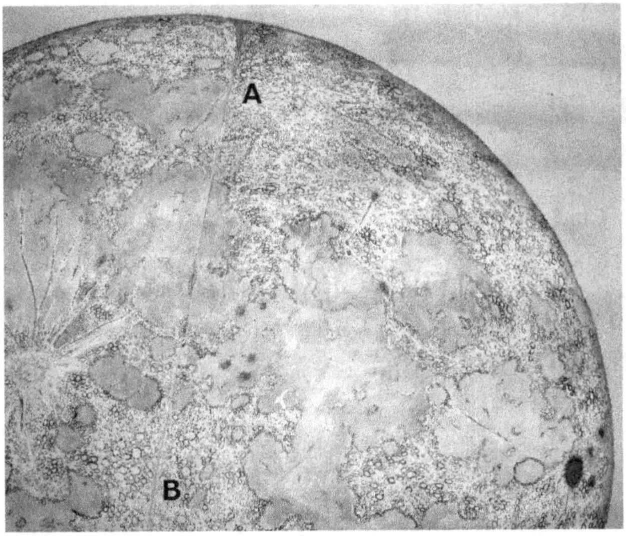

FIGURE 3.9
The upper half quarter of the Moon, showing the line (AB) jointing the two halves of the Moon.
COURTESY OF S. SHIMADA.

The Chinese explanation in figure 3.6 contains Masatami's signature and the date 1835, and states that he painted this lunar map during a single night's observation of the full moon in September 1813. However this statement is doubtful. Those who have some experience in telescopic observations of the Moon know that at full moon the lunar surfaces have least contrast, so it would be impossible to detect some of the detailed features painted by Masatami. Therefore, we infer that in reality he assembled this map from sketches made at multiple lunar phases.

Actually there is clear evidence that supports this inference. Upon scrutinizing the *Taiin no Zu*, one can see a line that connects two halves of the Moon (the line AB in Figure 3.9), and on both sides of this line the features drawn seem to be discontinuous.

This strongly suggests that each half-moon was observed at different phases and then the two were later combined into a full disk. In that sense, this diagram is surely not a simple sketch of the Moon but is a lunar map.

In Europe, it was M.F. van Langren, a Dutch astronomer and cartographer, who in 1645 published the earliest lunar map produced from observations at different phases, and the general scheme of feature depiction and nomenclature adopted in his map have basically been retained up to the present day.[19] Lunar map-making after van Langren, represented by astronomers like Jo-

19 Ewan A. Whitaker, "Selenography in the seventeenth century," in *The General History of Astronomy.* Volume. 2: Planetary Astronomy from the Renaissance to the Rise of Astro-

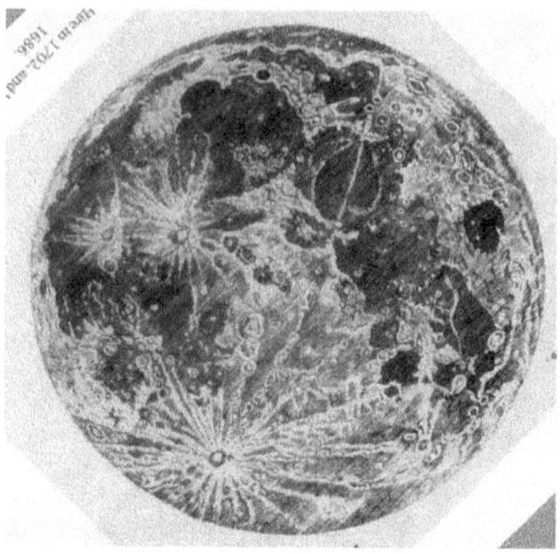

FIGURE 3.10
The lunar map by La Hire (1702). For easy comparison with other lunar maps, the figure shown here is rotated by 135 degrees from the original one.
Note: Whitaker 1989.

hannes Hevelius, Giovanni Battista Riccioli and Giovanni Domenico Cassini, gave birth to a new field of research called 'selenography' in the second half of the seventeenth century.[20] Since Masatami's lunar map was drawn nearly 150 years after these European activities, one may consider this map not so valuable from an historical point of view. However, we should also take into account that Masatami was totally ignorant of the astronomical situation in Europe – because of Japan's seclusion policy at this time – and he painted his lunar map independently and out of pure curiosity.

With this background in mind, we attempt here to compare the *Taiin no Zu* with some typical European lunar maps produced during the period 1650-1720. Figure 3.10 shows an example of a European lunar map with nicely-drawn ray craters.

This was produced by the French astronomer, Philippe de La Hire, and included in his Astronomical Tables, which were published in 1702. Upon comparing this lunar map with the painting by Masatami, it can be seen that the Masatami's lunar figure almost rivals the best of the European lunar maps, from both an artistic and a descriptive viewpoint. At the same time, however, it is worth noting that the bright ray craters were depicted differently by different observers even in Europe; for example, the lunar maps by Riccioli (1651) and Cassini (1679) did not show Tycho's ray system so clearly.[21]

Now we wonder how precisely Masatami's map is drawn from a scientific point of view. At this time in Japan there was a spiritual consensus among

physics. Part A: Tycho Brahe to Newton, ed. René Taton et al. (Cambridge, Cambridge University Press, 1989), 119-143.

20 Whitaker, "Selenography in the seventeenth century," 119-143.

21 e.g. see the photographs in Whitaker 1989.

Japanese intellectuals and scholars who were interested in so-called natural history, that they should describe natural things – like fauna and flora – by drawing them 'true to life'. Although this spirit may be somewhat different from the Western scientific attitude, there is no doubt that Masatami made efforts to record the Moon as realistic as possible.

In order to examine the positional accuracy of Masatami's lunar map, a comparison was made with an image generated by the planetarium software 'Stellar Navigator'[22] for 8 September 1813 (Gregorian calendar), which corresponds to the date of Masatami's alleged observations. Although we have had no chance yet to confirm how accurate the software is, it seemed that this software takes account of the lunar libration because it produces lunar images with different orientations for different dates.

It was found that Masatami's positions of ray craters like Tycho and Copernicus were fairly accurate, but the maria do not fit well. Also, the location of Plato crater was shifted to the left, and its size was exaggerated. However, the enlargement of specific features can also sometimes be seen in European lunar maps from the second half of the seventeenth century, probably because observers tended to be more affected by conspicuous features like Plato. As for the overall positional inaccuracy, it will not be fair for us to blame Masatami, because he was not a scientist. He was a painter after all, and he had no means of accurately measuring positions on the lunar disk.

Another possible cause of the geographical inconsistency in the Masatami's map could be that he actually made his observations not at full moon in September 1813 but on different days and at different phases, before combining them into a lunar map. In any case, the production of lunar maps like the one made by Masatami had never been attempted previously in Asia in pre-modern times, so we believe that *Taiin no Zu* represents an important achievement in the history of selenography.

4.2 *What Motivated Masatami to Draw His Lunar Map?*

It is well known during the latter half of the seventeenth century in Europe selenographical activities were motivated by two points.[23] The first was to accurately pinpoint the positions of minute but conspicuous lunar features, so they could serve as a 'common clock' and be used to determine the longitudes of places on the Earth. The second motivation emerged later, when the Moon was regarded as a habitable or creature-bearing world, so that astronomers needed a geographical map of our nearest celestial neighbor.

---

22 AstroArts, *Stellar Navigator,* ver. 6.0 (Tokyo: Astro Arts Inc, 1998).

23 Whitaker, "Selenography in the seventeenth century," 119-143.

Then, what motivated Masatami to draw such a detailed lunar map? Although we cannot of course be certain, there exist some clues. The existence of this map was first revealed to the Japanese astronomical community before World War II,[24] when relevant materials and references were unavailable. But our literature survey has revealed that Masatami's father (Masazane) also had made a vassal of a different feudal lord draw another lunar map for him, in 1799[25] – although its current whereabouts is unknown. According to Hukui, who recorded the explanation attached to the 1799 map, Masazane noted with dissatisfaction that he was then using a telescope that gave inverted images. This could be a reason why he made his son draw a new lunar map in 1813, but with an erect orientation. This episode also suggests that it was Masazane, rather than Masatami, who was more enthusiastic about producing lunar maps.

During the period from 1790 to 1840 Japanese interest in European science and arts experienced rapid growth. Some of the Dutch interpreters at Nagasaki (where the Japan branch of the Dutch East India Company was located) and pioneering local scholars imported Dutch books, read them and began writing and publishing books on novel Western things and affairs. Also affected by such developments were people who could not read Dutch but had a strong interest in Western learning. Among them, Shiba Kokan was a typical figure. He was a well-known painter at that time who mastered the technique of Western oil painting, and became famous for his invention of the copper-plate print. He also contributed to popularizing Copernicus' heliocentric system of the Solar System among the Japanese people.[26]

Around 1796 Siba published *Tenkyu Zenzu* (Figure of the Celestial Sphere), a series of colored copperplate engravings, which were widely welcomed by the Japanese of the Edo period. They included a few images of the Sun, the Moon and the celestial sphere, which were modified copies by Shiba, taken from the Western sources.

Among them, there was a plate of the Moon (Figure 3.11, left) entitled '*Getsurin Shinkei Zu*' (*True Moon*), which much impressed the Japanese because they had never before seen such a detailed figure of the Moon. In light of the

24 Kazuyuki Ikeda, "Bunka Nenkan" "*Taiin no Zu*" (The lunar map drawn in 1813, in Japanese), *Tenkai* 123 (1931): 337-340.

25 Kyuzo Hukui. 1937. *Shodaimyo no Bungei to Gakujutsu* (Literature and Scholastic Achievements by Daimyo Warlords, in Japanese) (Tokyo: unknown publisher, 1937): 336-337. Reprinted by Hara Shobo Inc. (1976).

26 Kokan Shiba, *Shiba Kokan Zenshu* (Collected Works of Shiba Kokan, in Japanese, four volumes, reproduced version) (Kyoto: Yasaka Shobo Inc, 1992-1993).

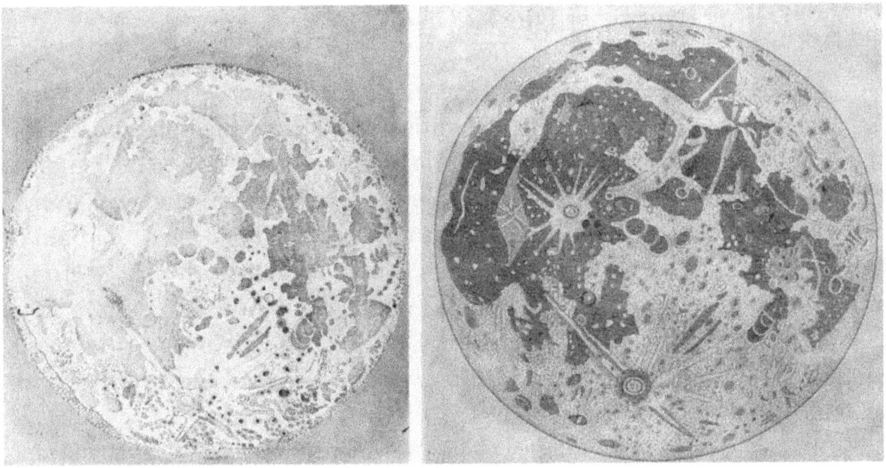

FIGURE 3.11 (left): Copperplate engraving '*Getsurin Shinkei Zu*' (True Moon) by Shiba Kokan (1796), and (right): The figure of the Moon cited in the book 'Underground World' published by Athanacius Kircher in 1664.

historical circumstances of that time, it is likely that Masazane was also influenced by this popular *True Moon,* because Masatami's first lunar map was drawn just three years after Shiba's publication.[27] We suggest that the *True Moon* provided the motivation for Masazane to confirm by himself the reality of the Shiba's print by observing the Moon with his own telescope.

It has recently been revealed[28] that Shiba's True Moon was actually based on an image in the book *Mundus Subterraneus* (Underground World) which was published in 1664 by Athanacius Kircher[29], a German Jesuit scholar and naturalist (see Figure 3.11. right). Also it has been well known that the Kircher's image was based on the lunar map made in 1655 by the eminent Jesuit astronomer Christoph Scheiner.

27 Kokan Shiba, *Tenkyu Zenzu* (Figure of the Celestial Sphere, in Japanese), 12 plates. (Tokyo: Shunparo, c.1796).

28 Kobe City Museum, *Shiba Kokan Hyakka Jiten* (Encyclopedia of Shiba Kokan, in Japanese). (Kobe: Kobe City Museum, 1998).

29 Athanacius Kircher, *Mundus Subterraneus, quo Universae Denique Naturae Divitiae* (Underground World, in Latin) (Amsterdam: Janssonio Waesbergiana, 1664-1678).

5 Early Photographs of the Moon

During the 1860s Japan experienced the civil war, conflicts between the Shogunal government and opposing feudal lords backed up by the Emperor. In 1868 the Shogunate finally surrendered and the political power went to the hands of the Meiji government, which is called the Meiji Restoration. The new government soon abandoned the national seclusion policy that had been maintained by the *Ancien Régime* for about 230 years, and rapidly introduced Western institutions in various fields of the society. As a result, the Western culture, educational system, science and technology flooded into Japan like a tsunami, along with European and American diplomats, teachers, and overseas merchants.

In 1874 a very rare astronomical event took place over the Far East including Japan. It is called the transit of Venus; a phenomenon that planet Venus passes across the solar disk, whose elaborate measurements allow us to determine a very precise solar parallax (a mean distance between the Sun and the Earth). Thus France, US and Mexico dispatched expedition teams to Japan for the observations. Thanks to this unparalleled fortune having received researchers from advanced countries in astronomy, the Japanese (both professional and non-professional) could learn and master three fundamental observation techniques in modern astronomy, namely the exact determination of astronomical latitude and longitude for a place, the time synchronization of two remote sites through the electric cable communication, and astronomical use of photography.[30]

Considering the above mentioned situations, it is likely that by around the end of the 1870s the Japanese had taken photographs of the solar and lunar disks through telescopes. Nevertheless, such materials are not known to exist. A possibility is that, for professional astronomers, simple photographs of the Sun and the Moon were not a target of serious science but no more than an observational testing, so they did not preserve them as astronomical archives. On the other hand, for ordinary photographers, astronomical images might not be a topic of interest after all.

In the case of the Moon, our literature survey found out a low quality lunar photograph shown in figure 3.12. This was included in the book *Seigaku Shokei*

30 e.g., Jules Janssen, "Mission du Japon pour l'observation du passage de Vénus," *l'Annuaire du Bureau des Longitudes de 1876* (1876): 572-588; Kuniji Saito and Shinozawa Shizuyo, "On the observations of the transit of Venus over the Sun with particular emphasis on the December 9, 1847 event observed in Japan: Part I," *Tokyo Astronomical Observatory Report*, 16/1 (1972): 72-162, 73; and Part II, 16/2 (1973): 259-385.

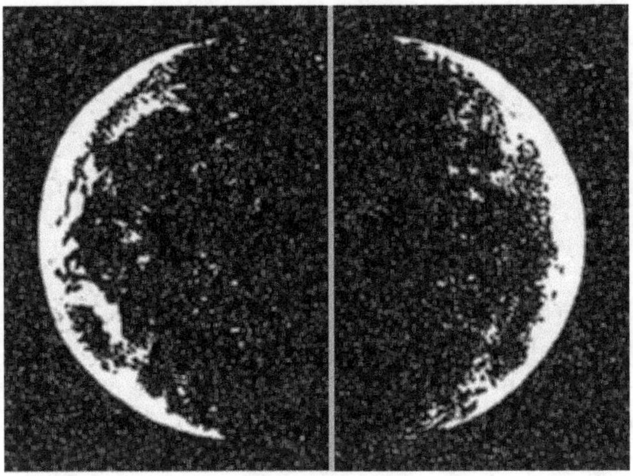

FIGURE 3.12
Early photographs of the Moon taken by the Japanese in or before 1875.

(Introduction to Astronomy) published by the Ministry of Education.[31] The explanation simply mentions that this photograph was taken in Japan, with neither the date and place of observation, nor information of photographic emulsion. Although the resolution of the photograph is fairly low, one can identify the Mare Crisium on the lower left and some other major features. This photograph could be an outcome learned from the experience in the Venus transit expedition of 1874.

6 Concluding Remarks

Finally we would like to add some comments based on a wider perspective than just the history of astronomy. As mentioned in the previous sections, during the period from the 1790s to 1840s in Japan researches in the field of the natural history became very popular among both specialists and dilettantes (amateurs), and at this time intellectual daimyos were also engaged in recording colorful and detailed drawings of plants, animals, insects, etc. Masatami, for instance, left us a beautifully-illustrated encyclopedia of butterflies and moths, '*Seicho hu*', which is now preserved in the National Archives of Japan.

Therefore, the enthusiasm of Masazane and Masatami to make lunar maps at this time can best be understood within this framework, which arose from a strong curiosity towards the wonders of Nature.

31 Naruo Sekidou, *Seigaku Shokei* (Introduction to Astronomy, in Japanese) (Tokyo: Ministry of Education, 1875), 76.

CHAPTER 4

Of Blurs, Maps and Portraits

Photography and the Moon

*Charlotte Bigg**

In the beginning was the blur
... this darkness I was looking at began to pale. I distinguished, what? Impossible to say. It was murky, fleeting, impalpable to the eye, so to speak. If nothing had a shape, this would be it.

VICTOR HUGO *on observing the Moon under Arago's guidance at the Paris Observatory in 1834*)[1]

•••

M. Daguerre has thrown the image of the Moon, formed at the focus of a mediocre lens, on one of his screens, and it left there a clear, white trace ... M. Daguerre will have been the first person to produce a sensible chemical modification with the help of our satellite's light rays.

FRANÇOIS ARAGO, *1839*[2]

•••

---

* This paper in part develops points made in Charlotte Bigg, «Sciences du portrait céleste,» *Tout ce que le ciel permet (en cinéma, peinture, photographie et vidéo)*, ed. Barbara Le Maitre and Bruno Nassim Aboudrar, collection Théorème nr. 24 (Paris: Presses de la Sorbonne Nouvelle, 2015), 119-126. For their feedback and suggestions, I am grateful to Carmen Pérez González as well as to the organisers and participants of events where earlier versions of this work were presented: *Images du ciel en cinéma, en peinture, en photographie,* conference at INHA Paris; *History of Amateur Astronomy: current research, future prospects*, workshop at the Swedish Royal Academy of Sciences/ Stockholm Observatory Museum; Oberseminar of the Abteilung für Wissenschaftsgeschichte, Ludwig Maximilian University, Munich.

1 "Cette noirceur que je regardais commença à blémir. Je distinguai, quoi? Impossible de le dire. C'était trouble, fugace, impalpable à l'oeil, pour ainsi parler. Si rien avait une forme, ce serait cela." Victor Hugo, *Le promontoire du Songe* (1863), pref. Annie Lebrun (Paris: Gallimard, 2013), 20. All translations are my own unless otherwise indicated.

2 "M. Daguerre a jeté l'image de la Lune, formée au foyer d'une médiocre lentille, sur un de ses écrans, et elle y a laissé une empreinte blanche évidente... M. Daguerre aura été le premier à produire une modification chimique sensible à l'aide des rayons lumineux de notre satellite." François Arago, "Fixation des images qui se forment au foyer d'une chambre obscure," *Comptes Rendus Hebdomadaires des Séances de l'Académie des Sciences* 8 (1839): 4-7, on 6.

 | DOI 10.1163/9789004298873_006

> On the very morning I left Paris, the ingenious man triumphantly brought to Arago's bedside (my radical! friend, one of the most amiable men on Earth, was unwell) the image of the Moon, its edges somewhat indistinct because the camera obscura had not been shifted fast enough to follow the motion of the Moon.
>
> ALEXANDER VON HUMBOLDT, *1839*[3]

•••

> Dr. Draper announced that he had succeeded in getting a representation of the Moon's surface by the daguerreotype. A portion of the figure was very distinct but owing to the motion of the Moon the greater part was confused. The time occupied was twenty minutes and the size of the figure was about one inch in diameter. Daguerre had attempted the same thing, but did not succeed. This is the first time that anything like a distinct representation of the Moon's surface has been obtained.
>
> WILLIAM DRAPER, *Report of a presentation, 1840*[4]

∵

Lost daguerreotypes figuring the Moon as a small white blur usually play the role of the first "firsts" in the history of astrophotography, the prelude to a succession of increasingly sharp, detailed and reliable representations of the Moon and other celestial bodies (see figures 4.1a and 4.1b).

Good arguments could be brought to support the view that historians of photography have already spent far too much attention to its beginnings, when so many of its later developments, especially with regards to the sciences, remain utterly virgin territory. (This fatigue is not new. Only a month

---

3 "Noch den Morgen, an dem ich Paris verliess, brachte der kunstreiche Mann vor Arago's Bette (mein radicaler! Freund, einer der liebenswürdigsten Menschen der Erde war unwohl) triumphirend das Mondbild, dessen Ränder darum etwas unbestimmt waren, weil die Camera obscura nicht schnell genug verschoben worden war, um der Bewegung des Mondes zu folgen." Letter from Alexander von Humboldt to the duchess Friederike von Anhalt Dessau, dated 7 February 1839, reproduced in Steffan Siegel, ed., *Neues Licht. Daguerre, Talbot und die Veröffentlichung der Fotografie im Jahr 1839* (Paderborn: Wilhelm Fink, 2014), 78-81, on 79.

4 *Minutes of the Lyceum of Natural History,* 23 March 1840, quoted in M. Susan Barger and William B. White, *The Daguerreotype. Nineteenth-Century Technology and Modern Science* (Baltimore and London: The Johns Hopkins University Press, 1991), 85.

after witnessing Daguerre's Moon, Alexander von Humboldt wrote to Francois Arago in a sigh that "M. Daguerre est d'ailleurs mon Chimborazo. Quarante fois par jour on me fait les mêmes questions."[5])

Yet the stakes of revisiting these narratives should not be underestimated, since, to quote Geoffrey Batchen, "any questioning of photography's beginnings therefore also represents a questioning of the trajectory of photography's history as a whole." Batchen sought to circumvent accounts focussing on discoveries and the disputes they gave rise to among actors (as the competing claims of Humboldt and William Draper in the quotations above already suggest) and historians after them. He proposed instead to situate the beginning of photography not in any individual acts but at the moment when "the discursive desire to photograph emerge(d)" and became "a demonstrably widespread, social imperative".[6] Though one might wonder whether this does not simply substitute a discursive determinism to a technological one, nonetheless this approach has the merit of opening up the history of early photography beyond a strict focus on technology. Taking a complementary stance, a number of recent publications have overhauled established stories and perspectives. Fine studies on Henry Fox Talbot and Daguerre, in particular, have helped put early photography back in its proper context, sometimes relativized its contemporary significance, and grasped its alternative meanings.[7]

Another approach might be to take blur seriously. Early daguerreotypes of the Moon are essentially defined by their indistinct, "unbestimmt" (Humboldt), "confused" (Draper) appearance.[8] Even though first photographs of the Moon had very few witnesses and were rapidly lost (often in fires), they have become an indispensable part of the story. Poor resolution, to use modern ter-

5 Alexander von Humboldt to François Arago, February 1839, in *Correspondance d'Alexandre de Humboldt avec François Arago (1809-1853)*, preface E.-T. Hamy (Paris: Guilmoto, 1908), 178.

6 Geoffrey Batchen, "Desiring Production," in *Each Wild Idea. Writing, Photography, History*, ed. Geoffrey Batchen (Cambridge, Mass. and London: The MIT Press, 2001), 3-24, on 3 and 5.

7 Larry Schaaf, *Out of the shadows: Herschel, Talbot and the invention of photography* (New Haven and London: Yale University Press, 1992); Stephen Pinson, *Speculating Daguerre. Art and Enterprise in the work of L.J.M. Daguerre* (Chicago: University of Chicago Press, 2012); Mirjam Brusius, Katrina Dean and Chitra Ramalingam, eds., *William Henry Fox Talbot. Beyond Photography* (New Haven: Yale University Press, 2013); Mirjam Brusius, *Fotografie und museales Wissen. William Henry Fox Talbot, das Altertum und die Absenz der Fotografie* (Berlin, Boston: De Gruyter, 2015); Tanya Sheehan and Andres Zervigon, eds., *Photography and its Origins* (New York: Routledge, 2015).

8 On Draper and his long-standing interest in astronomical photography, see Deborah Jean Warner, "The Draper family material: National Museum of American History," *History of Photography* 24 (2000), 16-23.

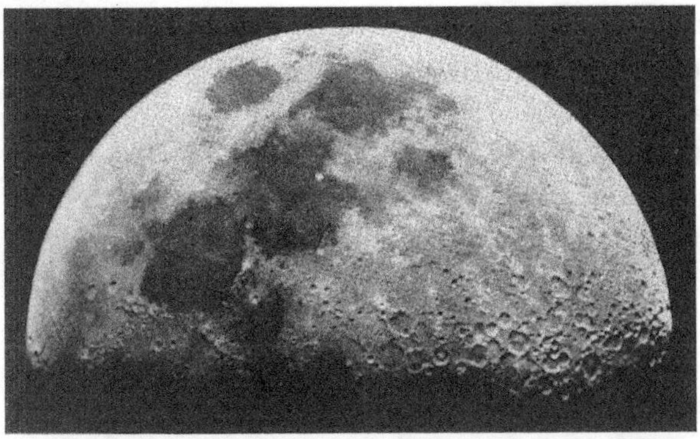

FIGURE 4.1
Fig. 4.1a "By A. A. Common on April 21, 1877, on a wet collodion plate with his 18-inch telescope." Gérard de Vaucouleurs, *Astronomical Photography. From the Daguerreotype to the Electron Camera* (London: Faber and Faber, 1961).

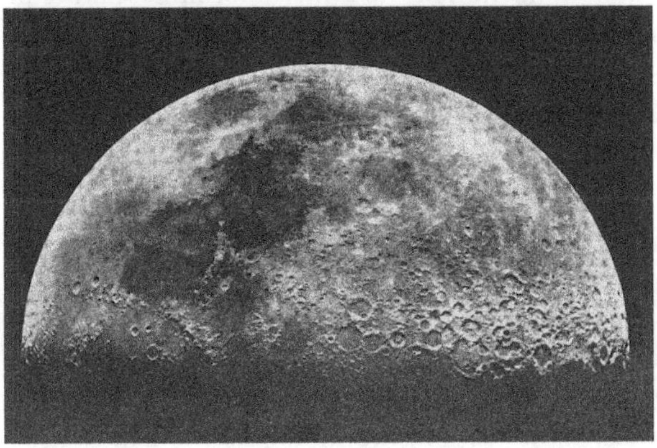

Fig 4.1b "By Loewy and Puiseux on February 12, 1894, on a dry plate with the great 60-cm (24-inch) equatorial coudé at the Paris. Observatory. A comparison of these two views reveals a significant new advance". Plate 4, Gérard de Vaucouleurs, *Astronomical Photography. From the Daguerreotype to the Electron Camera* (London: Faber and Faber, 1961).

minology, is their defining feature and is even often retrospectively exaggerated in the accounts of later practitioners and historians[9] (though it is never valued, as out of focus photography has sometimes been in artistic contexts).[10]

Blurriness is in fact a surprisingly widespread characteristic of "first" images in the realm of scientific imaging, viz. the early images of radioactivity, X-ray diffraction or electron microscopy. It may be taken as a fundamental property of visualizing technologies in the making. The indistinct image, at first decipherable only by its maker, functions as a testimony of a new technique and does not yet need to represent anything else than itself, its existence. Its imper-

9 Daniel Norman for instance wrongly attributes to Arago the claim that Daguerre's Moon plates "were so lacking in sensitivity that even a very long exposure gave only a faint image, completely lacking in detail." Daniel Norman, "The Development of Astronomical Photography," *Osiris* 5 (1938): 560-594, on 560.

10 See, e.g., Mirjam Brusius, "Impreciseness in Julia Margaret Cameron's Portrait Photographs," *History of Photography* 34 (2010): 342-355.

fect pictorial or representational qualities seem to entail the promise of technical improvement. The blurry image is seen to open a horizon of possibilities, all sharper and more definite; as if photography's indexical properties (Arago's "empreinte") preceded and anticipated its other functions.

This is not to say that subsequent photographs were never blurry, of course, but rather that, unlike "first blurs", these have usually been discarded and have vanished in the dustbins of history. As Peter Geimer has shown, photographic accidents and unintended effects are ubiquitous yet usually omitted from histories of photography. Taking them into account gives quite a different view on photography and its history.[11] Lost blurry daguerreotypes of the Moon may then be seen, perhaps, as a materialization of the underdetermined nature of photography itself. It was not a straightforward medium of representation but the locus of optical and chemical experimentation, and a surface of projection for competing conceptions of what the new medium was, was good for and was to become.

1 Photography and Astronomy

Mirjam Brusius makes this argument with respect to Talbot's photographic process as well as to two connected enterprises, the new science of archeology and the British Museum.[12] Likewise, in the case of astrophotography, we must keep in mind that photography also provoked a reflection on the nature of astronomy itself. Rather than studying how photography was *applied* to astronomy,[13] we then need to understand how photography and astronomy mutually made each other: in the elaboration of what came to be known as astrophysics, for instance; or in establishing the idea of photography as a reliable witness and scientific instrument.[14] The profound involvement of the most significant astronomers of their generation, beginning with John Herschel, François Arago

11 Peter Geimer, *Bilder aus Versehen. Eine Geschichte photographischer Erscheinungen* (Hamburg: Philo Fine arts, 2010). See also his historiographical survey on the indexicality of photography: Peter Geimer, *Theorien der Photografie zur Einführung* (Hamburg: Junius, 2009), 13-69.

12 Brusius, *Fotografie*, 11-15.

13 As in (the otherwise very competent) John Lankford, "The impact of photography on astronomy" in : *The General History of Astronomy. Vol. 4. Astrophysics and twentieth-century astronomy to 1950: Part A*, ed. Owen Gingerich (Cambridge: Cambridge University Press, 1984), 16-38.

14 On science as an important locus for establishing photography as a reliable tool see Kelley Wilder, *Photography and Science* (London: Reaktion Books, 2009).

and Alexander von Humboldt, working respectively with Talbot and Daguerre, confers a particular historical significance to the interactions of photographers and astronomers since the 1830s, and explains why, of all possible objects, the Moon was chosen as one of the first to figure on a daguerreotype.

For the French context, François Brunet has shown how decisive Arago's 1839 speeches to the French Academies were, not only in promoting and popularizing Daguerre's work but also in formulating an "idea of photography" as an instrument of scientific observation in place of the investigator's trained eye, an exact image because it was natural (thus erasing its technical nature),[15] along the lines of what was later described as "mechanical objectivity."[16] This idea (an exceedingly difficult one to realize in practice) was pursued among others by Jules Janssen, founder and director of the Meudon Observatory as well as for some years president of the Société Française de Photographie.[17]

Conversely, one cannot overestimate the transformative of power of photography on astronomy's aims, practices and social organisation. In 1922, Benjamin Baillaud, Arago's distant successor at the head of the Paris Observatory and president of the newly founded International Astronomical Union, claimed that photography counted among the most important tools of discovery of the previous fifty years in astronomy.[18]

15 François Brunet, *La naissance de l'idée de photographie* (Paris : Presses Universitaires de France, 2000). See also André Gunthert, "Spectres de la photographie. Arago et la divulgation du daguerréotype", in *Les Arago, acteurs de leur temps* (Perpignan: Archives départementales des Pyrénées-orientales, 2010), 441-451. Others have suggested that Arago's idea of photography was more complex, even contradictory. For John Tresch, the astronomer considered the daguerreotype as a "romantic machine." John Tresch, *The Romantic Machine: Utopian Science and Technology after Napoleon* (Chicago & London: The University of Chicago Press, 2012), 89-122.

16 On mechanical objectivity, see Lorraine Daston and Peter Galison, *Objectivity* (New York: Zone Books, 2007); on its limits in practice, see Alex Soojung-Kim Pang, "'Stars should henceforth register themselves'. Astrophotography at the Early Lick Observatory," *British Journal for the History of Science* 30 (1997): 177- 202.

17 André Gunthert, "La rétine du savant. La fonction heuristique de la photographie," *Etudes Photographiques* 7 (2000): 28-48; Françoise Launay, "Jules Janssen et la photographie," *Dans le champ des étoiles. Les photographes et le ciel, 1850-2000* (Paris: Editions de la Réunion des Musées Nationaux, 2000), 22-31.

18 "Address by M.M. Benjamin Baillaud, President of the International Astronomical Union," *Transactions of the International Astronomical Union* vol 1 (1922), 134. On the social reorganisation connected to the adoption of photography in observatories, see Charlotte Bigg, "Photography and the Labour History of Astronomy and Astrophysics," *The Role of Visual Representations in Astronomy: History and Research practice,* special issue ed. Klaus Hentschel and Axel Wittmann, *Acta Historica Astronomiae,* 9 (2000): 90-106.

This does not mean that the development of astrophotographic techniques was obvious or easy. Astronomy presented major technical challenges for photography: mobile objects, extreme distances, extraordinary low or strong luminosities, exceptional requirements of precision and reliability. But astronomy also supplied instrumental, theoretical and institutional resources for addressing these challenges. Whilst emulsion chemistry was usually quite foreign to astronomers before spectral analysis was adopted in observatories, astronomers were acknowledged experts in theoretical and practical optics. The heavens constituted for photography a field of experimentation and a source of technical invention for astronomers, photographers and the burgeoning photographic industry working together in close collaboration from the nineteenth century onwards.

At the same time, because astronomy long constituted a model of scientific observation for the empirical sciences, and because of the exceptional refinement of bodily and instrumental techniques involved in astronomy's precision work, in this field perhaps more than in others, photography had to prove that it could fulfil expectations, especially in the production of numerical data. This expectation was frequently disappointed, as witnessed by the very mixed success of the attempts to photograph eclipses or Venus transits between 1840 and the 1880s.[19] But even in later decades, when technical development had greatly improved the reliability of photography for most purposes, astrophotography was beset with difficulties. The Carte du Ciel project inaugurated in 1887 to capture photographically the state of the whole sky and constitute the largest star catalogue to date was so protracted in part because of the difficulties raised by astrophotographic technique: it was abandoned in 1970, or rather absorbed into other projects along very different technical and organisational lines.

---

19 e.g. in accounting for a particular type of blur known as the "black drop effect." For the observation of transits or eclipses, the traditional naked-eye, telescopic and drawing techniques were compared to a range of new photographic methods. See John Lankford, "Photography and the nineteenth-century transits of Venus," *Technology and Culture* 28 (1987): 648-657; Monique Sicard, "Passage de Vénus: Le révolver photographique de Jules Janssen," *Etudes Photographiques* 4 (1998): 45-63; Peter Hingley and Françoise Launay, "Expéditions scientifiques: éclipses et passages, 1860-19010," *Dans le champ des étoiles. Les photographes et le ciel, 1850-2000* (Paris: Editions de la Réunion des Musées Nationaux, 2000), 117-124; Alex Soojung-Kim Pang, *Empire and the Sun. Victorian Solar Eclipse Expeditions* (Stanford: Stanford University Press, 2002); Jessica Ratcliff, *The Transit of Venus Enterprise in Victorian Britain* (London: Pickering & Chatto, 2008). See also Bradley E. Schaefer, "The transit of Venus and the notorious black drop effect," *Journal for the History of Astronomy* 32 (2001): 325-336.

In light of these considerations, blurry daguerreotypes of the Moon may seem an appropriate starting point for a history of astrophotography considered as a history of the mutual elaboration of photography and astronomy, of high expectations frequently disappointed, of experimentation into photographic science, in other words a history in which blurs occupy a significant and legitimate place.[20] As the following sections suggest, drawing and painting proved to be essential resources to deal with early photographs, considered as blueprints, metaphors or predecessors of the new medium. The term "blur" seems particularly appropriate here, being a generalization of the original meaning of a smear on a surface as produced by ink on paper or paint on a canvas.

2 Photographic Maps of the Moon?

> There is reason to hope that photographic maps of our satellite can be made. That is to say that one of the most time-consuming, painstaking and delicate undertakings in astronomy can be executed in a few minutes.
>
> ARAGO, *1839*[21]

20 Kelley Wilder makes the distinction between scientific photography, the production of images in science using photography, and photographic science, research into photography itself. Kelley Wilder, *Photography and Science*, 7-17. On photographic techniques as scientific instruments for research into the nature of light for respectively Arago and Biot, and Brewster and Herschel, see Theresa Levitt, *The Shadow of Enlightenment. Optical and Political Transparency in France, 1789-1848* (Oxford: Oxford University Press, 2009), 129-159; Kelley Wilder, "A Note on the Invention of Photography: Reconsidering the Invention Story," in Sheehan and Zervigon, *Photography and its Origins*, 208-220. The work that come closest to a history of astrophotographic blur is Quentin Bajac, "1840-1875: les faux départs de la photographie astronomique," *Dans le champ des étoiles. Les photographes et le ciel, 1850-2000* (Paris: Editions de la Réunion des Musées Nationaux, 2000), 10-21. Other studies of early Moon photography include Barger and White, *Daguerreotype*, 84-91; Ann Thomas, "Capturing the Light: Photographing the Universe," in *Beauty of Another Order. Photography in Science*, ed. Ann Thomas et al. (New Haven: Yale University Press, 1997), 186-219.

21 "Il est permis d'espérer qu'on pourra faire des cartes photographiques de notre satellite. C'est dire qu'en quelques minutes on exécutera un des travaux les plus longs, les plus minutieux, les plus délicats de l'astronomie." François Arago, "Le daguerréotype," *Comptes Rendus Hebdomadaires des Séances de l'Académie des Sciences* 9 (19 August 1839), 263. All translations are mine.

One of the first expectations expressed with regards to astronomical photography was that it would supply something like an instant map of the Moon, replacing the extraordinarily lengthy and difficult business of observing, drawing and engraving. This was also the first expectation to be frustrated.

Arago had expressed this hope in 1839, explicitly comparing the daguerreotype to drawing and engraving:

> There are in M. *Daguerre's* pictures, in his copies, as in a pencil drawing, as in a burin engraving or better still (the comparison will be more exact), as in a mezzotint or *aquatint*, only white, black and grey, only light, darkness and half-tones. In a word, in M. *Daguerre*'s camera obscura light itself reproduces the shapes and the proportions of outside objects with near-mathematical precision; the photometric proportions of the different white, black and grey parts are exactly preserved; but half-tones represent red, yellow, green, etc., since the method creates drawings and not colour paintings.[22]

Arago later mentioned that painter Paul Delaroche described daguerreotypes as "photographic drawings."[23]

Ten years later, Harvard University president Jared Sparks held the same view. Upon receiving a daguerreotype of the Moon by daguerreotypist Samuel D. Humphrey, he wrote that

> Perhaps the art may at some future day be carried to such a degree of perfection, as to bring out the face of the Moon sufficiently large for a map. This would be a most desirable attainment; since a map of the Moon's surface can now be made only by the detached images presented to the eye through a telescope, and then transcribed by the hand. Such a map must necessarily be imperfect, and no two will precisely agree in all

22 "Il n'y a dans les tableaux, dans les copies de M. *Daguerre*, comme dans un dessin au crayon noir, comme dans une gravure au burin, ou mieux encore (l'assimilation sera plus exacte), comme dans une gravure à la manière noire ou à *l'aquatinta*, que du blanc, du noir et du gris, que de la lumière, de l'obscurité et des demi-teintes. En un mot, dans la chambre noire de M. *Daguerre*, la lumière reproduit elle-même les formes et les proportions des objets extérieurs, avec une précision presque mathématique ; les rapports photométriques des diverses parties blanches, noires, grises, sont exactement conservés ; mais des demi-teintes représentent le rouge, le jaune, le vert, etc., car la méthode crée des dessins et non des tableaux en couleur." Arago, "Fixation," 4-5.

23 "dessins photographiques," Arago, "Le daguerréotype," 260.

the parts. A Daguerreotype map would exhibit every feature with perfect exactness.[24]

Humphrey's daguerreotype and its popular success convinced Sparks to support the development of the method for astronomical uses at the recently founded Harvard College Observatory, and specifically the experiments made by its first directors William Cranch and his son George Phillips Bond together with John Adams Whipple in the 1840s and 1850s.[25]

Notwithstanding, thirty years later lunar photographic maps remained an elusive prospect. Berlin astronomer Johann von Mädler was reported saying in 1868 during a public lecture that

> Following the discovery of photography, hopes were voiced that can only be compared to those expressed by Descartes and his contemporaries after the discovery of the telescope. They pitied the unfortunate savants who had spent their lives observing, measuring and drawing. Not only would the same work be done without effort and in less time, but the results would be much better, more exact, more detailed than previously. What cost me seven years, the topographic map of the Moon, would be much better made in seven seconds. Today, thirty years have passed since Daguerre's invention; how have these ambitious hopes been realised?[26]

A teacher who had founded a private observatory in 1828 together with banker and astronomy enthusiast Wilhelm Beer (to whom he had been introduced by Humboldt), Mädler was eventually recruited as observer at the newly refounded Berlin observatory in 1835. The two men had spent seven years observing and drawing the Moon, culminating in the engraving and publication in 1834-

24 Samuel D. Humphrey, "Lunar Daguerreotypes," *Daguerreian journal* 1 (1 Nov. 1850), 14.

25 Barger and White, *Daguerreotype*, 87.

26 "Aussitôt après la découverte de la photographie, on entendit exprimer des espérances qui n'ont d'analogues que celles de Descartes et de ses contemporains à la suite de la découverte des lunettes astronomiques. On plaignait les malheureux savans qui avaient passé leur vie à observer, à mesurer et à dessiner. Non-seulement on devait faire la même chose sans peine et en moins de temps, mais on devait obtenir des résultats bien meilleurs, plus exacts, plus détaillés qu'auparavant. Ce qui m'a coûté sept années, la carte topographique de la lune, devait être bien mieux fait en sept secondes. Aujourd'hui trente années se sont écoulées depuis la découverte de Daguerre ; comment ces espérances ambitieuses ont-elles été réalisées?" Quoted in Rodolphe Radau, "Les Applications scientifiques de la photographie," *Revue des Deux Mondes* 25 (1878), 13.

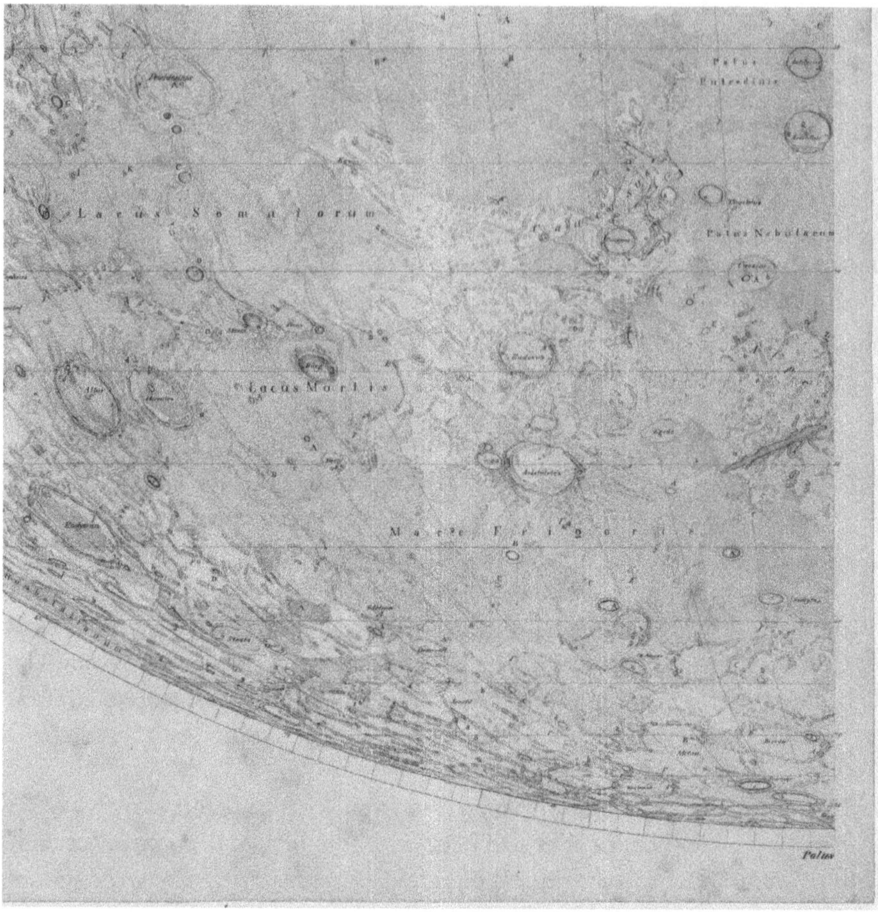

FIGURE 4.2 Wilhelm Beer, Johann Mädler, *Mappa selenographica totam Lunae hemisphaeram visibilem complectens observationibus propriis secundum projectionem orthographicam quartuor sectionibus constructa et delineata,* Berlin, 1834, detail of sheet 1. STAATSBIBLIOTHEK ZU BERLIN – PREUßISCHER KULTURBESITZ

1836 of their much acclaimed *Mappa Selenographica* (figure 4.2).[27] There was certainly more than a hint of *Schadenfreude* in Mädler's retrospective judgment on photography's ability to immediately outdate his painstaking work. But he was essentially right in claiming that photography had then, for all astronomical purposes, failed to replace older techniques.

27 Eino Eelsalu, Jürgen Hamel and Dieter Herrmann, *Johann Heinrich Mädler, 1794-1874. Eine Dokumentarische Biographie* (Berlin: Akademie-Verlag, 1985).

Not that Mädler was in principle against photography, to the contrary. He was among the very first commentators of Arago's 1839 speech in the German press – he is mainly remembered by historians of photography for having coined the term almost simultaneously with Herschel. He hailed the new technique, finding it much superior to drawing:

> Mr. Daguerre has succeeded in capturing it (the camera obscura image); albeit not in its natural colours, as some people wrongly think; but with its shadows and light, as the most skilled draughtsman would do or rather with a perfection exceeding the ability of any draughtsman and in every detail surpassing all expectations.

Mädler pointed out that, when fitted with an achromatic lens, the apparatus would yield sharp images; when observed with a magnifying glass, photographic lines were likely to be sharper than those produced by drawing or engraving.[28] But its practical use for imaging the Moon was perhaps simply unthinkable and Mädler did not broach the subject.

Mädler was not alone in his cautious appraisal of photography. In 1864, the British Association for the Advancement of Science set up a "Lunar Committee for mapping the surface of the Moon," that counted John Herschel among its members. In the instructions it issued to observers, photography was not entirely absent but occupied a minor position: "It is intended to collect in Tables various information appertaining to each object – as its selenographical coordinates, degree of reflective power, height or depth as the case may be, accompanied by references to more detailed descriptions, previous notices, drawings, and maps, including photographs, each separate Table being headed with one of the symbols in the annexed Table." The map, never completed, was to be drawn.[29]

---

28 "Hrn. Daguerre ist es gelungen, sie festzuhalten; zwar nicht in ihren natürlichen Farben, wie einige Personen mit Unrecht glauben; aber mit ihren Schatten und Lichtern, so wie nur der geschickte Zeichner vermöchte oder vielmehr mit einer Vollkommenheit, deren kein Zeichner fähig wäre und mit einer Vollendung in dem Einzelnen, die alle Erwartungen übersteigt." Anon. (probably Heinrich Johann Mädler), "Ueber die Erfindung des Hrn. Daguerre," *Königlich privilegirte Zeitung von Staats- und gelehrten Sache* (*Vossische Zeitung*), 16 (19 January 1839), 1-2. Reprinted in *Neues Licht. Daguerre, Talbot und die Veröffentlichung der Fotografie im Jahr 1839*, ed. Steffan Siegel (Paderborn: Wilhelm Fink, 2014), 57-60, on 58-59.

29 "On mapping the surface of the Moon, " *Report of the 35th meeting of the British Association for the Advancement of Science held in Birmingham in September 1865* (London: John Murray, 1866), 291. See also Edmund Neison, *The Moon and the Condition and Configurations*

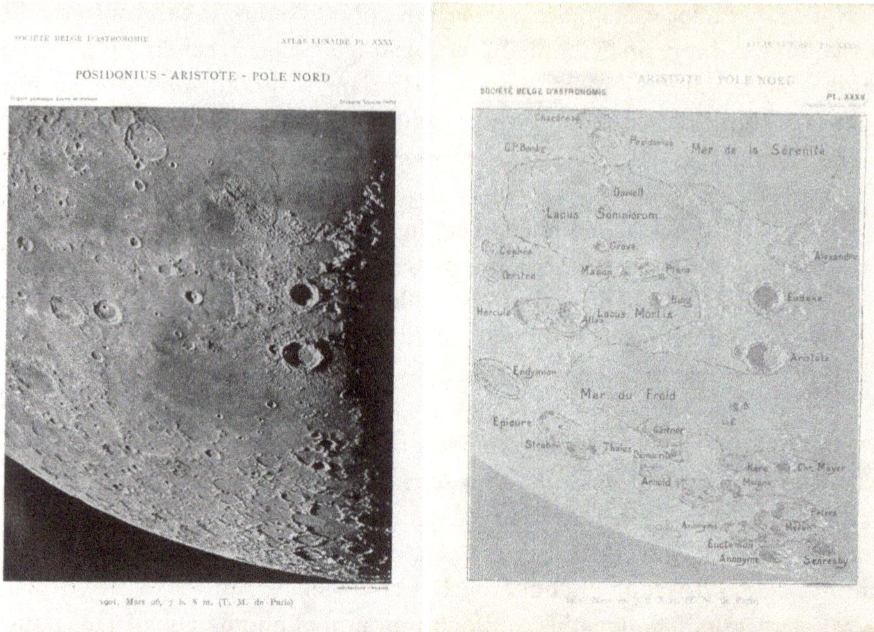

FIGURES 4.3a, b A small-size reproduction of a selection of Loewy and Puiseux's *Atlas Photographique de la Lune* was published by the Belgian Astronomical Society in 1899. The region of the moon represented here corresponds to the one shown in Beer and Mädler's map (figure 4.2). On the right, the heliogravure of the astronomers' photograph complete with the calque overlay featuring drawn contours and names. The region represented corresponds to the detail of Beer and Mädler's map in figure 4.2. *Atlas lunaire publié par la Société Belge d'Astronomie reproduisant à une échelle réduite aux 2/5 les agrandissements photographiques de M. Loewy et P. Puiseux*. Bruxelles, 1899. COURTESY OF STAATSBIBLIOTHEK ZU BERLIN – PREUßISCHER KULTURBESITZ

3 The *Atlas Photographique de la Lune*

It was not until the close of the century that the first Moon atlas based entirely on photographs appeared, prepared by astronomers Maurice Loewy and Henri Puiseux (1855-1928) of the Paris Observatory and published between 1896 and 1910.

Loewy and Puiseux wrote in the first volume that their aim had been to obtain

---

of its Surface (London: Longman, Green and co., 1876), a revision of previous Moon maps, including Mädler and Beer's, based on hundreds of "lunar sketches and drawings." (vi)

> of the present state of our satellite a representation as faithful and as precise as possible. In this respect the superiority of photography above all drawing methods no longer needs to be demonstrated… it stamps onto its images a character of undeniable authenticity, that the draughtsman's dexterity cannot live up to. In this way the uncertainties which have always prevented definitive conclusions to be made on the basis of older undertakings will be avoided.[30]

Between the 1860s and the 1890s, astronomers had gone from barely mentioning photography as a means of studying the Moon to off-handedly discarding drawing in a manner reminiscent of Arago and Sparks.

Looking more closely at their writings, however, it turns out that Loewy and Puiseux made important qualifications regarding the speed and reliability of photographic processes for imaging the Moon. Their striking photographs were made possible by the development of much faster, gelatine-based photographic emulsions in the previous decade. But the astronomers still had to put years of work, together with the Lumière photographic company and the Saint-Gobain glass manufacturers, into the improvement of emulsions, glass plates, as well as enlargement and reproduction techniques. These, together with sophisticated new tracking devices, helped to remedy, though only in part, the problem Daguerre and Draper had already faced, of the motion of the Moon causing blur during exposure.[31] Even then, of the circa 6000 photographs taken by Loewy and Puiseux during 500 nights, many were too blurry to be of any use and were immediately discarded. In other words, the beautiful *Atlas* photographs were anything but instant products, rather the best outcomes of dedicated and painstaking research and experimentation by many experts in a rather well-equipped institution, the Paris Observatory.

A major source of blur could never be entirely evacuated: *seeing*, as the astronomers call the fluctuations in the air due to atmospheric turbulence. Even when the atmosphere is calm, seeing causes oscillations in the apparent position of celestial features (twinkling), blurring details on long-exposure

30 "de l'état actuel de notre satellite une représentation aussi fidèle et aussi précise que possible. A cet égard, la supériorité de la photographie sur tous les procédés de dessin n'a plus besoin d'être démontrée… elle imprime à ses images un caractère d'authenticité indiscutable, auquel la dextérité du dessinateur ne peut suppléer. Ainsi seront évitées les incertitudes qui ont toujours empêché de dégager des travaux anciens des conclusions définitives." Maurice Loewy and Pierre-Henri Puiseux, *Atlas photographique de la lune publié par l'observatoire de Paris, 1er fascicule* (Paris: Imprimerie Nationale, 1896), 2.

31 Maurice Loewy and Pierre-Henri Puiseux, *Atlas photographique de la lune publié par l'observatoire de Paris, 2e fascicule* (Paris: Imprimerie Nationale, 1898), B2-B8.

photographs or altering their position on short- exposure plates. As businessman and astronomy enthusiast Warren de la Rue put it in 1859, reporting on his own experimentations with astrophotographic technique:

> When the telescope is employed optically, the mind can make out the proper figure of the object, although its image dances before the eye several times in a second and is able to select for remembrance only the states of most perfect definition; on the other hand, a photographic plate registers all the disturbances. The photographic picture will consequently never be so perfect as the optical image with the same telescope, until we can produce photographs of celestial objects instantaneously: we are still a long way from this desirable end.[32]

Until the twentieth century, the human eye, able to an extent to filter the seeing, usually remained more competent than the photographic plate for the determination of positions.[33] This problem was only completely solved decades later with the use of space telescopes, operating entirely outside the atmosphere; or with "Lucky Imaging" techniques that improve resolution by digitally combining the best photographs chosen from a series taken in quick succession.[34]

But there were more fundamental objections to lunar photographic maps. For the *Atlas* was not, strictly speaking, a map. It was a collection of enlarged photographs reproduced by heliogravure, a source of headaches in its own right). Between each thick cardboard page, a calque overlay displayed the drawn contours and names of the Moon features visible by transparency on the photograph underneath (Figures 4.3a and 4.3b). Loewy and Puiseux freely admitted that their *Atlas* was not cartographic:

> representations of the Moon made to this day belong to one of two classes: those aiming to reproduce faithfully the appearance of a limited portion of the Moon under a given angle of lighting; and those summarizing a long series of observations and researches in graphic and conventional

32 Warren de la Rue, "Report on Celestial Photography in England," *Reports of the British Association for the Advancement of Science for 1859* (London: Taylor and Francis, 1860), 131-153, on 133.

33 Whitaker, *Mapping*, 142.

34 personal communication of Craig Mackay, emeritus Professor of Image Science in the Institute of Astronomy, Cambridge university. For publications references and details on Lucky Imaging techniques, see the Institute's comprehensive webpages <http://www.ast.cam.ac.uk/research/lucky/> (last accessed 12 February 2016).

> form; it is only to the images in the first category that photographic exposures can be fairly compared... A single object cannot be considered to be known unless it has been observed as illuminated successively at least from the East, the West and from the Meridian.[35]

For Loewy and Puiseux, photographs could profitably replace drawings of individual observations, but they could never be maps, which were necessarily *graphic* and *conventional* representations synthesizing many observations. Here again, drawing remained indispensable.

The picture that emerges from this nineteenth-century work on moon cartography is then not one of photography immediately or obviously displacing drawing as the basis and product of Moon cartography but rather of photography being incorporated and experimented upon in combination with existing techniques (and in the process giving rise to many blurry plates). As in the case of nineteenth-century research into nebulae, visual observation and drawing long remained much superior and sometimes irreplaceable for many purposes.[36]

4 Albedo and Its Discontents

> I had just seen the sun rising in the Moon. The flash met something, perhaps a peak, and hit it, a snake of fire materialized in this darkness, traced a circle and became motionless; a crater had appeared... Meanwhile the circles grew, widened, joined at the edges, exaggerated until they all conflated; valleys deepened, abysses deepened, rifts opened their lips overflowing with shadowy froth, deep spirals sunk, slopes frightening to the eye, immense cones of obscurity were projected, shadows stirred, stripes

35 "Les représentations de la Lune exécutées jusqu'à ce jour se partagent en deux classes: les unes visant à reproduire fidèlement l'aspect d'une portion limitée de la Lune sous un éclairage donné; les autres résumant sous une forme graphique et conventionnelle une longue série d'observations et de recherches; c'est seulement aux images de la première catégorie que les épreuves photographiques peuvent être équitablement comparées... Un objet unique ne saurait être considéré comme connu si l'on a pu, au moins, le voir éclairé successivement par l'est, par l'ouest et du côté du Méridien." Maurice Loewy and Henri Puiseux, "Sur les photographies de la Lune obtenues au grand équatorial Coudé de l'Observatoire de Paris," *Comptes Rendus Hebdomadaires de l'Académie des Sciences* 119 (1894): 254-259, on 256-267.

36 Omar Nasim, *Observing by hand. Sketching the nebulae in the nineteenth century* (Chicago: Chicago University Press, 2013).

> of rays settled as would an architrave from one outcrop to the next, knots of craters created wrinkles around the peaks, all kinds of furnace profiles appeared haphazardly, some made of smoke, others of light; capes, points, gorges, passes, inclined planes, escarpments, chiasms mingled, entangled their curves and angles; the figure of mountains could be seen. It existed magnificently.
>
> VICTOR HUGO, *1863*[37]

Victor Hugo's eloquent testimony serves as a reminder that even before photography, observing the Moon was an uncertain business. Successive Moon mappers frequently disagreed about the existence, shape and name of features, albeit in a manner certainly less virulent than those involved in the disputes about the existence of canals on Mars.[38] One of the major difficulties faced by Moon observers is the variable appearance of the lunar landscape. The contours, height or depth of mountains, valley and craters seem to change under different angles of solar illumination as the Moon goes through its different phases. It was not for instance until after Beer and Mädler's observations that a consensus emerged on the absence of a lunar atmosphere. Loewy and Puiseux's proposed photographic solution was to make observations of the Moon under at least three different angles of illumination. Only then could reliable information be obtained on the shape and size of landscape features.

Here the astronomers' approach echoed contemporary developments in photogrammetry. In the 1850s geodesists, cartographers and architects had

37 "Je venais de voir le soleil se lever dans la lune. L'éclair fit une rencontre, quelque chose comme une cime peut-être, et s'y heurta, une sorte de serpent de feu se dessina dans cette noirceur, se roula en cercle et resta immobile; c'était un cratère qui apparaissait.... Cependant les cercles grandissaient, s'élargissaient, se mêlaient par les bords, s'exagéraient jusqu'à se confondre tous ensemble ; des vallées se creusaient, des précipices s'ouvraient, des hiatus écartaient leurs lèvres que débordait une écume d'ombre, des spirales profondes s'enfonçaient, descentes effrayantes pour le regard, d'immenses cônes d'obscurité se projetaient, les ombres remuaient, des bandes de rayon se posaient comme des architraves d'un piton à l'autre, des noeuds de cratères faisaient des froncements autour des pics, toutes sortes de profils de fournaise surgissaient pêle-mêle, les uns fumée, les autres clarté; des caps, des promontoires, des gorges, des cols, des plans inclinés, des escarpements, des coupures, s'enchevêtraient mêlant leurs courbes et leurs angles ; on voyait la figure des montagnes. Cela existait magnifiquement." Hugo, *Promontoire*, 27-29.

38 On the history of lunar nomenclature see Whitaker, *Mapping*; Scott L. Montgomery, *The Moon and the Western Imagination* (Tucson: University of Arizona Press, 1999). On the disputes about the topography of Mars, see, e.g., K. Maria D. Lane, *Geographies of Mars*, Chicago, 2011; Joshua Nall, *News from Mars. Transatlantic Mass Media and the Practice of New Astronomy, 1870-1910* (Ph.D. Dissertation, University of Cambridge, 2013).

begun experimenting with photography for the measurement and representation of volumes, from buildings to geographical features. These techniques were increasingly used with the development of aerial photography, especially during and after the First World War. Like the astronomers, specialists warned against any spontaneous tendency to equate photographs with maps:

> One might be tempted to believe, given the objective's accuracy, that a vertical aerial photograph gives an accurate image of the terrain, and that once the scale is corrected, one would have a map of the photographed zone. Unfortunately this is not the case, and multiple causes of error complicate the problem.[39]

To the contrary: photogrammetry was defined in 1891 by one practitioner as the "inverse of perspective," as "this part of the graphic arts that has as its purpose the re-establishment, with the help of exact perspectives, of the orthogonal projections of external objects."[40] Photogrammetry aimed to substitute a number of visual and instrumental techniques used by military engineers, cartographers and architects with photographs, but this implied subjecting the information these supplied to a considerable number of transformations. Mapping terrestrial features by photogrammetry required taking at least two photographs from different standpoints; moon studies required a series of observations made under different angles of illumination. One suspects that Nasmyth and Carpenter's multimedia project of photographing plaster-cast models of moon reliefs was intended not only to study the geology of the moon but also simultaneously, through a reverse-engineering of moon photographs, to explore photography's ability to represent relief, to learn to read moon photographs and make them useful for deciphering lunar geography.[41]

39 "On serait tenté de croire, vu la fidélité de l'objectif, que la photographie aérienne verticale est une image fidèle du terrain, et qu'une fois remise à l'échelle nous aurons une carte de la zone photographiée. Malheureusement il n'en est rien, et de multipes causes d'erreur compliquent le problème." A. H. Carlier, *La photographie aérienne pendant la guerre* (Paris: Delagrave, 1921), 44.

40 "réciproque de la perspective", "cette partie des arts graphiques qui a pour objet le rétablissement, à l'aide de perspectives exactes, des projections orthogonales des objets extérieurs." Victor Legros, *Sommaire de Photogrammétrie. Application élémentaire de la photographie à l'architecture, à la topographie, aux observations scientifiques et aux opérations militaires* (Paris: Sociétés d'éditions scientifiques, 1891), 2, 1.

41 James Nasmyth and James Carpenter, *The Moon, considered as a Planet, a World, a Satellite* (London: John Murray, 1874). For a careful study of this book and undertaking, see Omar W. Nasim's chapter in this volume.

When it came to drawing maps, whether terrestrial or lunar, whether using drawings or photographs, a convention for representing the third dimension had to be chosen. Lunar cartography closely parallels the history of terrestrial cartography with a progressive standardization of the use of hachuring for representing the third dimension, landscapes being increasingly represented using a simulated uniform lateral illumination.[42]

However, not all maps obeyed these conventions. As Albert van Helden, among others, has insisted, there was a continuing, cartographically-aberrant, habit of representing albedo on moon maps, that is, the differential reflection of light on different parts of the moon, as distinct from relief (see figure 4). Van Helden distinguishes between *maps* strictly speaking and "images, pictures of the moon" which he calls *likenesses*. This points to an important yet frequently omitted dimension of the history of moon imaging: its uses and users, that both changed considerably over time. Until the early eighteenth century and the development of chronographic methods, moon maps were important for determining longitude at sea by the lunar distance method; and they became again vital in the Space Age for preparing moon landings. In between, van Helden argues, representations of the moon were driven either by the "not very pressing" need to keep up with terrestrial cartography or by aesthetic motivations.[43]

Van Helden's remark provides an important corrective to histories that take technological improvement as the driving force of the historical development of astronomical imaging: different intended uses obviously give rise to very different kinds of representations and different conceptions of accuracy. The important development of amateur astronomy over the course of the nineteenth century provided a market for moon maps featuring albedo for those observers focussed not or not only on cartography but on the identification and study of lunar features. For instance, the map of the moon (figure 4.4) prepared by popular astronomer Casimir-Marie Gaudibert under the direction of Camille Flammarion in 1887 was explicitly aimed at "diverse social conditions, the taste for astronomy being nowadays widespread among all classes of society."

It featured albedo, with the North of the moon at the bottom as visible through a telescope, and was published in a sufficiently small format for convenient use by amateurs in the field, unlike, Mädler's or, later, Loewy and Puiseux's unwieldy and expensive volumes (these, as figure 4.3. shows, were also

42 Whitaker, *Mapping*, 116.

43 Albert van Helden, "Moon Maps in History" (review of Ewin Whitaker's *Mapping and Naming the Moon*), *Journal for the History of Astronomy* 33 (2002): 285-289.

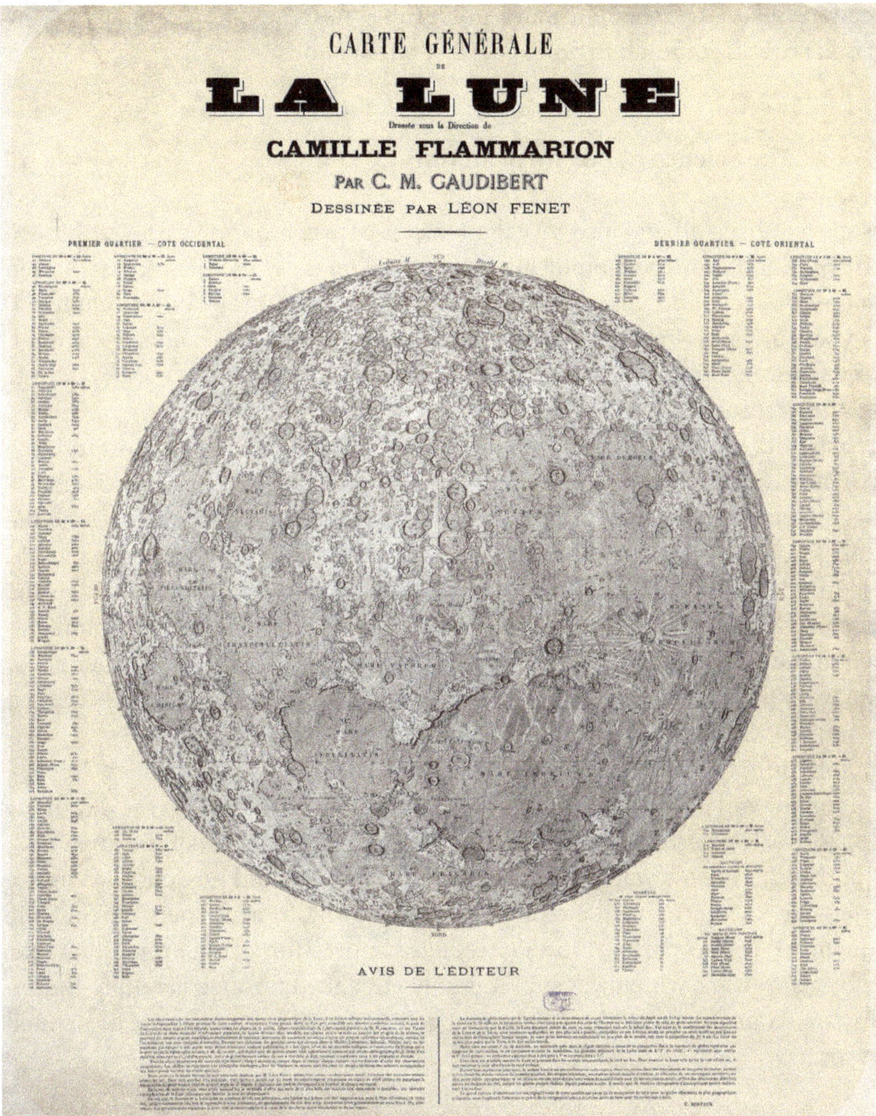

FIGURE 4.4 "Les observateurs du ciel réclamaient depuis longtemps une bonne carte géographique de la lune, d'un format suffisant mais commode, contenant tous les détails indispensables à l'étude pratique de notre satellite, et cependant d'une grande clarté et d'un prix accessible aux diverses conditions sociales, le goût de l'astronomie étant aujourd'hui répandu parmi toutes les classes de la société." Gaudibert also prepared a globe of the moon for Flammarion. *Carte générale de la lune, dressée sous la direction de Camille Flammarion, par C.-M. Gaudibert. Dessinée par Léon Fénet* (Paris: E. Bertaux, 1887).
COURTESY OF BIBLIOTHÈQUE NATIONALE DE FRANCE.

made available in smaller, more practicable formats).[44] Such maps of the moon are still common in non-specialist publications.

5 The Science of Lunar Portraiture

Maps featuring albedo also reveal a tension between purely cartographic enterprises and more geographical or geological ones, where the representation not only of relief but also of texture plays an important role. Where photography was involved in this type of Moon imaging, the preferred analogy and reference was no longer drawing, but often *painting*.

The common reference to early astronomical photographs as "portraits" here points to the cultural status and scientific meanings of lunar portraiture. The story of Moon photography unfolded on the margins of the vivid debates that flared from the late 1830s about the artistic merits of photography and the threat it potentially represented for established artistic values and livelihoods. Figure 4.5b, the German translation of a Grandville engraving in which the moon was shown daguerreotyping itself (1847) while it had merely painted itself in the original French edition of 1843 (Figure 4.5a) is a pointed example of the ways in which scientific and cultural debates intermingled and were put before a wide public.

Early moon photographs were frequently described as portraits.[45] In 1839, Humboldt was perhaps the first to do so when he wrote "that on the day of my departure, Daguerre brought us at the observatory the image of the lunar disc, a portrait of Luna brought forth by herself."[46] One should no doubt see here a reference to the ancient literary and artistic tradition of personifying the moon;[47] but also perhaps to the aesthetic impression of moon photographs upon its viewers, which was subsequently enhanced through carefully chosen display strategies, including elaborate golden frames reminiscent of painted portraits.

44 For a study of a nineteenth-century French amateur Moon observer who owned many such maps, see Laurence Guignard, "Le `tourment lunaire' de Jules Pierrot-Desseilligny. Pratiques amateur d'observations de la lune," *Romantisme* 166 (2014): 65-78.

45 see Thomas, "Capturing the Light," Barjac, "faux départs".

46 "dass am Tage meiner Abreise Daguerre uns nach dem Observatorium das Bild der Mondscheibe brachte, ein portrait von Luna selbst hervorgebracht." Letter from Alexander von Humboldt to Carl Gustav Carus dated 25 February 1839, reproduced in Siegel, *Neues Licht*, 85-87, on 86.

47 On the old literary-scientific tradition of engaging with the Moon, see Frédérique Aït-Touati, *Fictions of the Cosmos* (Chicago: Chicago University Press, 2011).

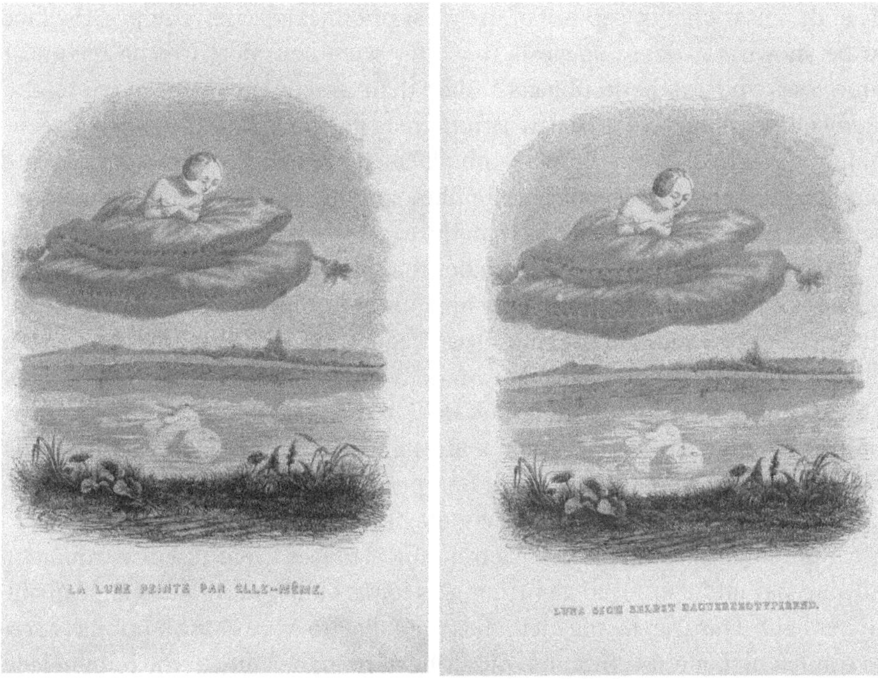

FIGURES 4.5a, b Autoportraits of the moon. While in the original edition of 1843 Granville's engraving bears the legend "The moon painting itself," in the German translation of 1847 one reads, with no further comment, "the moon daguerreotyping itself." J. J. Grandville, *Un autre monde...*, Paris, 1843, opposite page 144 and Plinius dem Jüngsten, *Eine andere Welt*, illustriert von J. J. Grandville, Leipzig, 1847, opposite page 128. COURTESY OF STAATSBIBLIOTHEK ZU BERLIN – PREUẞISCHER KULTURBESITZ

Astronomical daguerreotypes were unique, fragile and beautiful objects that exerted a strong attraction beyond the scientific community. The moon daguerreotypes that resulted from the experimentations of the Bonds and Whipple at Harvard College Observatory were taken on a triumphant tour of Europe by George Phillips Bond in 1851 (Figure 4.6). They were shown at the Crystal Palace Exhibition where one won a gold medal, the Royal Astronomical Society, the French Academy of Sciences where they produced a sensation "scarcely ever before witnessed", the British Association for the Advancement of Science's annual meeting in Ipswich, Hamburg, Pulkowo, Berlin where they were compared by Humboldt to photographs on paper and a daguerreotype of the 1851 eclipse, Halle, Göttingen, Geneva, Bonn, Parsonstown, etc.[48] The pub-

48 "Diaries of George Bond" in *Memorials of William Cranch Bond and of his son George Phillips Bond*, ed. Edward S. Holden (San Francisco: C.A. Murdock and Co./ New York City: Lehmke and Buechner, 1897), 87-122.

lic and "sensational" reception of these astronomical daguerreotypes, the first to be shown so widely, suggests that they were considered to be technical prowesses and aesthetic objects rather than practical contributions to the study of the moon; even though astronomers paid close attention to the techniques developed by the Bonds to obtain such a level of detail, especially the modification of the telescope lens for photographic use (emulsions being more sensitive to shorter wavelengths than the eye).

Lunar "portraits" are also a reminder that portraiture was a major concern of the early photographers and later an important source of income for the industry and profession. In the mid-1830s, Daguerre and Nicéphore Niépce were already working on emulsions that would be fast enough for making portraits. The "daguerreotypomania" that took hold in the 1840s was largely focused on portraits, replacing the cheap silhouettes and miniatures bought in previous decades by a growing proportion of European populations – including astronomers (Figure 4.7).[49] Conversely, technologies developed for portrait photography were sometimes adapted for astronomical use, as when Yerkes astronomer Edward Barnard employed a portrait lens for its ability to collect a large quantity of light. This use, he pleaded, "has not fallen to a lower level but has risen to a much higher one – from the picturing of human vanity in the human face to the picturing of the sublime features of the face of the heavens."[50]

Portraiture was not a scientific term. Hervé Faye, disciple of Arago, astronomer at the Paris Observatory and a long-standing advocate of astrophotography, lamented in the 1850s that "images of the moon, the sun, the planets remain, despite the considerable importance I myself give to them, mere portraits."[51] For Faye, these portraits such as, presumably, Whipple and Bond's, had little to do with science and constituted mere likenesses, in van Helden's terminology, or what astronomers would today call "pretty pictures." Faye regretted that observatories had not developed photography to function as a recording instrument while in physics it was employed for the observation of thermic, electric and magnetic phenomena. This was not quite true since Faye himself was developing photographic methods for assisting meridian observa-

49 In Paris in 1868, portraits were the second most frequent subject of commercial photography, after stereoscopic views. Elizabeth Anne McCauley, *Industrial Madness. Commercial Photography in Paris, 1848-1871* (New Haven and London: Yale University Press, 1994), 97.

50 Edward E. Barnard, "Some of the results of astronomical photography pertaining specially to the work with a portrait lens," *Proceedings of the American Philosophical Society* vol. 46/187 (1907), 422.

51 "les images de la lune, du soleil, des planètes ne sont encore, malgré l'importance considérable que je leur attribue moi-même, que de simples portraits." Quoted in Bajac, "Faux départs," 15.

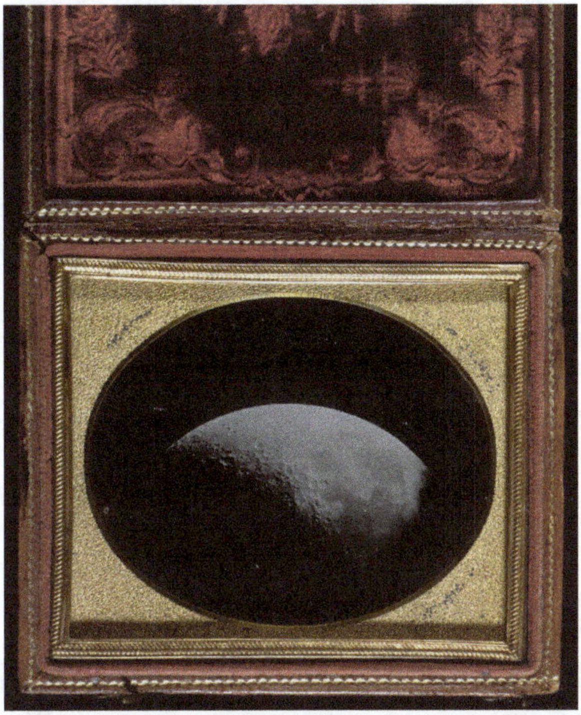

FIGURE 4.6
Daguerreotype of the moon made at the Harvard College Observatory in 1851 by the astronomer William Cranch Bond and the photographer John Adams Whipple.

tions and the measurement of time, and at Greenwich Observatory, George Biddell Airy had, since the 1840s, used self-recording photographic techniques for registering magnetic and meteorological observations, but admittedly hardly any quantitative or systematic study of the celestial bodies had at the time been undertaken involving photography.[52]

This changed in 1858 when John Herschel and Warren de la Rue obtained the support of the British Association for the Advancement of Science to launch a programme of regular photographic documentation of the solar surface and of sunspots at Kew Observatory using devices developed by de la Rue, that ran until 1872. De la Rue had originally been encouraged by Nasmyth to begin astronomical observations and was later inspired by Bond's 1851 visit to London to take up photography. He claimed to have been first to apply the new wet-plate collodion process to photograph the moon in 1852 and spent the following years experimenting with the technique, a difficult undertaking that had to be abandoned, hampered especially by the lack of a driving mechanism

52 See Holly Rothermel, "Images of the Sun: Warren de la Rue, George Biddell Airy, and Celestial Photography," *British Journal for the History of Science* 26 (1993): 137-169.

to follow the motions of the moon.[53] One ambition at the time had been, not unlike the later solar programme, to document the evolution of the moon's surface over time: "it will be evident that selenological disturbances, extending over two or three miles, would not escape detection, if such occur, provided photographs continue to be taken for a sufficiently long period."[54]

6 Physiognomies of Landscape

This approach, defined by Herschel as a "history"[55] of celestial bodies' appearances and by Faye as "descriptive astronomy"[56] involved the use of photography for observing solar eclipses and the surface of the planets in order to study their physical, geological make-up and its evolution over time.

In 1872, Faye presented new photographs of the moon obtained by the American chemist and photographer Lewis Rutherford using gelatine-based emulsions:

> A single look at these magnificent prints makes one appreciate the services they could render to the study of Lunar Geology. The great bright lignes or fractures outlining the arcs of large circles intersect at angles that can be measured with some precision … the cirques, the craters and the smallest circular depressions presented by the surface of the Moon in such numbers are represented on a large scale and with a striking accuracy unmatched by any topographical map. It will be possible to study step by step the variety of these diverse types, so similar at first sight to

53 "I could not take photographs of the Moon in this way alone, but required always the aid of an experienced coadjutor, willing to lose the greater portion of a night's rest, often to be disappointed by failures resulting from the state of the weather, and numberless impediments sufficient to damp the ardour of the most enthusiastic… but the difficulties we had to encounter were so great that it was at last resolved to discontinue the experiments until such time as a driving motion could be applied to the telescope." Warren de la Rue, "Celestial Photography," 131-132.

54 De la Rue, "Celestial Photography," 140.

55 Quoted in Rothermel, "Images of the Sun," 151. This approach was inspired by William Herschel's project for a "natural history of the heavens." See Simon Schaffer, "Herschel in Bedlam: Natural History and Stellar Astronomy," *British Journal for the History of Science* 13 (1980): 211-239.

56 "astronomie descriptive," Hervé Faye, "Sur l'état de la photographie astronomique en France," *Comptes Rendus Hebdomadaires des Séances de l'Académie des Sciences* 50 (1860): 965-967.

> our extinct volcanoes yet so different, in certain respects, to their terrestrial counterparts. Here photography supplies the heights (in the region of their projected shadows) as well as linear dimensions in the horizontal plane. One is struck by the appearance of (the seas), as through the direct inspection of the Moon, of the idea of having under one's eyes the product of vast oupourings of fluid matter that erased the ancient surface accidents, leaving here and there on the edges a few vestiges of primitive cirques.[57]

As Faye had pointed out in 1860, this work was especially developed by British and American astronomers, a situation that would become a major justification for the creation of the Meudon Observatory fifteen years later under Jules Janssen and funding astrophysics in France.

The origins of this type of work were older, however. In his address to the academies in August 1839, Arago had explained that, with regards to photography's application in science, with the daguerreotype, as with the telescope,

> we discover a myriad of new worlds; that, penetrating into the constitution of the six planets of the Ancients, we find it analogous to that of our Earth, with mountains whose height is measured, with atmospheres

57 "Il suffit d'un coup d'œil sur ces magnifiques épreuves pour faire apprécier les services qu'elles pourraient rendre à l'étude de la Géologie lunaire. Les grandes lignes lumineuses, sortes de cassures dessinant des arcs de grands cercles, se croisent suivant des angles qu'il est possible de mesurer avec une certaine exactitude... Les cirques, les cratères et jusqu'au moindres fosses circulaires que la surface de la Lune nous présente en si grand nombre y sont représentés à grande échelle, avec une fidélité saisissante qu'aucune carte topographique ne saurait reproduire. On pourra y étudier pas à pas les variétés nombreuses de ces types divers, si semblables de prime abord à nos volcans éteints et si différents toutefois, à certains égards, de leurs analogues terrestres. Ici la photographie donne les hauteurs (dans la région des ombres portées) aussi bien que les dimensions linéaires dans le sens horizontal.... On est frappé par leur aspect (des mers), tout aussi vivement qu'à l'inspection directe de la Lune, de l'idée qu'on a sous les yeux le produit de vastes épanchements d'une matière fluide, qui serait venue effacer les accidents antérieurs de la surface, en laissant subsister çà et là sur les bords quelques vestiges des cirques primitifs." Hervé Faye, "Sur les photographies de la lune de M. Lewis Rutherfurd," *CRAS* 75 (1872), 1072. Faye writes prudently that "Ces belles photographies ne dispensent naturellement ni du secours d'une carte bien faite, comme celle que nous devons à celle de MM. Beer et Mädler, ni de l'étude directe de la Lune elle-même au moyen de télescopes plus ou moins puissants."

whose upheavals are studied, with phenomena of polar ice formation and fusion that are analogous to those of the terrestrial poles.[58]

What Arago described here went far beyond cartography. It corresponded rather to the "idea of a physics of the world" proposed by his friend Humboldt, that included the study of botany, climate and geology. Humboldt took inspiration from the eighteenth-century naturalist project that rested both on precise and quantitative observations and on the "coup d'oeil", the panoramic view that made it possible to make sense of underlying structures, such as the geological makeup of the Alps as revealed from a summit to Horace-Bénédicte de Saussure.[59] Humboldt's science of the cosmos was also inspired by Kaspar Lavater's physiognomy of facial expressions defined as

> The Science which teaches the knowledge of the relationship of the exterior to the interior, of the visible surface to the invisible it embraces, of animate and perceptible matter to the non-perceptible principle that imprints upon it the character of life, of the manifest effect with the hidden force that produced it.[60]

Humboldt had proposed in his *Ansichten der Natur* (1807) to establish a physiognomy of Nature "capable of surveying nature with a comprehensive glance" in order to grasp the character of each landscape, as spontaneously characterized by painters:

> That which the painter designates by the expressions "Swiss scenery" or "Italian sky" is based on a vague feeling of the local natural character. The

58 "on découvre des myriades de nouveaux mondes ; que, pénétrant dans la constitution des six planètes des anciens, on la trouve analogue à celle de notre terre, par les montagnes dont on mesure les hauteurs, par les atmosphères dont on suit les bouleversements, par des phénomènes de formation et de fusion de glaces polaires, analogues à ceux des pôles terrestres." Arago, "Le daguerréotype," 264.

59 See Charlotte Bigg, "The panorama, or la nature à coup d'œil," in *Observing Nature – Representing Experience. The Osmotic Dynamics of Romanticism 1800-1850*, ed. Erna Fiorentini (Berlin : Reimer Verlag, 2007), 73-95.

60 "la Science qui enseigne à connoître le rapport de l'extérieur avec l'intérieur, de la surface visible avec ce qu'elle embrasse d'invisible, de la matière animée & perceptible avec le principe non perceptible qui lui imprime ce caractère de vie, de l'effet manifesté avec la force cachée qui le produit." Johann Casper Lavater, *Essai sur la physiognomonie, destiné à faire connoître l'homme et à le faire aimer* (The Hague : Jaques Van Karnebeek, 1781) (first edition 1772), 22.

> azure of the sky, the effects of light and shade, the haze floating on the distant horizon, the forms of animals, the succulence of plants, the bright glossy surface of the leaves, the outline of mountains, all combine to produce the elements on which depends the impression of any one region.[61]

Such a general impression was to be made up of visual observation and measurement, of experience and experiment, they involved both aesthetic pleasure and rigorous investigation, an approach which could be expressed and communicated through landscape painting.

Humboldt's project became one of the foundations for the new science of geography in the nineteenth century. Without exaggerating the influence of Lavaterian physiognomy on Humboldt's approach, one might see in Humboldt's "portrait of Luna" and in Arago's vision an extension of the Humboldtian physiognomy of Nature to the moon.

Entrusting photography with the ability not only to record likeness but also to speak of physiognomic identity proved tricky, however: photographic portraits were often found unresembling and uncanny, photographic evidence frequently useless for identification, and moon photographs ambiguous, as we've seen.[62] Nevertheless, Arago and Humboldt, the first exegetes of Daguerre's invention, shaped its perception and the horizon of possibility of its later applications, including the physiognomy of landscapes. Photographic portraits of the moon were forebears of the physical study of the moon and the other planets by analogy with the study of terrestrial landscapes, an analogy already made explicit in Nasmyth and Carpenter's idiosyncratic project, or in the da-

61 "Was der Maler mit den Ausdrücken Schweizer Natur, italienischer Himmel, bezeichnet, gründet sich auf das dunkle Gefühl dieses lokalen Naturcharacters. Himmelsbläue, Beleuchtung, Duft, der auf der Ferne ruht, Gestalt der Thiere, Saftfülle der Kräuter, Glanz des Laubes, Umriß der Berge – Alle diese Eindrücke bestimmen den Totaleindruck einer Gegend." Alexander von Humboldt, "Ideen zu einer Physiognomik der Gewächse" in *Ansichten der Natur* 1 (Tübingen: J.G. Cotta, 1808), 173-174. This translation is quoted from Alexander von Humboldt, *Views of Nature: or Contemplations on the sublime Phenomena of Creation*, trans. E.C. Otté and Henry G. Bohn (London: Henry G. Bohn, 1850), 217-218.

62 On likeness, identity and resemblance in portrait photography, see Alan Trachtenberg, "Likeness as Identity: Reflections on the Daguerrean Mystique," in *The Portrait in Photography*, ed. Graham Clarke (London: Reaktion Books, 1992), 173-192; Jan von Brevern, "Resemblance after photography," *Representations* 123 (2013): 1-22. On the difficulty of recognizing persons on photographs, see Jennifer Tucker, *Nature Exposed: Photography as Eyewitness in Victorian Science* (Baltimore and London: Johns Hopkins, 2005). One of the proposed ways of making photographic portraits resemblant was to make them blurry: see Brusius, "Impreciseness."

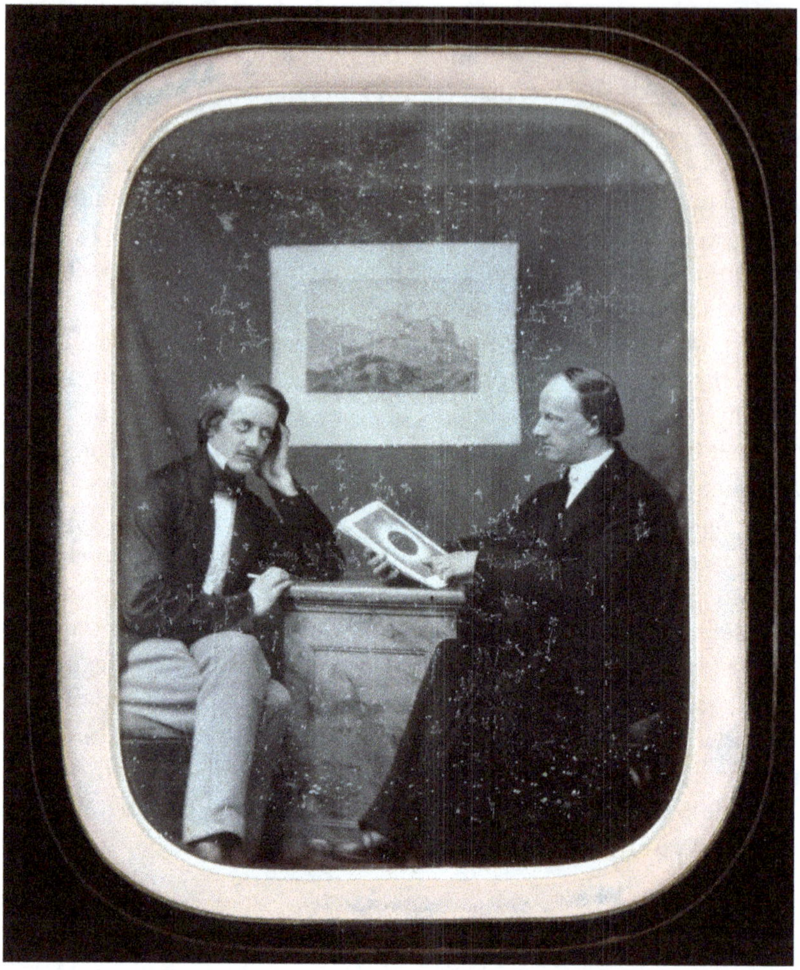

FIGURE 4.7 Dr. Bartels and astronomer Julius Schmidt, Daguerreotype by Wilhelm Breuning, 1856. COURTESY OF MUSEUM FÜR KUNST UND GEWERBE HAMBURG. Johann Friedrich Julius Schmidt (1825-1884) was at the time of the photograph director of the privately-owned Olmütz observatory before he took over the Athens observatory in 1858. He spent three decades drawing the moon, preparing a map more detailed than Beer and Mädlers, published in 1874 and he also studied the physical geography of the earth and of other planets. The study of eclipses of the moon and of terrestrial geography are showcased in this portrait of the astronomer, making explicit the connection between terrestrial and planetary geography. Photography here is purely a means of portraying a scientist who relied on drawing in his research. Ferrari d'Occhieppo, Konradin, "Schmidt, Johann Friedrich Julius" in: Neue Deutsche Biographie 23 (2007), S. 199

guerreotype portrait of Julius Schmidt, and that would much later take the name of planetary science.

7 Conclusion: Whose Amateurs?

> As always with the introduction of new, epoch-making inventions into scientific problems, at the beginning celestial photography delivered important results effortlessly; today still, a single look at the first photograph of a nebula can for instance lead to an important scientific result that could not have been obtained in earlier times, even with a great amount of work. But this easy harvest of results has led to negligence in the accuracy of research, to the domination of fantasy over sober reason, and the first signs of such corruption can already be seen in the history of astrophotography. … Above all, every individual must strive to help celestial photography steer clear from the downright offensively unscientific character of much of photography's specialist literature, as betrayed for instance by the frequent improper use of technical terms.[63]

At the close of the nineteenth century, Julius Scheiner, astronomer at the Potsdam Observatory, took to task the dabblers and unserious practitioners of astrophotography. Against "fantasy", "corruption", "negligence" and "improper" usage of terminology, he recommended the cultivation of "reason", "precision," "accuracy" and hard work. He granted that amateurs and photographers, to name them, had made contributions to the burgeoning field, but now was the

63 "Wie stets nach Einführung neuer, epochemachender Erfindungen in die naturwissenschaftlichen Probleme, so hat auch die Himmelsphotographie anfangs mit spielender Leichtigkeit wichtige Ergebnisse geliefert; selbst heute noch vermag z.B. ein einziger Blick auf eine zum ersten Male erhaltene Photographie eines Nebelfleckes ein wichtiges wissenschaftliches Resultat zu liefern, welches vielleicht früher trotz Aufwendung vieler Arbeit nicht hatte gewonnen werden können. Diese Leichtigkeit der Ausbeute hat aber noch immer zu einem gewissen Nachlassen der Exactheit der Forschungen geführt, zur Herrschaft der Phantasie an Stelle des nüchternen Verstandes, und die ersten Anzeichen eines derartigen Verfalles sind bereits deutlich in der Geschichte der Himmelsphotographie zu erkennen.… Vor allem muss auch jeder sich bemühen, die theilweise geradezu beleidigende Unwissenschaftlichkeit, welche einen grossen Theil der photographischen Fachliteratur beherrst, und die sich z. B. in einer Menge falsch gebrauchter technischer Ausdrücke verräth, von der cölestischen Photographie fern zu halten." Julius Scheiner, *Photographie der Gestirne* (Leipzig: W. Engelmann, 1897), 2-3.

time for the expert practitioners to take over. One hundred years later, distinguished astronomer Ewin Whitaker who considered himself one of the "very few professional astronomers interested in lunar research" felt compelled, in his *History of Lunar Cartography and Nomenclature* to insist, citing the necessities of navigation for Lunar spacecrafts that, that "putting names on lunar features is not just an idle pastime for dilettantes trying to ensure their place in history but is a serious and essential scientific pursuit akin to naming terrestrial mountains, seas, rivers, etc."[64]

Both the moon and photography have been favourite hunting grounds for amateurs and it should come as no surprise that a remarkable number of the characters in the history of nineteenth-century moon studies were neither professional astronomers nor professional photographers. Businessmen (de la Rue), bankers (Beer), chemists (Draper), industrialists (Nasmyth) instrument-makers or inventors (Whipple) contributed crucially to the production of lunar maps, to the development of astrophotography and to the study of the physical appearance and constitution of the moon and other planets. Labelling these figures as outsiders to astronomy (or to photography) is however not particular helpful and does little more than serving the purposes of disciplinary and professional boundary-work. A cross-comparison of photographers' and astronomers' histories indeed suggests that the amateur is a relative rather than an absolute category, and in any case an historically-situated one. A more symmetrical approach such as the one proposed here considers that both photography and astronomy (and in particular astrophysics) were not fixed entities but fields in the making. The fact that individuals such as Mädler, Janssen or Whipple eventually became professional astronomers or professional photographers testifies to their ability to transform these fields as much as to the sharp increase of institutional and commercial opportunities over the course of the nineteenth century. The history of astrophotography, considered as a history of experimentation and the production of many blurs, took place in many different settings and involved the active contribution of a range of skilled participants, some from inside and some from outside the observatory and photographic studio. Here the notion of "expert amateurs" introduced by photohistorian Clément Chéreau is useful, as distinguished from the "users" of black-boxed equipment such as the Kodak camera. Resonating with Alan Chapman's "Grand Amateurs" in Victorian astronomy, it focuses on the high

64 Whitaker, *Naming*, 189.

level of competence and commitment of individuals to their chosen pursuit rather than on their source of income.[65]

That this should still be the case today may perhaps seem more surprising, despite the wishes of the likes of Scheiner or Whitaker. The development of CCD cameras for astronomical uses since the 1970s provides a case in point, which saw an influx of physicists and electronics specialists into astronomical departments and observatories, bringing in new and important skills but who had not necessarily been previously conversant with astronomy or, for that matter, with analogue photography.[66]

Non-scientists also continue to play a role in the production of lunar and planetary iconography and its study, as Pedro Ré and Carmen Pérez González show in this volume. In the nineteenth century, lunar and planetary geography was a strong focus in popular observatories and publications, where the Mars canal discussions also took place, giving rise to a rich, sometimes fantastic iconography that sometimes echoed earlier explorations of moon photography (Figure 4.8).[67]

In today's age of planetary science, communities of "enthusiasts" study and create digital visualizations using the raw – and blurry – data from space probes that space agencies have been putting online since the mid-2000s, including from the missions to Mars and Titan.[68]

Space agencies have also pursued the old tradition of participatory astronomy that Arago had particularly valued,[69] witness Nasa's clickworkers pro-

65 Clément Chéreau, "Le jeu des amateurs," in *L'Art de la Photographie, des Origines à nos Jours*, ed. André Gunthert and Michel Poivert (Paris: Citadelles & Mazenod, 2007), 256-271. Alan Chapman, *The Victorian Amateur Astronomer: Independent Astronomical Research in Britain, 1820-1920* (Chichester: Wiley, 1998).

66 On the making of the Hubble images, see Elizabeth Kessler, *Picturing the Cosmos. Hubble Space Telescope Images and the Astronomical Sublime* (Minneapolis: University of Minnesota Press, 2012)

67 see Lane, *Geographies;* Nall *News from Mars.*

68 E.g. the "Mosaic of river channel and ridge area on Titan" made of three frames from the Huygens Descent Imager/ Spectral Radiometer (DISR) instrument provides unprecedented detail of the high ridge area including the flow down into a major river channel from different sources. See <http://www.esa.int/spaceinimages/Images/2005/01/Mosaic_of_river_channel_and_ridge_area_on_Titan> (last accessed on 5 April 2018). See also Charlotte Bigg, "In weiter Ferne so nah. Bilder des Titans," *Bildwelten des Wissens*, 5 (2007): 9-19.

69 On Arago's enrollment of the public to observe eclipses, see Theresa Levitt, "I thought this might be of interest… The Observatory as a Public Enterprise" in *The Heavens on Earth. Observatories and Astronomy in Nineteenth-Century Science and Culture*, ed. David Aubin, Charlotte Bigg and H. Otto Sibum (Durham, NC: Duke University Press, 2010), 285-304.

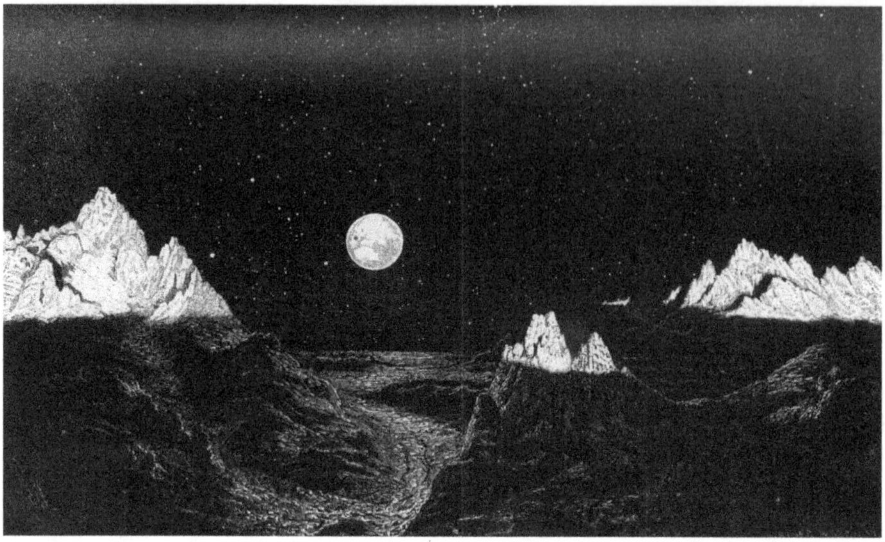

FIGURE 4.8 La Earth seen from the Moon (theatre set) for the Urania, Berlin's popular observatory. Max Wilhelm Meyer, *Illustrierter Leitfaden der Astronomie, Physik und Mikroskopie in Form eines Führers durch die Urania zu Berlin* (Berlin: Hermann Patel, 1892), 153.

gramme, an experimental online project "where you can do volunteer image analysis to help planetary science" by marking and classifying Mars craters or cataloguing its landforms.[70]

70 www.nasaclickworkers.com (last accessed 16 February 2016)

CHAPTER 5

James Nasmyth on the Moon

Or on Becoming a Lunar Being, without the Lunacy

Omar W. Nasim

In the spring of 1889, Sir Robert Ball's publisher contacted their competitor, John Murray, requesting permission to use four plates from James Nasmyth and James Carpenter's *The Moon: Considered as a Planet, a World, a Satellite* (1874). The requested plates were meant for a brand-new edition of the Irish astronomer's best-selling work, *The Story of the Heavens*. Nasmyth's publisher responded, at least at first, in the negative, writing: "You are doubtless aware that these [illustrations] are unique & form one of the special features of Mr. [Nasmyth's] book ... [and] still maintain a steady sale."[1] In the end Ball got his wish, and later editions of his book included three plates from *The Moon*, each of which was engraved and designated "after Nasmyth."[2] From their first appearance in 1874, and well into our own day, the photographs printed in *The Moon* (ex. Figures 5.1 and 5.2) have garnered a good deal of attention. The fact that many are actually prints of photographs of plaster and wood models of the moon has only increased their appeal, drawing scholarly attention from many fronts, including historians of science, art, literature, and photography. Among other things, the printed images in *The Moon* have been appreciated for their realism as well as their positivism; for their uncanny resemblance to science fiction effects and their relation to mechanical objectivity; and for the ways in which they straddle science and art.[3] It is due to such diversity of treatments that the images have often been treated in fragmented and

1 Cassell and Co. Ltd. to John Murray, with copy of response from the latter, May 11, 1889, John Murray Archive, National Library of Scotland, Edinburgh, MS. 40200, f.26-27.

2 See for example the new and revised edition, Sir Robert S. Ball, *The Story of the Heavens* (London: Cassell and Co., 1897), plates VII, VIII, IX.

3 F. Robertson, "Science and Fiction: James Nasmyth's Photographic Images of the Moon," *Victorian Studies* 48 (2006): 595-623; C. Armstrong, *Scenes in a Library: Reading the Photograph in the Book, 1843-1875* (Cambridge: MIT Press, 1998); C. Armstrong, "Der Mond als Fotografie," in *Diskurse der Fotografie. Fotokritik am Ende des fotografisches Zeitalters*, ed. Herta Wolf (Frankfurt am Main: Suhrkamp Verlag, 2003), 359-383; Ben Carver, Ben, "'A Gleaming and Glorious Star': Rethinking History in the Plurality-of-Worlds Debate," *Journal of Victorian Culture* 18 (2013): 429-452; Boris Jardine, "Made Real: Artifice and Accuracy in Nineteenth-Century Scientific Illustration," *Science Museum Group Journal* 1 (2014), accessed November

 | DOI 10.1163/9789004298873_007

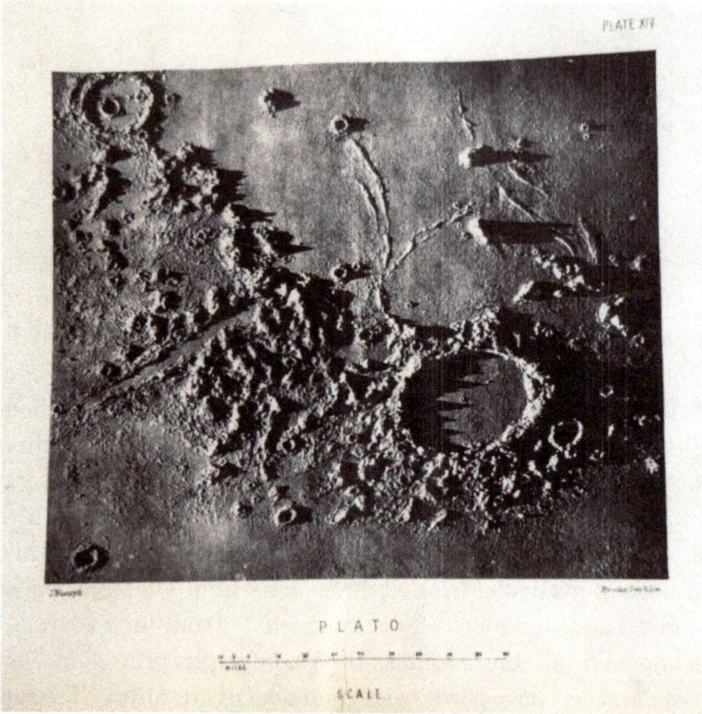

FIGURE 5.1 "Plato," James Nasmyth and James Carpenter's *The Moon: Considered as a Planet, a World, a Satellite* (1874), 1st edition, plate 14.

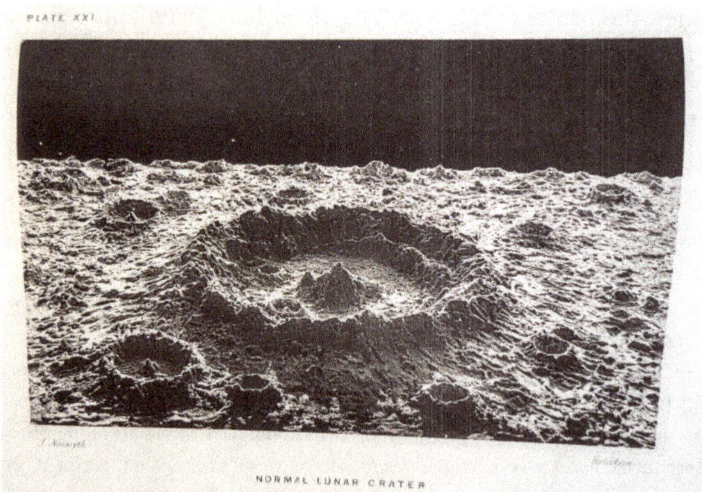

FIGURE 5.2 "Normal Lunar Crater," James Nasmyth and James Carpenter's *The Moon: Considered as a Planet, a World, a Satellite* (1874), 1st edition, heliotype, plate 21.

decontextualized ways. It is time that we add more historical depth to these pictures.

One notable exception is the article "Science and Fiction" by Francine Robertson, who has made inroads by providing a relatively well-rounded account of Nasmyth's photographs. Building on her work, I take the next step to an even richer milieu for Nasmyth's work on the moon. Beginning in the early 1840s when Nasmyth first started to make systematic observations of the moon, I embed his drawings, models and photographs back into a history that spans at least thirty-five years. The history is a robust one, which connects the products of Nasmyth's long lunar engagement to major geological, astronomical, and ethnographic expeditions; to museums, exhibitions, and lecture displays; to forgotten photographs, photographers, and models; and even to racialized studies of the human hand and empire. Though I cannot claim to be exhaustive, in the first two sections I attempt to plot out a pre-history to *The Moon*, and set up along the way numerous signposts for those wanting to follow-up.

The second way in which I wish to add historical depth is to use the richer contexts developed in this paper to challenge what is more often than not an ocular-centric approach to Nasmyth's printed photographs. Carol Armstrong's work is culpable as much as more recent work that describes Nasmyth's photographs as "replicating what he saw through the telescope with his own eyes;" a process that the author goes on to describe as imaginatively "extending beyond what is seen with the eyes alone."[4] As we shall see, without at least the inclusion of the hand, this is inadequate to a proper understanding of Nasmyth's photographs. Robertson, to be sure, has drawn our attention to the importance of the hand in Nasmyth; but it still stands in need of significant supplementation and integration. So, for instance, she does not develop the fundamental role played by the mind in Nasmyth's work, particularly as it relates to the hand. It is when we contextualize the hands, eyes, and mind into Nasmyth's longstanding interest in forms and their materialization, that we will be in a much better position to delineate his distinctive mode of observation, a mode that anchors the resulting photographs into "the actual."

Indeed, it is in appreciating the ways in which Nasmyth's otherworldly photographs are grounded in a number of concrete and tangible *handlings* that will, in turn, help us to comprehend his insistence that such photographs are "legitimate" as scientific images. In other words, however much they might seem to

---

24, 2015, <http://www.dx.doi.org/10.15180/140208>; Martin Kemp, *Seen/Unseen: Art, Science, and Intuition from Leonardo to the Hubble Telescope* (Oxford: Oxford University Press, 2006).

4 Colleen Boyle, "You Saw the Whole of the Moon: The Role of the Imagination in the Perceptual Construction of the Moon," *Leonardo* 46 (2013): 249, 250.

us today to be exemplary illustrations of fiction or fantasy – a standpoint normally assumed without much discussion in recent literature – Nasmyth insists that they are not. By taking him seriously, and on his own terms, we will see in the third section of this essay that it is precisely because Nasmyth's hands, mind and eyes *are* implicated in the production of each photograph, that he deems them to be "well-directed." It is in this way that the following is also an attempt to add layers, texture and depth to the use of photography in science, so that what follows might also act as an antidote to the widespread treatment of photographs as slick, flat-surfaces, that stand alone and untouched; for without a deeper treatment we will fail to capture what Nasmyth took to be the scientific value of photographs of plaster models, loaded with mind, hands and eyes.[5]

1 Drawings and Models, 1840-1851

The idea to photograph the moon harkens back to the early days of photography. When the daguerreotype was first publically announced in the early parts of January 1839, the director of the Paris Observatory, Dominique François Jean Arago, described multiple scientific applications for the new device. Among these was the application of the daguerreotype to the moon. Since the French Academy of Sciences had already appointed Arago to head a committee for the purpose of photographing the moon, it is clear that Arago's suggestion was more than just a vision for the future – it was something he was already engaged with. Arago includes a short report of the committee's preliminary findings already on 7 January 1839, recommending ways that the inventor of the new device, Louis-Jacques-Mandé Daguerre, might improve his optics and chemistry. Arago also reports that Daguerre, upon the committee's request, "obtained, in twenty minutes, on his dark ground, a white image of the moon, with a lens far less powerful than ours." But these early attempts appear to have been defective: a reporter described Daguerre's picture of the moon as appearing more "like the tail of a comet."[6] This was due to the fact that the motion of the moon across the sky was too quick for the camera's exposure time, resulting in a trail of moonlight left behind on the sensitive copperplate. In any case,

5 Also see Nasim, "Handling the Heavens: Things and Photo-Objects of Astronomy," *Photo-Objects: On the Materiality of Photographs and Photo-Archives in the Humanities and Sciences*, edited by Julia Bärnighausen, Costanza Caraffa, et. al. (in press, 2018).

6 In Hippolyte Gaucheraud, "Fine Arts – The Daguerotype," *The Literary Gazette; and Journal of the Belles Lettres, Arts, Sciences, &c.*, January 19 (1839): 43, 44. See also, François Arago, "Fixation des images qui se forment au foyer d'une chambre obscure," *Comptes Rendus* 8 (1839): 4-7.

Daguerre's daguerreotypes of the moon were destroyed when his laboratory caught fire and burnt to the ground on 8 March 1839. It is the American physician, John William Draper, who is thus typically credited with the first daguerreotypes of the moon, made in the spring of 1840.[7]

Well aware of the French experiments, William Henry Fox Talbot attempted a series of his own photogenic experiments with moonlight. Writing to John F.W. Herschel on 15 November 1839, Talbot reported that he took advantage of a full moon in September "to try the effect of the lunar rays upon my paper, and I obtained a distinct impression in 10 minutes."[8] Nearly a month later, Herschel too was reporting to Talbot his own experiments using salt paper to capture the lunar rays.[9] Despite these glimmers of hope for the photogenic process, in the face of the moon, it turned out to have been much too insensitive. But even when a year later Talbot developed the more sensitive calotype process, it did not seem to have obtained any notable results for selenography either. As late as 1848, to be sure, the director of the Harvard College Observatory recounts that he, along with his son George Bond, and the photographer John A. Whipple, tried out the calotype for astronomical purposes, but to no avail. Having set aside the calotype, it was in the spring of 1851 that Whipple, employing a faulty clock-drive for the Harvard equatorial telescope that barely permitted him to track the lunar orb as it traveled across the sky, successfully obtained daguerreotypes of the moon. Whipple's daguerreotypes were exhibited at the Great Exhibition of 1851, winning him the Gold Medal.

Whipple's visual representations of the moon, however, were not the only ones on display at the Great Exhibition in London. There were at least three others, each representing the celestial orb in its own distinct manner. One was a globe of the moon, done in relief by Charlotte Readhouse. Another was by the chemist, amateur astronomer, and painter, Henry Blunt of Shewsbury, who presented, in a hand-held wooden case, an electrotype copy of a plaster model of the lunar surface around the crater Eratosthenes (14.47°N, 11.32°W, diameter:

7 See, Sarah K. Gillespie, "John William Draper and the Reception of Early Scientific Photography," *History of Photography* 36 (2012): 241-254. Also see: Jon Darius, *Beyond Vision* (Oxford: Oxford University Press, 1984); and K. Hentschel, "Why not one more Imponderable?: John William Draper and his 'Tithonic rays'," *Foundations of Chemistry* 4 (2002): 5-59.

8 William H. Fox Talbot to John F. Herschel, 15 November 1839, *The Correspondence of William Henry Fox Talbot,* collection number: HS 17:298, last accessed November 15, 2015, <http://www.foxtalbot.dmu.ac.uk>.

9 John F. Herschel to William H. Fox Talbot, 4 December 1839, *The Correspondence of William Henry Fox Talbot,* collection number: 1937-4854, last accessed November 15, 2015, <http://foxtalbot.dmu.ac.uk/letters>.

59 km, depth: 3.6 km).[10] A short report on Blunt's model, presented at the meeting of the British Association for the Advancement of Science (BAAS) in Birmingham, points out that "the model ought to be held in an oblique light in order to view it to advantage."[11] Perhaps it was due the spectators' difficulties in obtaining just the right angles to see the appropriate effect, that Blunt's model did not win much recognition. And finally, the third lunar representation exhibited was by the engineer and inventor of the steam-hammer, James Nasmyth, who entered a painted map of the moon that was over six-feet in extent. The catalogue description for the map seems to have been impressed by the fact that it showed "the relative positions and character of the most striking features of [the moon's] surface, as they appear when seen under the most favorable circumstances in respect to light and shade."[12] Nasmyth's large map of the moon won the exhibition's Prize Medal and attracted much attention, including that of Prince Albert.[13]

On the evening of October 9, 1851, Nasmyth was granted an audience with the Royal Family, who, while visiting Manchester, were guests of Lord Ellesmere. The Prince was keen to learn more about Nasmyth's "graphic lunar studies," and requested that the inventor be invited over after dinner.[14] Queen Victoria records this meeting in her diary:

> After dinner we saw some very interesting drawings of the moon by Mr Nasmyth, a Scotchman, & an Engineer, a most intelligent man, who has entirely raised himself by his own efforts, & has a large factory at Paticroft [sic], where he has a steam hammer. He has made many wonderful & useful inventions, & is very simple, modest & unaffected. He is quite an enthusiast about the moon & the researches he has made by means of a telescope of his own construction are very wonderful. He made the map of the moon, which is in the great Exhibition. By these drawings he shows that there are immense volcanic mountains & craters in the moon, that they are perfectly round, proving the entire absence of all atmo-

10 *Great exhibition of the works of industry of all nations, 1851: official descriptive and illustrated catalogue*, vol. 1 (London: Spicer Brothers and W. Clowes, 1851), 468-69, 450.

11 Henry Blunt, "On a Model of the Moon's Surface," *Notices and Abstracts of Communication to the British Association for the Advancement of Science, at the Birmingham Meeting, September 1849* (London: John Murray, 1850): 1. The model is now a part of the Science Museum in London's collection, object number: 1858-43.

12 *Great exhibition of the works of industry of all nations, 1851*, 471.

13 It was at the Great Exhibition of 1851 that Nasmyth also won the Council Medal for his steam hammer.

14 James Nasmyth, *James Nasmyth: Engineer, An Autobiography*, ed. Samuel Smiles (London: John Murray, 1912), 332.

> sphere & of all wind.... He also has made drawings representing the effect he supposes the sun to have on the moon, – a perfectly black sky, from the absence of atmosphere, – no vegetation, – nothing but the roughest, boldest rocks, & the most intense sunshine. It was extraordinarily interesting to hear what he had to say & he explained all so delightfully. What fields science opens to the human mind! How much more of God's visible (& invisible) world may we not possibly soon be led to admire & comprehend! We saw also some other very pretty sketches by Mr Nasmyth.[15]

For his part, what Nasmyth recalled most about that evening was the Prince's reaction, who "said that the drawings opened up quite a new subject to him, which he had not before had the opportunity to consider."[16] As we shall see, this was not the last time Nasmyth's sketches, not to mention his enthusiasm, effected in others an interest in the study of the moon; nor was this the first time that the sketches of the moon were shared, put on display or used by Nasmyth to explain what he had seen on the moon.

Over seven years earlier, Nasmyth read a paper to the Royal Astronomical Society (RAS) that included seven small pen and ink sketches, and one large painting of a prominent lunar crater Maurolycus (41.8°S, 14.0°S; diameter: 114 km, depth: 4.7 km).[17] This lecture of 1844 was the result of nearly four years of work using two Newtonian reflecting telescopes, constructed by Nasmyth for use at his home in Patricroft, near Manchester: one with an 8¾ inch aperture and a focal length of 9 feet; and the other had a 12 inch aperture and a 13 feet focal length. It was also during this period that Nasmyth constructed what would be his most effective reflecting telescope, one with a mirror of 20 inches in diameter. Accompanying the drawings and painting presented at the lecture, Nasmyth also displayed to the RAS a plaster model, supported on wood, and done to scale, of the crater Maurolycus and its surrounds. In fact, the drawings and painting shown at the meeting were only a selection of a series of drawings that would have contributed to the production of the three-dimensional model of the crater (Figure 5.3), which was presented by Nasmyth at the meeting as a gift to the RAS.

15 Entry for 9 October 1851 (Princess Beatrice's copies), vol. 32 (1st July 1851-31st December 1851), 139-141, RA VIC/MAIN/QVJ (W); retrieved: November 18, 2015.

16 Nasmyth, *James Nasmyth, Engineer*, 333.

17 James Nasmyth, "Some Remarks on the Telescopic Appearance of the Moon," *Monthly Notices of the Royal Astronomical Society* 6 (1844): 79-82.

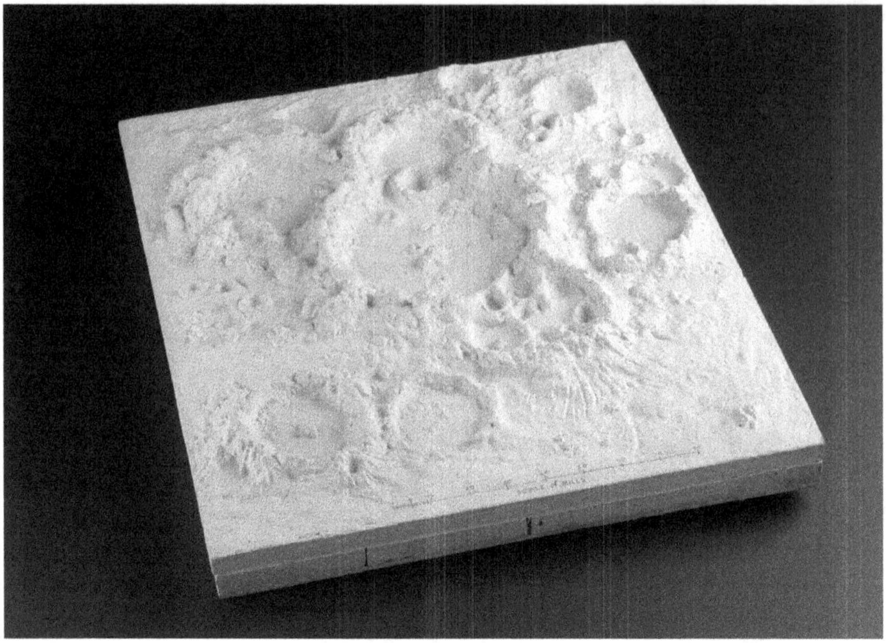

FIGURE 5.3 Nasmyth's early plaster and wood model of the crater Maurolycus, originally presented as a gift to the Royal Astronomical Society in 1844, now at the SCIENCE MUSEUM IN LONDON, RAS INV. 1931-682. REPRODUCED WITH PERMISSION.

Nasmyth was making models of the moon as soon as he began making observations in the early part of the 1840s. Rather than remaining content with flat, two-dimensional representations of the moon, that is, he went the extra-mile to produce three-dimensional relief models in plaster and wood. Although I will explore in more detail why this was the case, it is important to note that already in the 1840s Nasmyth provides us with two interesting hints. In the first case, Nasmyth writes: "With a desire to become *minutely acquainted* with the nature and structure of the surface of the moon, I have almost exclusively confined my attention, in such matters, to a *close inspection* of the lunar surface during the last four years.... It was with a view to *communicate* more clear and just ideas on this subject, that I have constructed the model which I have the honour to submit to the attention of the Society."[18] As we shall see, close inspection, attention, and the minute acquaintance gained thereby, are related to the tangible work of the eyes, hands and the mind in the construction of plaster models. And secondly, by providing the third dimension

18 James Nasmyth, "On the Telescopic Appearance of the Moon," *Memoirs of the Royal Astronomical Society* 15 (1846): 147; my italic.

Nasmyth believed he could make the moon relevant for geology as well: "The model was constructed with a view of illustrating the close relationship which appears to exist between the structure of the lunar surface and that of a considerable portion of the earth."[19] When one closely examines Nasmyth's early plaster model of Maurolycus (see the bottom right-hand corner of the model in Fig. 5.3), one finds on the same surface a tiny, scaled model of Mount Vesuvius. Finding a small terrestrial volcano next to a lunar crater not only expresses an interest in utilizing analogies based on forms, but it also highlights the performative character of the models; in other words, the plaster models were primarily used as center pieces of explanation and public display, and not something made to be photographed – at least not yet.

Apart from the plaster model of Maurolycus, Nasmyth built others, which were not photographed until a decade later. In the meantime, he continued to exhibit the models and drawings at public lectures. In the spring of 1845, Nasmyth wrote to an old Edinburgh friend, that one of his recent lectures in London was received "with 'Eclat' as they say. The subject appeared to interest a most full audience of authoritative persons very deeply. I had a very large and fine model and a lot of very large and effective drawings done in Distemper ... and brought the subject before them in good understandable style ... My theory of the contraction of the surface as the *cause* of the volcanic Era alike on the moon as on the Earth has taken Root in the minds of the knowing men."[20] Considering the doubly controversial nature of Nasmyth's theory, he had good reason to be pleased at this time. The theory about the earth's original molten state was just regaining ground thanks to the work of geologists like George Poulett Scrope (1797-1876). But the molten state theory was also loosely connected with the controversial Nebular Hypothesis – a connection that was soon to be fortified in the work of Herbert Spencer, and others.[21] Nasmyth was at the cutting-edge of the science of the day, one that employed theories of cosmological significance to explain geological features of the earth.

At around this time, Nasmyth presented lunar drawings and another, newly constructed plaster model of the moon to the Museum of Economic Geology at Craig's Court, Whitehall in London.[22] This was a museum connected to the

---

19 Nasmyth, "Some remarks," 79.

20 Nasmyth to D.O. Hill, April 30, 1845, A9.42B, Royal Observatory of Edinburgh; italics in original.

21 For more on Scrope, see Martin J. Rudwick, "Poulett Scrope on the Volcanoes of Auvergne: Lyellian Time and Political Economy," *British Journal for the History of Science* 7 (1974): 205-42. Also see the wide-ranging review by Herbert Spencer, "Recent Astronomy, and the Nebular Hypothesis," *The Westminster Review* 14 (1858): 185-225.

22 Nasmyth had already submitted other items for display at the Museum and seems to have suggested ways of displaying things to the director, see letter from Nasmyth to De la

Geological Ordinance Survey directed by Sir Henry De la Beche, and curated, from 1839 onwards, by the chemist Richard Phillips. From its very inception in 1837, the museum was conceived as a space where members of the public might enter for free and acquaint themselves with all sorts of geological specimens ranging from ores, soil samples, minerals, fossils, and many other materials brought back from ongoing geological surveys around the empire.[23] Visitors would have seen Nasmyth's model and drawings alongside survey tools, precious metals, large maps, and crystals, all meant to give a picture of the empire, and the practicable roles of geology and economics in it. Though seemingly out of place, Nasmyth's lunar representations were seen as usefully contributing to the geological sciences as well.

We learn about this connection between the Museum of Economic Geology and Nasmyth by means of the first letter by the engineer to John F.W. Herschel, sent in the summer of 1845. Having heard that Herschel was preparing a presentation on the moon for the November meeting of the Royal Astronomical Society, Nasmyth ceased the opportunity to make Herschel an offer:

> I have taken the liberty to address you in order to offer to you my humble services in respect to furnishing you for that occasion with some large drawings which I have made from time to time of the various and more remarkable portions of the Lunar surface – some of these drawings are now under the charge of Sir H. De La Beche at the Museum of Economic Geology, where there is also a large model of a remarkable portion of the moon's surface, which should you think suitable as an illustration of your subject I doubt not Sir H. would make you most welcome to the use of it.

Nasmyth goes as far as to offer Herschel custom made models or drawings of any part of the moon he required, all to be made directly "*from Nature*."[24]

Herschel, however, had already made other arrangements for his talk on the moon, which was to be centered on an exquisite globe of the moon constructed in 1838-9 by Frau Hofrätin Wilhelmine von Witte of Hanover, instead.[25] She constructed the lunar globe using wax and mastic. It stands at 26 inches high

---

Beche, July 2, 1842 in Tom Sharpe and P.J. McCartney, *The Papers of H.T. De la Beche (1796-1855) in the National Museum of Wales* (Cardiff: National Museum Wales Books, 1998), 78.

23 Thomas Sopwith, *Account of the Museum of Economic Geology, and Mining Records Office* (London: John Murray, 1843).

24 Nasmyth to Herschel, August 1, 1845, Royal Society Archives, London, 13:76.

25 It was on Herschel's way back from the Cape of Good Hope 1838 that he visited his Aunt in Hanover, where he also first encountered von Witte's lunar globe. John Herschel to Margaret Herschel, July 22, 1838, TxU:H/L-0538.5.

and is about 12 inches in diameter.[26] The many minute details on the globe, executed in relief, were laid down on the basis of the latitudes and longitudes provided in the meticulous triangulation surveys of the lunar surface by Wilhelm W. Beer and Johann H. Mädler. The micrometric details were checked and rechecked through von Witte's own Fraunhofer telescope and inscribed onto the globe's surface using a magnifying glass. Despite the accuracy obtained in laying down two-dimensional details onto the surface, due to the small scale of the model, the three-dimensional heights of the lunar craters or mountains had to be doubled in proportion to the rest in order for the details in relief to be noticeable to the eye. After Herschel's lecture on the globe, von Witte's "admirable work" was "submitted to the closer inspection of the members" of the RAS.[27] Soon afterwards, De la Beche made attempts through Admiral William Henry Smyth and John Herschel to purchase her model for the Museum of Economic Geology – perhaps in order to exhibit it next to Nasmyth's drawings and model of the moon.[28]

Utilizing nearly nine years' worth of observations, in 1849 Nasmyth completed a brand-new map of the moon, one that was formulated – just as von Witte had done – on the basis of Beer and Mädler's celebrated *Mappa Selenographica* (1834). Beer and Mädler's map is made up of four separately lithographed quadrants – each printed on an elephant folio sized page – which when taken together represent the moon with a total diameter of about 37 inches. Nasmyth's painted map of the moon is in fact a scaled-up version of Beer and Mädler's already giant map. In the case of the latter, however, the map relies on topographical conventions for depicting thousands of surface-features, and thus lacks the pictorial interplay of light and shadow on the moon's surface. In contrast, Nasmyth's large "picture-map" incorporates the "appropriate shading," not only because he wished to depict the moon with the effect of relief, but because the shadows are essential to the observation of the moon: "[an observer] would be struck by the shadows by which the lunar objects reveal themselves: he would get to know them mostly by their shadows, since it is mainly by those that their forms are revealed to a terrestrial observer."[29]

---

26 Herschel must have seen von Witte's model that now resides at the Historisches Museum Hannover. For more information on von Witte's and other relief models of the moon from the period see, Juergen Blunck, "Deutsche Pionierarbeiten der Monddarstellung mit erhabenem Relief," *Der Globusfreund* 47/48 (2000): 293-303.

27 John Herschel, "Description of model of the surface of the moon by Frau H. Witte," *Monthly Notices of the Royal Astronomical Society* 7 (1845): 15-16.

28 Herschel to De la Beche, June 11, 1846, *The Papers of H.T. De la Beche*, 55-56.

29 Nasmyth and Carpenter, *The Moon*, 68-69.

Considering the picture-map's size (197 × 199cm) – truly meant to engulf, if not overwhelm, the viewer – it is no wonder that it was this very map of the moon that Nasmyth entered for display at the Great Exhibition of 1851 – it is the same that would so impress the Prince and which would go on to win the Prize Medal at the exhibition.[30] But before he submitted the picture-map, Nasmyth had the opportunity in 1850 to show it at the BAAS meeting in Edinburgh. At first, it was presented to the elite Physical Section A; but because it was so well received there, the council asked Nasmyth to re-present the material to an assembly of all the members of the various sections. "The novelty of the subject," he later wrote, "and the graphic illustrations helped me very much. I was quite full of the Moon."[31] It was also this same picture-map that was in later years to be photographed, reduced in scale, and reproduced as plate IV for Nasmyth's *The Moon*.

2 Photographing Models, 1853-1865

We have seen that at least until the beginning of the 1850s, Nasmyth was busy lecturing, making many different sorts of lunar drawings based on a series of telescopic observations, and on their basis, constructing plaster models of the moon. What we do not find in this period is Nasmyth busy with photography – it is conspicuously missing in action. Given the iconic status of *The Moon* for the history of photography, this might come as a surprise to some. But it was only in the summer of 1853 that the suggestion was first made to Nasmyth by David Roberts, the Scottish painter of the Orient. After praising Nasmyth's depictions of "The Landscape of other worlds," which Nasmyth shared with him, Roberts recommended photographing "those wonderful drawings" – no mention of photographing the models is made, however.[32]

This is not to suggest that Nasmyth had no previous interest in photography. In letters to an old family friend, the renowned Scottish calotype photographer, David Octavius Hill, Nasmyth referred to photography in the most reverent of terms. When asking after Hill's artistic collaborator Robert Adamson, and their female assistant, Nasmyth wrote: "How goes on the divine solar art? And how does that worthy artist Mr Adamson the authentic contriver & manipulator in the art of light and darkness? And thrice worthy Miss Mann that

30 This large picture-map of the Moon can be found in Science Museum's collection in London, 1956-153/3.

31 Nasmyth, *James Nasmyth, Engineer*, 328, 324.

32 In Nasmyth, *James Nasmyth, Engineer*, 351.

most skillfull [sic] and zealous assistant ... I have the remembrance of All on Em so clearly calotyped in my minds [sic] Eye as last I saw them in full manipulation of the divine art of light."[33] In another letter, when referring to a set of calotypes by Hill and Adamson, Nasmyth declared, "it is nature reduced to an artistical standard and abounding in the most noble suggestion."[34] A leading authority on Hill has suggested that it was none other than Nasmyth that was Hill's "first conduit to" photography.[35]

Bearing in mind Nasmyth's admiration for what he called "that most Divine & magic art," it might be asked: Why do we not find Nasmyth taking photographs – whether of drawings or models – during his early observations of the moon? It turns out that he did not quite know how to use photography.[36] As late as 1847, when referring to the "progress" of the calotype, Nasmyth admitted that, "I wish I had time and knowledge to practice so fascinating an art as it is..."[37] And yet, in 1855 the first photographs of Nasmyth's lunar models were published (Figure 5.4).[38] However, it was not Nasmyth who took these photographs of two plaster models (one of the crater Copernicus and the other of Walter).[39] Instead, it was his Mancunian friend, Joseph Sidebotham who made these using the recently invented collodion process. Sidebotham was, among other things, a calico printer at the Strines printing company in Manchester, a fellow of both the Royal Astronomical Society and the Society of Antiquarians, a member of the Manchester Literary and Philosophical Society, a naturalist, a founding member of the Photographic Society of Manchester, and a respected photographer – Nasmyth's models were certainly in good hands.

Sidebotham's two collodion pictures of the two lunar models were first printed and published in 1855 for a volume of the *The Strines Journal: A Monthly Magazine of Literature, Science and Art*; the company's own in-house journal, edited by two of its employees. Sidebotham, who would remain the company's director for thirty years, often contributed drawings, photographs and articles to the journal. The two salt-paper prints of Sidebotham's collodions were a

33 Nasmyth to Hill, March 27, 1847, A9.42B, Royal Observatory of Edinburgh.

34 Quoted in Sara Stevenson, *The Personal Art of David Octavius Hill* (New Haven: Yale University Press, 2002), 33.

35 Stevenson, *The Personal Art*, 22.

36 Nasmyth to Hill, April 30, 1845, A9.42B, Royal Observatory of Edinburgh.

37 Nasmyth to Hill, March 27, 1847, A9.42B, Royal Observatory of Edinburgh.

38 James Nasmyth, "The Moon and its Surface Inequalities," *The Strines Journal: A Monthly Magazine of Literature, Science and Art* 4 (1855): 110-115. See also, Michael Hallett, "The Strines Journal and the Nasmyth Steam Hammer," *History of Photography: An International Quaterly*, 13 (1989): 221-222.

39 33.0°S, 0.7°E, diameter: 132 × 140 km, depth: 4.1 km

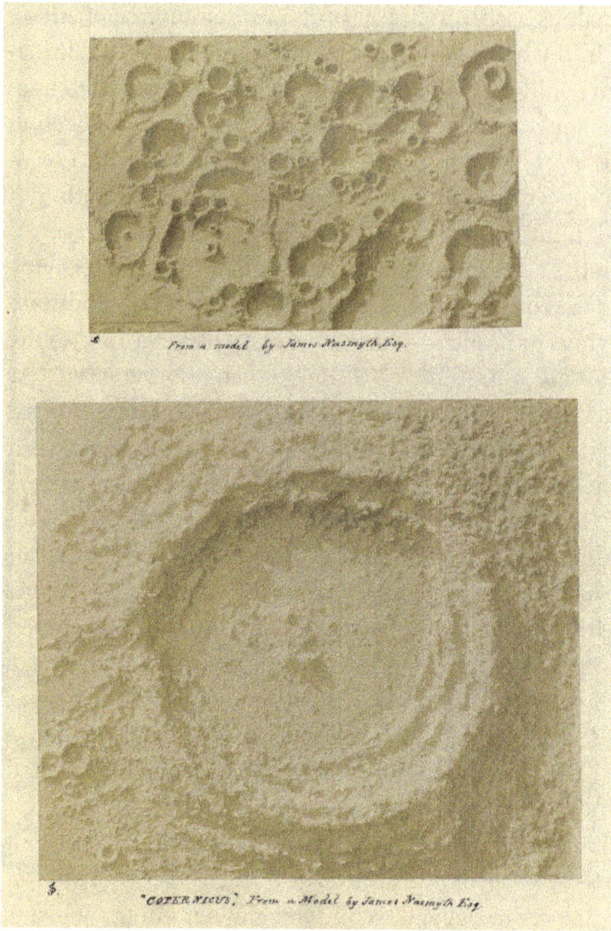

FIGURE 5.4
Two salt-paper prints of Sidebotham's collodions of Nasmyth's models. Printed for James Nasmyth, "The Moon and its Surface Inequalities," *The Strines Journal: A Monthly Magazine of Literature, Science and Art* 4 (1855): 110-115.

part of Nasmyth's article on the moon, which also included a reproduction of Whipple's famous lunar photograph. A year later, both of Sidebotham's collodion pictures (along with seventeen of his waxed paper photographs) were exhibited at the exhibition of the Manchester Photographic Society.[40] With a total of 739 photographs exhibited, it was a relatively large exhibition; and among those showing were Roger Fenton, David Octavius Hill, Adolphe Braun, and Oscar Gustav Rejlander.[41] Sidebotham's photographs of Nasmyth's lunar

40 [Thomas Sutton], "Photographic Exhibition of the Manchester Photographic Society," *Photographic Notes: Journal of the Photographic Society of Scotland and of the Manchester Photographic Society* 1 (1856), 204; and page 126 in same volume.

41 Information retrieved from "Photographic Exhibitions in Britain 1839-1865", last accessed December 17, 2015, <http://peib.dmu.ac.uk/index.php>. It should also be noted that

models, in other words, were seen in the same venue as some of the most celebrated photographers of the nineteenth century. Considering the importance of Sidebotham's collodions for the history of photography, it is both a surprise and unfortunate that they continue to be neglected or incorrectly attributed, even today.[42] At any rate, it was at this high point that, at the very end of 1856, Nasmyth retired at the age of 48 from his Bridgewater Foundry in Manchester, and with his wife moved to a home in Kent, where he could focus exclusively on his "hobbies," and perhaps find some time for learning photography too.

Besides his usual drawings and plaster models, Sidebotham's photographs were now a part of Nasmyth's amplified repertoire. Although he could certainly marshal them in subsequent lectures, there remained challenges posed by the day's technology in reproducing the photographs for print. Sidebotham's photographs were reproduced for *The Strines Journal* as salt paper prints, but they diminished the stark contrasts in tone so important to Nasmyth's interest in forms shaped by shadows and light. But there were ways around these technical challenges. Among the most striking was Edwin M. Williams exquisite chronolithograph of the dense surroundings of the crater Walter, done for Nasmyth's article in the first volume of the *Quarterly Journal of Science* (1864): it was, in fact, made after a drawing by Nasmyth of Sidebotham's photograph of the crater.[43] But despite these challenges, the photographs traveled as reproductions and were used by others in a variety of ways. So, for instance, just a year after William's chronolithograph was published, we find both of Sidebotham's photographs reproduced as engravings by E. Salle in the first edition of the gorgeous, abundantly illustrated, and popular volume, *Le Ciel* (1865) by Amédée Guillemin. Significant for us is how Guillemin employs the two figures. Recognizing Nasmyth's interests in conceptual and physical analogies,

---

Rejlander was associated with Darwin, see the wonderful account in Phillip Prodger, *Darwin's Camera: Art and Photography in the Theory of Evolution* (Oxford: Oxford University Press, 2009). As Robertson has pointed out, it was Darwin's success, especially in John Murray's eyes, in using new photomechanical reproduction methods that paved the way for James Nasmyth's daring reproductions (Robertson, "Science and Fiction," 610, 611, 613).

42 Unfortunately Sidebotham's photograph of Nasmyth's model of the crater Copernicus, for example is still incorrectly attributed to John Herschel, like in Robertson, "Science and Fiction," 609; Frances Terpak, "Imaging the Moon," in *Devices of Wonder*, ed. Barbara Maria Stafford and Frances Terpak (Los Angeles: Getty Research Institute, 2001), 200-203; and Michael Benson, *Cosmigraphics: Picturing Space Through Time* (New York: Abrams, 2014), 89. Even in Guillemin, one of the engravings of Sidebotham's collodions is attributed to Admiral W.H. Smyth.

43 James Nasmyth, "On the Physical Aspects of the Moon's Surface," *The Quarterly Journal of Science* 1 (1864): 395-401, plate between pages 394-95.

the French writer asks his readers to compare Nasmyth's lunar craters with another illustration, a topographical map of a terrestrial peak by Charles Piazzi Smyth.[44] This map was one of the many results of an expedition by the Smyths to Tenerife in the Canary Islands in 1856, the primary goal of which was to determine the practicality of fulfilling Isaac Newton's recommendation to build astronomical observatories on mountain tops. But there is more to this connection between Piazzi Smyth and Nasmyth.

Charles Piazzi Smyth, the Astronomer Royal for Scotland and a Professor of Practical Astronomy at the University of Edinburgh, was also a mutual friend of both Nasmyth and Sidebotham; indeed, the biographies of all three overlap in many interesting ways. So, for example, Smyth turned to Sidebotham for suggestions on how to employ photography in the dark tombs of the Egyptian pyramids – Sidebotham recommended flash which Smyth used to some effect. And sometime in 1857, Nasmyth constructed for Smyth, upon the latter's return from the Canary Islands, a coloured, plaster relief model of the peak of Tenerife. The model was built according to the specifications provided by Smyth's survey of the region, and was subsequently photographed in stereo by Smyth and used as the frontispiece to his *Report on the Teneriffe Astronomical Experiment of 1856, Addressed to the Lords Commissioners of the Admiralty* (1858) (Figure 5.5).[45] This was a summary of the scientific results of his Tenerife expedition that included a batch of stereo-photographs he was not permitted to print in an earlier version of the report published for the *Philosophical Transactions*.[46] The results of the expedition secured Smyth's reputation in scientific circles, and formed the basis of a series of popular lectures. At these lectures, along with lanternslides made by Sidebotham, Smyth exhibited Nasmyth's plaster model of the peak of Tenerife.[47] He did this, for instance on 5 March 1858 at the Royal Institution of Great Britain in London. A week later, Smyth again lectured on the expedition but this time to the Royal Astronomical Society, where once again he displayed Nasmyth's model of the terrestrial mountain. At some point during this lecture, the model was inundated with limelight so as to bring out variations in form and elevation depicted in the

44 Amédée Guillemin, *Le Ciel*, 228, fig. 83.

45 And in turn, Smyth let Nasmyth use two stereo-photographs from his Tenerife expedition for *The Moon*, where the two are combined to figure a line drawing, see figure 40 in Nasmyth and Carpenter, *The Moon*, 129.

46 For details see Larry Schaaf, "Piazzi Smyth at Teneriffe: Part 1 The Expedition and the Resulting Book," *History of Photography* 4 (1980): 289-307.

47 See for instance one he gave at the Royal Institution on 5 March 1858. See full account in *Notices of the Proceedings, Royal Institution of Great Britain*, Vol. 2, No. 28 (5th March, 1858), 493-97.

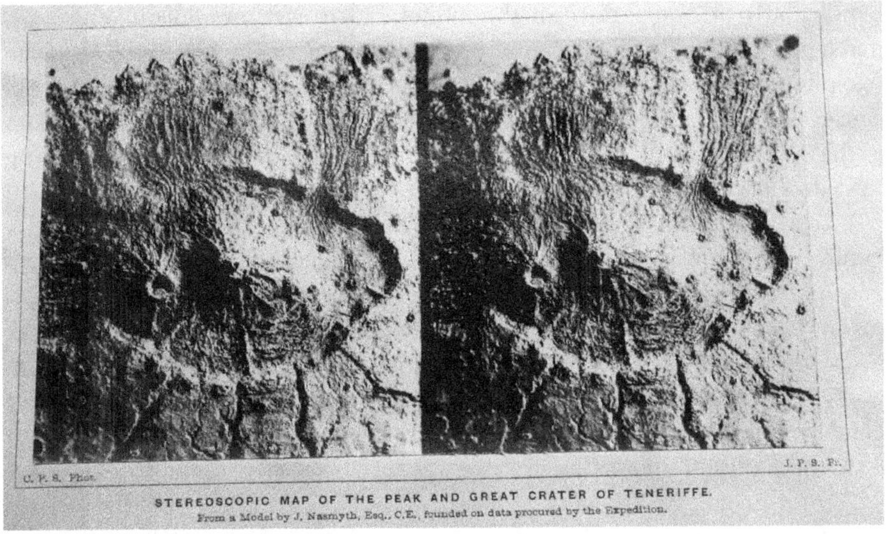

FIGURE 5.5 Stereo-photograph of a coloured plaster and wood model of the peak of Tenerife by James Nasmyth, and printed as the frontispiece to Charles Piazzi Smyth's *Report on the Teneriffe Astronomical Experiment of 1856, Addressed to the Lords Commissioners of the Admiralty* (1858).

coloured plaster relief. In addition to employing Nasmyth's model for his talk and six large drawings, also by Nasmyth, Smyth presented eighteen photographs, projected onto the wall by limelight. After Smyth's talk, George Biddell Airy asked Nasmyth to say a few words on the connection between lunar and terrestrial craters and volcanoes, and his drawings.[48]

Smyth's stereo-photograph of Nasmyth's plaster model of a terrestrial peak is reminiscent of another set of photographs made just a few years earlier. These too were photographs of plaster models of terrestrial peaks (i.e., the Alpine mountains of Monte Rosa and the Zugspitze), published in 1854 by the brothers Adolphe and Hermann Schlagintweit. The artist F. Warnstedt built two relief models of Monte Rosa for the Schlagintweits, in strict accordance with the data acquired by the brothers' extensive surveys of the mountains, with special reference to their hypsometric results. One of these models was left white and the other was coloured. The Berlin photographer, Wilhelm Halffter took collodion images of each of the models as they stood vertically, and with a source of light coming in at an angle of 40° or 50° from the north-west part of the model (Figure 5.6). Attempts were also made to have light fall per-

48 Charles P. Smyth, "On Certain Connecting Points between Lunar and Terrestrial Volcanoes," *Monthly Notices of the Royal Astronomical Society* 18 (1858): 156-162.

pendicularly on the surface of the model, but it cast "unrealistic" shadows, making certain parts too dark. The brothers, in fact, go on to stress that of the two photographs it was the *uniformly white model* that was to be preferred because it showed features "in the highest precision;" while the photograph taken of the colored model presented only a "natural and general picture" of Monte Rosa and its surroundings.[49]

Like Adolphe and Hermann Schlagintweits' white plaster models, Nasmyth's lunar models too were left uncoloured. To understand why this might have been preferred or even deemed to be more precise, especially for photographical purposes, it is worthwhile recalling two salted paper prints from paper negatives of a white plaster cast of a bust, published in William Henry Fox Talbot's *The Pencil of Nature*. The accompanying text to one of these two plates details photography's ability to capture a variety of effects of light and dark. It is worth quoting Talbot at length:

> Statues, busts, and other specimens of sculpture, are generally well represented by the Photographic Art; and also very rapidly, in consequence of their whiteness. These delineations are susceptible of an almost unlimited variety; since in the first place, a statue may be placed in any position with regard to the sun, either directly opposite to it, or at any angle: the directness or obliquity of the illumination causing of course an immense difference in the effect. And when a choice has been made of the direction in which the sun's rays shall fall, the statue may be turned round on its pedestal, which produces a second set of variations no less considerable than the first. And when to this is added the change of size which is produced in the image by bringing the *camera obscura* nearer to the statue or removing it further off, it becomes evident how very great a number of different effects may be obtained from a single specimen of sculpture.[50]

49 Adolphe and Hermann Schlagintweit, *Épreuves de cartes géographiques produites par la photographie d'après les reliefs du Mont Rose et de la Zugspitze* (Leipzig, J.A. Barth, 1854). Electrotype reproductions of the plaster models were made in order that they might be sold, along with stereo-photographs of the relief models (see: Otto Funke, *Bericht ueber die Schlagintweit'schen Reliefe des Monte-Rosa in der Schweiz und der Gruppe der Zugspitze und des Wettersteines in den bayerischen Voralpen* (Leipzig, J.A. Barth, 1855)). I was first made aware of these photographs in a paper presented by Sigrid Schulze, "Seen from Above: Photographs of Terrain Models by Hermann und Adolph Schlagintweit 1854," (paper presented at the workshop, Hybrid Photography, Berlin, Germany, February 19-21, 2015).

50 William Henry Fox Talbot, *The Pencil of Nature* (London: Longman, Brown, Green, & Longman, 1844-6), text to plate V. A good analysis is provided in Joel Snyder,

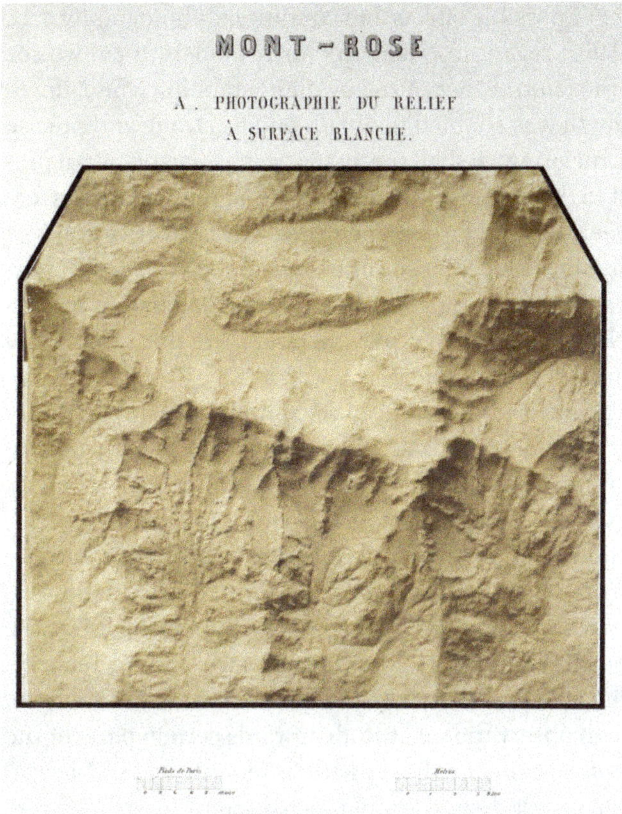

FIGURE 5.6
A print from a collodion by Wilhelm Halffter of a white plaster model for Adolphe and Hermann Schlagintweit, *Épreuves de cartes géographiques produites par la photographie d'après les reliefs du Mont Rose et de la Zugspitze* (Leipzig, J. A. Barth, 1854). COURTESY OF THE ROYAL GEOGRAPHICAL SOCIETY.

Just as in Pliny's legend of the discovery of drawing and from it sculpture, shadows have immense significance for photography too.[51] In effect, Talbot's statement pinpoints the practical elements that make photographs of white plaster models so much more effective than drawings or models alone. This is mainly because of the variety of effects attainable by multiple configurations of a white model's position in relation to light-source and camera, which in turn points to more control in obtaining the right light or dark effects. And in contrast to walking around and looking at a relief model on display, where the best conditions can be hard to acquire (recall Blunt's model), photography can be used to arrest and show just the right forms, chiseled out of shadows and light.

"Nineteenth-Century Photography of Sculpture and the Rhetoric of Substitution," in *Sculpture and Photograph: Envisioning the Third Dimension*, ed. Geraldine A. Johnson (Cambridge: Cambridge University Press, 1998), 21-34.

51 Hagi Kenaan, "Photography and Its Shadow," *Critical Inquiry* 41 (2015): 541-72.

In the face of the number of possible effects and conditions, photography acts, therefore, as a control. With regard to Nasmyth's lunar models, then, we see that it was by means of the controls permitted by photographing white, uncoloured models that Nasmyth was enabled to obtain and fix strong and precise photographic forms, so crucial, as we shall see, to his overall observational program. Finally, in contrast to the contingently unexpected in photography, a feature that has recently been regarded to be one of the most effective traits of "explorative photography" in science, these controls actually limit – if not extinguish – the unexpected or the contingent that might remain latent in Nasmyth's photographs.[52] Nasmyth's photographs, therefore, are not exploratory as much as they are domesticating.

Getting the least unexpected photographs was central to Nasmyth's lunar project, not only for purposes of visualization, but also analogy. Returning to the stereo-photograph of Nasmyth's model of the peak of Tenerife for Smyth, we find a kind of interchangeability between the lunar and the terrestrial, which is evident in methods and techniques, but also in conceptions and nomenclature; while Nasmyth encounters *terrestrial* forms on the Moon (i.e., volcanoes), Smyth sees *lunar* forms on Earth – such as the "Lunar Rocks" at the bottom of the "Great Crater" (Figure 5.7) in the volcanic surrounds of Tenerife. But the interchangeability is reflected in more than just the look of the terrain, it is explicit in Smyth's own observational attitude towards certain parts of the earth. Some of the volcanic terrain Smyth met with in Tenerife was so "savage" that no horse or foot could access them without much discomfort; so how were these dangers to be overcome? Smyth offers an intriguing methodological answer. He writes that it was thanks to the "bird's-eye view commanded by our astronomical observatory on Mount Guajara, assisted by the hypsometric shading of morning and evening suns, enabled us, with comparative ease, to arrive at these broad results, of what may be called *telescopic petrology*, and in a manner much more applicable to questions of the lunar surface than the ordinary observations and measures of the geologist."[53] In other words, on Smyth's sponsored mission to assess the practicality of mountaintop observatories, he was not just concerned with how well telescopes might function in making celestial observations, but with how well they might function in making terrestrial observations from such heights. Indeed, at his lecture to the RAS, Smyth went as far as to suggest that geologists should use mountaintop

52 Jan von Brevern, "Counting on the Unexpected: Aime Civiale's Mountain Photography," *Science in Context* 22 (2009): 409-437.

53 Charles P. Smyth, *Report on the Teneriffe Astronomical Experiment of 1856, Addressed to the Lords Commissioners of the Admiralty* (London and Edinburgh: 1858), 542; my italics.

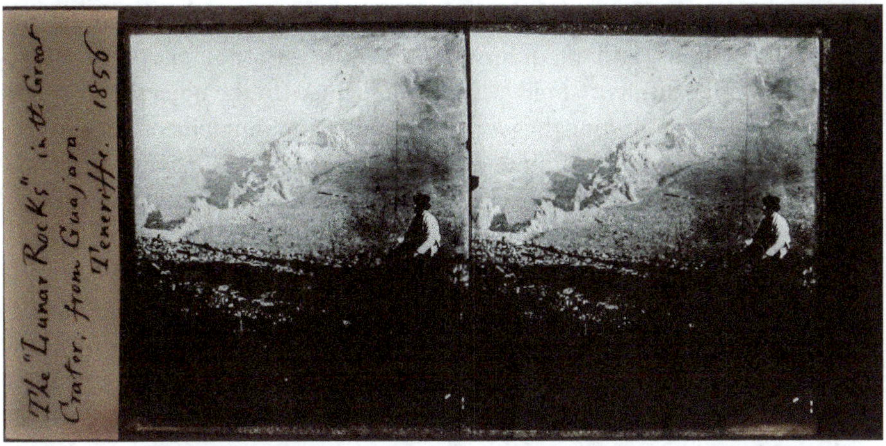

FIGURE 5.7 Charles Piazzi Smyth, "The 'Lunar Rocks' in the Great Crater; from Guajara." Stereo-photographic glass plate, box 1, "Guajara 1856", COURTESY OF THE ROYAL OBSERVATORY OF EDINBURGH.

observatories in order to take advantage of the bird's eye view to see what they might otherwise miss; in this case, lunar-like craters right here on earth. The suggestion was explicitly made in order to link "methods of observation" between geology and astronomy, so that the former might have a chance to view the earth's surface and its features "from such heights and distances as in the naked eye to subtend something like the same angles that those of the moon do in the telescope."[54]

Smyth's telescopic petrology is clearly reminiscent of Nasmyth's telescopic studies of the moon; Smyth, in other words, comports himself while observing terrestrial features as if he were observing the moon. In 1864 Nasmyth must have had Smyth's remarks in mind when he argued that *unlike* geological observations that often demanded physically trying or dangerous crossings to access sites of interest, the astronomer's tranquil and steady position secured views that mitigate against "hasty" observations. "In these respects," Nasmyth writes, "telescopic visits to the surface of the moon yield more correct and reliable results than would many a visit to portions of our world where the scenery

54 Smyth, "On Certain Connecting Points between Lunar and Terrestrial Volcanoes," 157. It should be noted that Smyth's published Report to the Royal Society did not received such great internal reviews from geologists like Charles Lyell and Charles Darwin, see Schaaf, "Piazzi Smyth, Part 1", 298-299.

to be surveyed is not, perhaps, conveniently accessible."[55] As we shall see, for Nasmyth the stability and tranquility granted by the astronomer's position also provided an opportunity to hold a steady dialogue with the object of observation, essential for grounding the work in actuality and tangibility.

Smyth had become acquainted with Nasmyth's observations of the moon a few years before, not just through articles, letters or lectures, but by means of a personal visit. Thanks to a delightful watercolour that Smyth painted of James Nasmyth observing at his 20-inch telescope, we know that Scotland's Astronomer Royal paid the engineer a visit at his home sometime in December 1850.[56] It was during this visit that Nasmyth must have shown Smyth his drawings and models of the moon. These glimpses into Nasmyth's work induced a lunar enthusiasm in the astronomer, for not even two months pass and we find Smyth producing two of his own Nasmythian moonscapes (Figures 5.8 and 5.9). When Smyth's two watercolours are taken together, they clearly reflect the kind of conceptual and pictorial work that Nasmyth was already involved in: the transformation of the telescopic view of a lunar crater into a landed view on the basis of the geometry of shadows.

The transformation from bird's eye view to moonscape is a fundamental one for Nasmyth, a fact that is brought into relief in a letter to Josiah Crampton, the author of *The Lunar World* (1853). Upon receiving a gift-copy of Crampton's book, and seeing therein the ways in which he depicted hypothetical views of someone standing upon the moon's surface (Figure 5.10), Nasmyth wrote:

> I am glad to observe you have made an attempt to realize lunar landscape and scenery, as it would appear were we 'there to see.' I also have done something in this way. I cannot imagine any subject more glorious to feast the mind's eye upon, than to wander, in thought, amid the fearfully grand scenery in the moon; and any one who is artist enough can do so, from what is revealed to us by the telescope; in fact, this is to me the chief charm of the pursuit---not an indulgence of 'mere wild fancy', but a most legitimate exercise of the reason, and most legitimate powers of the imagination.[57]

---

55 James Nasmyth, "On the Physical Aspects of the Moon's Surface," *Notices of the Proceedings at the Meeting of the Members of the Royal Institution of Great Britain* 4 (1866): 395-6.

56 Charles P. Smyth, December 1850, "Sketches Chiefly in Scotland 1849-", vol. 2, 77, Royal Observatory of Edinburgh

57 Nasmyth to Josiah Crampton, November 5, 1853, quoted in J. Crampton, *The Lunar World: Its Scenery, Motions, Etc., Considered with a view to Design* (Edinburgh: Adam and Charles Black, 1863), 130.

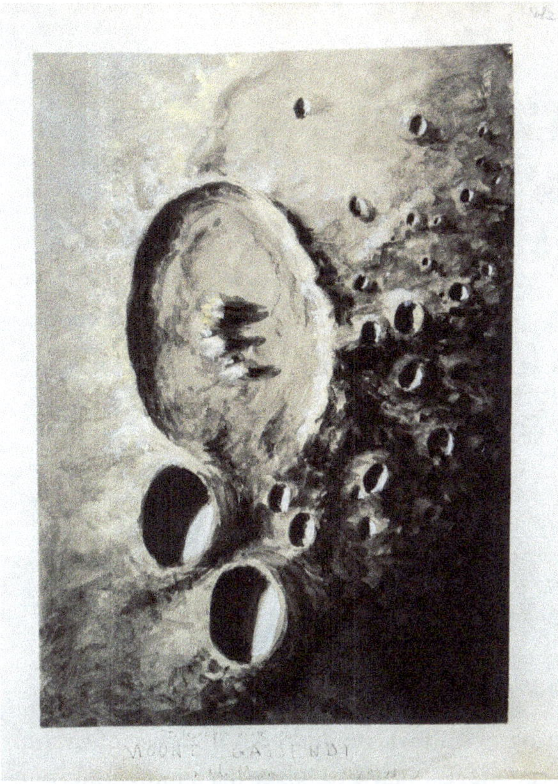

FIGURE 5.8
Charles Piazzi Smyth, "Telescopic View of Mount Gassendi in the Moon, February 1851," watercolour in "Sketch Book, 1849-1850: Sketches Chiefly in Scotland," volume 2, COURTESY OF THE ROYAL OBSERVATORY OF EDINBURGH.

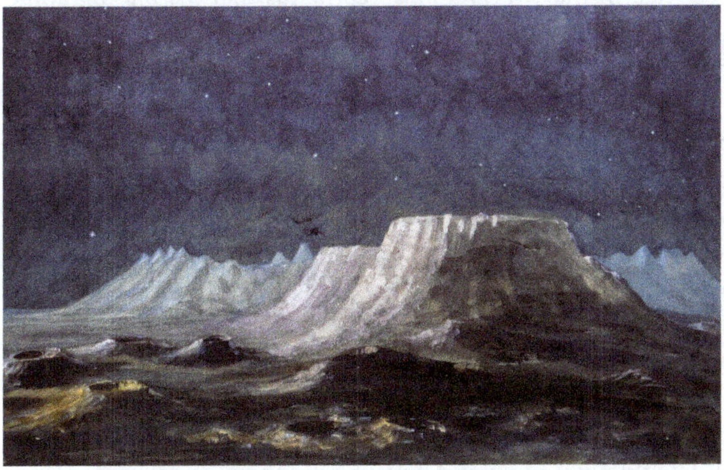

FIGURE 5.9 Charles Piazzi Smyth, "Lunar Landscape, Mount Gassendi, March 1851," watercolour in "Sketch Book, 1849-1850: Sketches Chiefly in Scotland," volume 2, COURTESY OF THE ROYAL OBSERVATORY OF EDINBURGH.

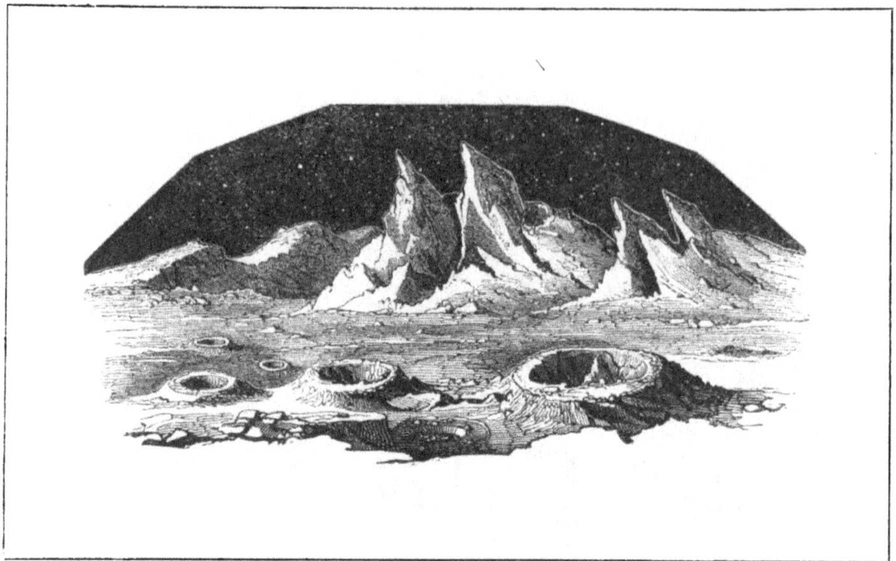

FIGURE 5.10 Josiah Crampton, *The Lunar World: Its Scenery, Motions, etc. Considered with a View to Design*, Edinburgh, 1863, Plate IV.

Although Nasmyth is referring to his own moonscapes in praise of Crampton's, Nasmyth's insistence on their legitimacy extends to his photographs published in 1874, as well. In that work, he also contends that his photographic depictions of lunar landscapes are not mere fantasy, or what we might today call, science fiction. Especially in the case of photography, I want to show that this insistence is warranted by his views on the role of tangibility and actuality for astronomical observations – a view that might be described as the mechanic's mode of observation.

3 Forms in Hand, Eye and Mind

When the painter Philip H. Calderon received his copy of *The Moon* in the mail, he could not put it down: "I forgot everything," he enthused in a letter to Nasmyth, "and, indeed, I have been up in the Moon ... The photographs took my breath away. I could not understand how you did them, and your explanation of how you *built* the models from your drawings only changed the wonder into admiration. Only an artist could have said what you say about the education of the eye and of the hand. You may well understand how it went home to me."[58] The Keeper of the Royal Academy in London was expressing more than just a resonance with the dynamics of the eye and hand expressed in the book's pictures, but he was doing so as a painter and as one who had studied engineering as a young man.[59] And as both these craftsmen would have admitted, the production of models and drawings demanded more than the skilled use of hands and eyes, they also required the activities of the mind. In this section I will detail the function of hands, eyes and mind in Nasmyth's work on the moon. I will argue that it is precisely because these are built into and layer his sensuous photographs for *The Moon*, that Nasmyth believes he can maintain their scientific value as opposed to their fictional one. As he insists in the book and elsewhere, "we are invoking the conception of things that actually exist; and that we are not, like some imaginary voyagers to the moon, indulging in mere flights of fancy."[60] I want to show that for Nasmyth, what made his staged photographs "legitimate" and "well directed," as he says, was the ways in which the mind processed *actual* tangibles received through eyes and hands, in order to ground materialized forms.

Form and matter are key, then, to unraveling what Nasmyth understood to be the intricate relationships between the hands, eyes and mind, but also drawings, models and photography.[61] Forms are the products of the artists' or engineers' prowess, but are also givens in the world that require explanation. Nasmyth spent a good part of his life engineering forms *and* pursuing their origins; after all, who better to interpret forms than one trained in materializing or embodying them – this, at least, is the implicit rhetoric underpinning much of his work. By clarifying these aspects of Nasmyth's vocation and thought, I hope to be able to tease out a number of ways in which Nasmyth supposed he was anchoring his photographs into tangibility and actuality. As Martin Kemp points out, Nasmyth's photographs "involve both appearance

58 Quoted in Nasmyth, *James Nasmyth, Engineer*, 381.

59 G.A. Storey, "Philip Hermogenes Calderon, A.R.A. (1833-98)," *Magazine of Art* 22 (1897-1898): 446-52.

60 Nasmyth and Carpenter, *The Moon*, 157.

61 Contrast this to Oliver Wendell Holmes famous proclamation that with photography, "Form is henceforth divorced from matter" and that photography expresses humankinds "triumph over earthly conditions, the divorce of form and substance," in "The Age of Photography," *The Atlantic* 3 (1859): 738, 748. Nasmyth's images force us to bring the earthly conditions back into play and reconnect form with matter, Holmes, notwithstanding.

and process."[62] But this is not just at the level of natural appearances and processes but also of appearances and processes formed by means of Nasmyth's hands, eyes, and mind in the form of drawings, models and photographs. In a word, *The Moon* is the result of manual observations characterized by the dexterity of a mechanic.

3.1 *Origins of Forms*

Upon his return to England from the Ascension Islands of Saint Helena, a stranger paid a visit to Josiah Crampton. When the stranger entered Crampton's drawing room and saw the pictures of lunar craters hung on its walls, he "exclaimed ... that they were so like the scenery of that remarkable volcanic island in which he had resided some years, that nothing could more perfectly represent it. He said that the circular form and appearance of the craters was *identical*."[63] The drawings that hung in the room included not just those by Crampton but also a few by Nasmyth's own hand. The analogy between the surface features of the earth and the moon was a powerful, visual one.[64]

In a manner of speaking, Nasmyth's selenographic work was grounded in the earth, constantly informing his observations of the moon; for the earth, in the broadest sense, is "the great field from which we must glean our experiences and form our analogies."[65] One of the things that ties his constructive, visual work on the moon to actuality is the earth, in other words, with its concrete and actual forms that, in turn, underpin analogies and inferences in tangible ways: "Anyone who is familiar with terrestrial volcanic craters must," declares Nasmyth, "at the first glance at those which are scattered in such infinite numbers over the Moon's surface, detect this well-known analogous feature, the central cone, and at once reasonably infer that these similar forms arose from a common cause, that cause being no other than volcanic action,

---

62 Martin Kemp, *Visualizations: The Nature Book of Art and Science* (Berkeley: University of California Press, 2000), 63.

63 Crampton, *The Lunar World*, 110.

64 Nasmyth was a part of the BAAS's Lunar Committee. One of its leading members, the geologist, John Phillips went as far as to claim that "it may be boldly affirmed that a competent theory of volcanic action [on earth] can hardly be regarded as having been adequately tested, much less completed without a careful study of the magnificent volcanic surface of the moon ... [where] a long train of igneous eruptions are exhibited as clearly as in the celebrated region of Auvergne." John Phillips, "Suggestions for the Attainment of a Systematic Representation of the Physical Aspect of the Moon," *Proceedings of the Royal Society of London* 12 (1862-63): 31-32.

65 Nasmyth and Carpenter, *The Moon*, 1.

accompanied by all its most marked characteristics."[66] Nasmyth's firsthand experience of seeing the transformations in Mt. Vesuvius' form between his first and second trips to Italy was fundamental in this regard.[67] The formation and translation of form is what anchors many of Nasmyth's wide-ranging scientific interests.

Nasmyth's analogies, moreover, should not be taken in isolation from what he calls his "very favorite subject of inquiry, namely the *origins of forms*."[68] In fact, *The Moon* can be seen as Nasmyth's most developed expression on the subject, but it is certainly not the only example. Throughout his career, the mechanical engineer pursued the origin of forms in a number of other ways. Take, for example, his long-standing interest in Assyrian cuneiforms, an interest that arose during his many, regular visits to the British Museum, a habit begun when he first moved to London in 1829. After making repeated examinations of this ancient system of writing, Nasmyth was able to extract from these enigmatic inscriptions on clay tablets the "elementary," "wedge-like form" fundamental to the whole system. Comporting himself, as he puts it, a "mechanic," Nasmyth reconstructs the shape of the stylus required to inscribe the elementary forms found in soft clay.[69] He then goes on to derive the origin of the stylus itself in clay bricks formed in the clay-rich Tigris and Euphrates valley. And finally, according to Nasmyth, this wedge-like form is the original source of specific features of certain capital letters in the Greek, Roman, and English alphabets.[70] Nasmyth's comportment as a mechanic, taken when examining

66 Nasmyth, "On the Physical Aspects of the Moon's Surface," 397. For a brief history of the study of craters, see Christian Koeberl, "Craters on the Moon from Galileo to Wegener: A short history of the impact hypothesis, and implications for the study of terrestrial impact craters," in *Earth-Moon Relationships*, ed. Cesare Barbieri, et al. (New York: Springer, 2001), 209-24.

67 And besides Nasmyth's trips and sketches of Vesuvius, he would have had access to the many published images of this and other volcanoes, like those in the Auvergne region, by one of the lead volcanologists of the day, George Poulett Scrope. Despite the resemblance of some of Scrope's images to lunar craters, it should be noted that when Scrope's reviewed Nasmyth's *The Moon*, he disagreed with the latter's analogies, claiming the craters were not formed by lunar volcanoes, per se, but by means more akin to the bubbling-action of molten, see George P. Scrope, "Review of The Moon," *Geological Magazine* 1 (1874): 272-79. But Scrope actually makes this remark about the lunar craters already in 1859, see George P. Scrope, *On the Mode of Formation of Volcanic Cones and Craters* (London: Taylor and Francis, 1859), 37 fn.

68 Nasmyth, *James Nasmyth, Engineer*, 432.

69 Nasmyth, *James Nasmyth, Engineer*, 436, 437.

70 Nasmyth gave two lectures on the subject, one at the Royal Institution in London in 1839, and James Nasmyth, "On the Plastic Origins of the Cuneiform Characters and its Relation

the materials and forms presented at the British Museum, is what also characterizes his mode of astronomical observation.

There are many other examples of Nasmyth's interest in "the etymology of form."[71] One of the most well-known is surely Nasmyth's discovery of the forms taken by sunspots resembling "willow-leaves"; a form that demanded an explanation, one that Nasmyth was only too happy to provide.[72] Another example stands out in Nasmyth's archives, namely, his unpublished studies on various forms given to classical urns (some are drawn as pencil outlines on the reverse of a photograph of the moon), which all originate, he maintained, in different but well-defined geometrical constructions of the ellipse.[73] Also consider, to take yet another example, Nasmyth's peculiar conclusion that the origin of the form given to the pyramids is due to the triangular shape taken by Sun's rays as they descend through clouds and down to the Earth.[74] Let us not forget to also include Nasmyth's take on the origin of the peculiar form taken by the swarm of asteroids and their paths between Mars and Jupiter, which he explains on the basis of the shattering action of a Prince Rupert drop and other actions on masses of brittle metal observed at his own foundry.[75] There is also his short but important abstract, "On the Form of Lightning," which distinguishes between natural or true forms of lightning in visual representations from artistic ones, the latter originating in Greek sculptures of Zeus.[76]

---

to our own Alphabet," *Notices and Abstracts of Communication to the British Association for the Advancement of Science, at Cheltenham in August 1856* (London: John Murray, 1857), 118-19

71 Nasmyth, *James Nasmyth, Engineer*, 443.

72 Nasmyth's first remarks on the structure of sunspots are in a letter to Sidebotham that was published as: James Nasmyth, "On the structure of the luminous envelope of the Sun," *Memoirs of the Literary and Philosophical Society of Manchester*, 1 (1862): 407-411. Also James Nasmyth, "New Theory of the Nature and Source of the Sun's Light and Heat," in *The Year-Book of Facts in Science and Art*, ed. John Times (London: David Bogue, 1852), 152-54; C.E. Bartholomew, "Discovery of the Solar Granulation," *Quarterly Journal of the Royal Astronomical Society* 17 (1976): 263-289; and Holly Rothermel, "Images of the Sun: Warren De la Rue, George Biddell Airy and Celestial Photography," *The British Journal for the History of Science* 26 (1993): 137-169.

73 See, MS.42502 (438 B1); MS.3241, 205, V.229; MS.3242, 10-12, V.38,176, James Nasmyth Archive, National Library of Scotland, Edinburgh.

74 James Nasmyth, *James Nasmyth, Engineer*, 432-34; see also, MS.3242, 217, James Nasmyth Archive, National Library of Scotland, Edinburgh.

75 James Nasmyth, "Proposed Theory of the Origin of the Asteroids," *Notices and Abstracts of Miscellaneous Communications to the Sections* (London: John Murray, 1853), 21-22.

76 James Nasmyth, "On the Form of Lightning," *Notices and Abstracts of Miscellaneous Communications to the Sections of BAAS* (London: John Murray, 1857), 14. See also, Jennifer

Another example clearly links the originator of forms to the origin of forms, and that by means of Nasmyth's commercial interests in constructing "self-acting" tools for machines precise enough for the requirements of the industrial age. On the basis of years of experience, he reduced the many parts composing modern machines into six elementary forms: line, plane, circle, cylinder, cone, and sphere. "That the more near to absolute mathematical truth we can have the forms of those parts [of any machine]," explains Nasmyth, "the more perfectly will the machine perform its duties."[77] It is the job of self-acting tools, such as those manufactured at his own foundry, to achieve these perfect forms for the service of modern machines – a feat that he thought, ironically enough, could only be achieved by "iron hands."[78] But by far the most daring of his speculations is one intended to explain the spiral form of some nebulae discovered by William Parsons, the third Earl of Rosse.[79] The whirling motion assumed by water when allowed to escape from the bottom of its container is the basis of the analogy used to explain the origin of the spiral nebulae.[80] Nasmyth reveals himself to be a lively interpreter of what Robert Chamber refers to as "Sky-Hieroglyphics."[81]

On the basis of our brief excursion into Nasmyth's long-standing interest in the origins of forms, a few fundamental points arise. The first relates to the fact that for Nasmyth analogy is used to reveal the source of forms, even those that

---

Tucker, *Nature Exposed: Photography as Eyewitness in Victorian Science* (Baltimore: Johns Hopkins University Press, 2005), 127-130.

77 Nasmyth, "Appendix B. Remarks on the Introduction of the Slide Principle in Tools and Machines Employed in the Production of Machinery", in *Practical Essays on Mill Work and other Machinery with notes and additional articles by Thomas Tredgold and revised into a third edition with additions by George Rennie* (London: John Weale, 1841), 394, 95.

78 Nasmyth, "Appendix B," 399, and also 394, 395, 410. For more context on the hand and labour, see Peter J. Capuano, *Changing Hands: Industry, Evolution, and the Reconfiguration of the Victorian Body* (University of Michigan Press, 2015).

79 See, Omar W. Nasim, "Observation, Working Images and Procedure: the "Great Spiral" in Lord Rosse's Astronomical Record Books and Beyond", *British Journal for the History of Science* 43 (2010): 353-389.

80 James Nasmyth, "Suggestions respecting the Origin of the Rotatory Movements of the Celestial Bodies and the Spiral Forms of the Nebulae as seen in Lord Rosse's Telescopes," *Monthly Notices of the Royal Astronomical Society* 15 (1855): 220-21.

81 [Robert Chambers], "Sky-Hieroglyphics," *Chamber's Journal of Popular Literature, Science, and Arts*, 180 (1857): 372-374. Writing to Hill, Nasmyth recommended "an Enchanting book called 'Vestiges of Creation' ... It is all the talk. It is a noble and extraordinary work." (Nasmyth to Hill, April 30, 1845, A9.42B, Royal Observatory of Edinburgh). Since the work was published anonymously, it is unlikely that either of these men could have known that it was written by their mutual acquaintance, Robert Chambers.

remain hidden or inaccessible. Nasmyth, indeed, assumes that there are deeper things at work than those given, surface-forms, but that the latter might intimate the former. As such, analogous forms often ground what is inaccessible or unfamiliar into the mundane, accessible, and wholly familiar; and that the causes which give rise to specific forms at the level of the mundane might be similar to those unfolding at the level of the grand, sublime and distant. This point relates, then, to Nasmyth's tethering the process of observation to the earth and the mundane, the familiar and the concrete, the tangible and the actual – it is what makes his imaginary wanderings legitimate or "well-directed." But it also relates to Nasmyth's views on the relationship between the microcosm and the macrocosm, that there exists a "wonderful integrity" between the two, a harmony that can cut across huge differences in distances, scales, and beings.[82]

Many of the examples above point to a common feature of Nasmyth's approach, one that was recognized by a reviewer of *The Moon*, who put it thus: "Mr. Nasmyth brings his mechanical training and experiences to bear in dealing with the moon's physical history."[83] There is a clear sense that Nasmyth is manually active and experimenting with materials rather than composing, for instance, thought experiments. Many of the explanations provided for the origin of forms, in fact, arise straight from such tangible, hands-on experiences, including ones Nasmyth would have had at his own foundry or workshop. One gets the distinct sense that as an artist, craftsman, and mechanical engineer, he must have had a good handle on forms, their origins, and structural anlogies. In this connection, we might pay heed to Pierre Duhem's characterization of French or German scientists in contrast to British ones: while in the case of the former we expect to inhabit "the abode of reason," when it comes to British scientists we can be rest assured that "we [will] find ourselves in a factory."[84] The case of Nasmyth shows that this generalization might have applied to more than just experimental physics, as Duhem had initially meant it to be applied, but also observational astronomy.

It is with these points at the forefront that we should approach the pictures in *The Moon*, including these well-known heliotypes (Figure 5.11): one of

82 Nasmyth, "Suggestions respecting the Origin of the Rotatory Movements", 220; and also see Nasmyth and Carpenter, *The Moon*, 1.

83 [Anonymous], Review of "The Moon: Considered as a Planet, a World, and a Satellite, by James Nasmyth, C.E., and James Carpenter, FRAS," *The Edinburgh Review or Critical Journal* 140 (1874): 78.

84 In Eugene S. Ferguson, *Engineering and the Mind's Eye* (Cambridge, MIT Press, 1994), 45. Duhem's remarks have been nicely contextualized in Bruce J. Hunt, *The Maxwellians* (Cornell University Press, 2005), 87-95.

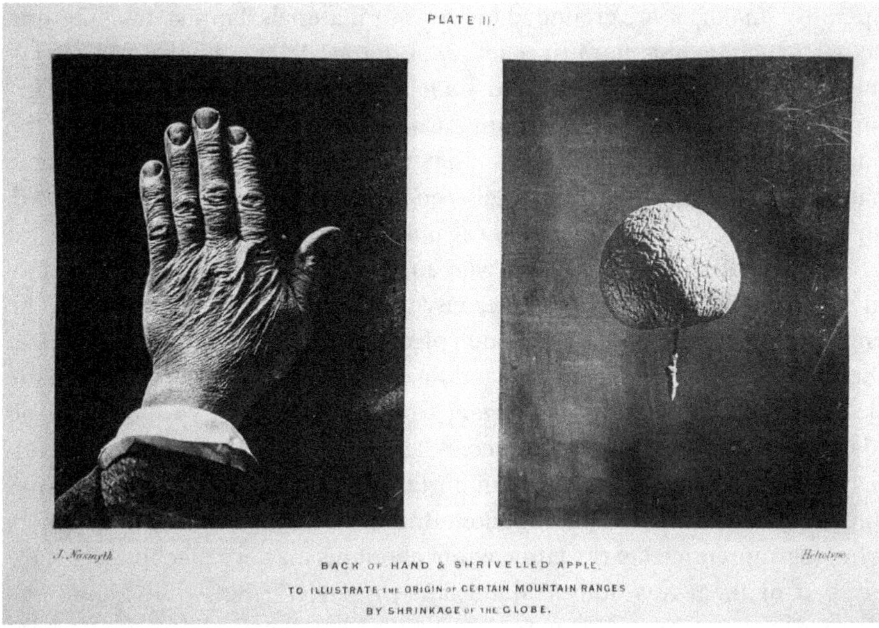

FIGURE 5.11 "Back of Hand & Shrivelled Apple," James Nasmyth and James Carpenter's *The Moon: Considered as a Planet, a World, a Satellite* (1874), 1st edition, heliotype, plate 2.

Nasmyth's own aged, wrinkled hand and the other of a rotten, shriveled-up apple. The two forms are part of an extended argument about cooling igneous bodies, which over time contract so that deeper strata beneath the surface shrink resulting in excess upper strata that "wrinkle" to form lunar mountain ranges and volcanoes. And when we look to maps of terrestrial mountain ranges, we also see, according to Nasmyth, "pictures of wrinkles," whose forms indicate analogous actions occurring on the earth as well. Similar actions occur, albeit at different scales and rates, in the human hand or an apple, as each wither away into old-age. Nasmyth concludes that the "wrinkle theory" used to explain the existence of certain appearances, "opens up a chapter in geology and physiology well worthy of consideration."[85]

Notwithstanding the use made of Nasmyth's photograph of his own hand, it might still be strange to encounter this image in the context of a book on the moon. I want to show that, given certain visual codes, Nasmyth's hand presented layers of meaning to the Victorian audiences that saw it. To begin with, the context into which these photographs are presented make clear that the

85 Nasmyth and Carpenter, *The Moon*, 29; Nasmyth, *James Nasmyth, Engineer*, 320.

operative analogies are grounded in forms or materials familiar, tangible, and actual, which include not just relief maps, terrestrial mountains, and old apples; but also, Nasmyth's own hand – a hand, to be sure, implicated and ever-present in different forms throughout his work on the moon. The picture of the hand in *The Moon* presented to its nineteenth-century audiences the reliable hand of an accomplished industrialist, engineer and inventor – it was, indeed, an emblem of assurance for viewers against fantasies or grandiose claims.

Nasmyth's ambidextrous hands were present, even when ostensibly absent, in a pen and ink drawing sent to the Viscount Duncan, who wrote back to the engineer declaring: "I shall have much pleasure in being reminded as I look at the drawings that the same hand and head that executed them invented the steam hammer, and many other gigantic pieces of machinery which will tend to immortalise the Anglo-Saxon race."[86] There were, to be sure, others of Nasmyth's ilk who too were seen to immortalize the "race" with their machines and hands, such as Isambard Kingdom Brunel and Henry Maudslay. Nasmyth, who was apprenticed to the latter, wrote about his master's machines, as if every part of them was "full of Maudslay's presence."[87] But while Maudslay's presence may permeate his many mechanical handiworks, Brunel's assistant and biographer, Richard Beamish, turned his attention to the *form* of Brunel's hand; something that Nasmyth's own biographer and friend Samuel Smiles would do for Nasmyth later.

Indeed, in a footnote to a reprint of the photograph of Nasmyth's wrinkled hand, Smiles refers the reader of Nasmyth's autobiography, which he edited, to Richard Beamish's *The Psychonomy of the Hand* (1864).[88] Based largely on the works of French authors like Casimir Stanislaus D'Arpentingy and A.A. Desbarrolles, Beamish's book is about the diversity of human hands and how their many forms correspond to specific national, racial, moral, vocational and psychological traits.[89] Beamish distinguishes psychonomy into two parts: there is chiromancy, the study of lines on the palm, where we find aspects like the mound of Moon that manifests the levels of imaginative power in a person. But he dedicates the bulk of the book to chirognomy, which studies the diver-

86 In Nasmyth, *James Nasmyth, Engineer*, 311.

87 In Samuel Smiles, *Industrial Biography: Iron-Workers and Tool-Makers* (Boston, 1864), 279.

88 The footnote only occurs in later editions of Nasmyth, *James Nasmyth, Engineer: An Autobiography Edited by Samuel Smiles*, London, 1905, fn. 320.

89 Casimir S. D'Arpentigny, *La chirognomie, ou l'art de reconnaitre les tendances de l'intelligence d'apres les formes de la main* (Paris: C. le Clere, 1843); C.G. Carus, *Ueber Grund und Bedeutung der venchiedenen Formen der Hand in verschiedencn Personen*, Stuttgart, 1846; Adolphe Desbarrolles, *Les mystéres de la main révélés et expliqués* (Paris: E. Dentu, Libraire-Éditeur, 1859).

sity of outline forms of the human hand. Beamish classifies at least eight different kinds of hand-forms, providing visual examples for each: the elementary, pre-human hand, the labour hand, the useful hand, the philosophic hand, the artistic hand, the psychical hand, the mixed hand, and the female hand. Given the form of Nasmyth's hand in the printed photograph, Smiles claims that it fits best under the "labour hand" or the "mechanical hand."[90] This class, which in contrast to the delicacy of the French and Spanish, is primarily British, self-confident, and includes the "Great labourers, great navigators, great hunters – in a word, those who devote their lives to action and locomotion, possess strong hands and spatulous fingers. They have made rough places smooth, and are even now brining the ends of the earth together."[91] It is under this same head that Beamish includes the hands of Brunel.

Beamish saw the study of the psychonomy of the hand to be of general benefit to the empire. The differences established between the eight classes and their subclasses are used throughout to explain many kinds of things, including national characteristics and how these manifest in industry, creativity, and the character of those under colonial rule. "To the geographer," Beamish continues, "the Psychonomy of the Hand will give a new interest. Every portion of the habitable globe, in place of offering a dry nomenclature conveying no definite ideas, will become a reflection of the greatest wonder of creation, opening up the inner life of a people, and the relation subsisting between nations. It will point out the physiological boundaries of races, and, to a certain extent, the unequal destiny which they are called upon to fulfill."[92] Indeed, it is under the class of "psychic hands" that Beamish includes two tracings taken from casts made available to him at the museum of the East India House in London, one from a Himalayan, and another from a person of central India (Figure 5.12).[93]

The casts at the East India House, to be sure, were reproductions of plaster casts taken by the Schlagintweit brothers on their famed scientific expeditions of 1854-57, through the Indian subcontinent, and sponsored by the East India Company and the King of Prussia, Frederick William IV. Among the thousands of drawings, measurements, photographs, specimens and samples

90 Nasmyth, *James Nasmyth, Engineer*, fn. 320.

91 Richard Beamish, *The Psychonomy of the Hand,* 2nd edition (London: Frederick Pitman, 1865), 52.

92 Beamish, *The Psychonomy of the Hand*, 91.

93 Beamish, *The Psychonomy of the Hand*, plates 29 and 30. For more on this museum, see: Jessica Ratcliff, "The East India Company, the Company's Museum, and the Political Economy of Natural History in the Early Nineteenth Century," *Isis* 107 (2016): 495-517.

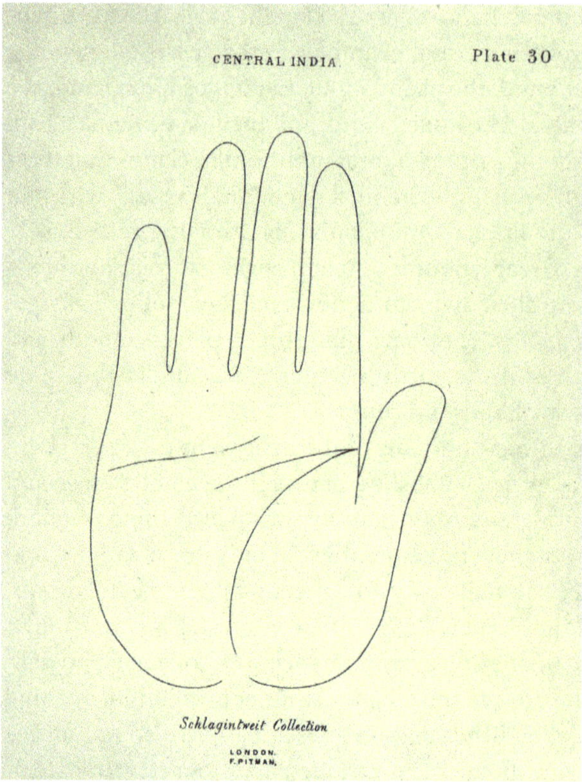

FIGURE 5.12
Hand from central India, traced at the East Indian House from a cast originally made by the Schlanginweit brother's expedition. Robert Beamish, *The Psychonomy of the Hand*, 2nd edition, London, 1865, Plate 30.

that they brought back with them to Europe, there were 275 plaster casts of faces, and 37 of hands and feet, all made directly from living indigenous persons that they encountered on their travels.[94] Once brought back to Germany, the Schlangintweits had electrotype reproductions made of the plaster casts so that they could be sold to scientists and institutions throughout Europe. Joseph Barnard Davis, the co-author of *Crania Britannica* (1856) praised these casts, writing, "I regard them as of the utmost importance to the British Government and people, who have never before had any adequate means of be-

94 [Anoymous], *Prospectus of Messrs. De Schlagintwetis' Collection of Ethnographical Heads from India and High Asia*, 2nd edition (Leipzig: J.A. Barth, 1859). See also: A. Zimmerman, "Die Gipsmasken der Brüder Schlagintweit: Verkörperung kolonialer Macht," in *Über den Himalaya: Die Expedition der Brüder Schlagintweit nach Indien und Zentralasien 1854-1858*, eds. M. von Brescius, et al. (Cologne: Böhlau, 2015), 241-49; G. Armitage, "The Schlagintweit Collections," *Indian Journal of History of Science* 24 (1989): 67-83; G. Finkelstein, "'Conquerors of the Kuenluen'? The Schlagintweit Mission to High Asia, 1854-57," *History of Science* 38 (2000): 179-218. For a broader context, see Aviva Briefel, *The Racial Hand in the Victorian Imagination* (Cambridge: Cambridge University Press, 2015).

coming acquainted with our fellow-subjects in India. Indeed, I may safely say, there never was before such an admirable exemplification of the Ethnology of any people."[95] Another to sing the praises of the Schlangintweits and their casts was Sir Roderick Impey Murchison, who by this time had succeeded De la Beche as the director of the Geological Ordinance Survey and as the head of the Museum of Practical Geology in Jermyn Street, a direct descendant of the Museum for Economic Geology. One cannot help wonder where in the museum Nasmyth's plaster model of the lunar surface was in relation to some of Schlangintweits' casts of the faces, feet and hands of *Naturvölker*. Whatever the case may have been, by appealing to the work of Beamish, Smiles further grounds Nasmyth's drawings, models, and photographs into hands inscribed with a civilizing force, hands with nationality, race, and an empire – a fact that might have been obvious to readers at the time, but is very much worth recalling today.[96]

It was with these hands, finally, that Nasmyth traced, at the level of the microcosm, the formative actions wrought at the level of the macrocosm: "We trust that we, on our part, have shown that the study of the moon may be a benefit not merely to the astronomer, but to the geologist; for we behold in it a mighty 'medal of creation' doubtless formed of the same material and struck with the same die that moulded our earth; but while the dust of countless ages and the action of powerful disintegrating and denuding elements have eroded and obliterated the earthly impression, the superscriptions on the lunar surface have remained with their pristine clearness unsullied, every vestige sharp and bright as when it left the Almighty Maker's hands."[97] The clarity of the "superscriptions" on the moon's surface, as opposed to the sullied surface of the earth, reveal God's hand; and when these are moulded in graphite or ink, gypsum or light they reveal the hand of our mechanic in originating these forms anew, at the level of the human. We have seen ways in which Nasmyth's abiding interest in the origin of forms was grounded in analogies that related the inaccessible and the unfamiliar to the tangible and concrete – in ways that might even be traced back to his factory or workshop, to assumptions about character and empire, and to retracing the steps of the Almighty Hand with his own. As such, we now turn to Nasmyth as the originator of forms.

---

95 In [Anonymous], *Prospectus of Messrs. De Schlagintwetis' Collection*, iv.

96 This might a good place to record that Nasmyth designed weapons, see, James Nasmyth, "On War Rockets," *Memoirs of the Literary and Philosophical Society of Manchester* 4 (1871): 206-10; and "Submarine Mortar Frigate," *Scientific American* 8 (1853): 178.

97 Nasmyth, *The Moon*, 179.

3.2 *An Originator of Forms*

Nasmyth's archives are filled with all kinds of drawings. There are hundreds of technical drawings for engineering works and mechanical plans found in his "Scheme Book," including one that is supposed to be the first drawing of the steam-hammer – the invention that made his name a household one.[98] At another archive, three large albums contain hundreds of non-technical sketches which include pencil, pen and ink, and watercolor drawings of landscapes, buildings, ruins, "a very bad dream," geological and volcanic scenes, alongside fantastical ones, and multiple pictorial variations of antiquarians in their studies. Among these one also finds many sketches on loose paper of various sizes and media of astronomical objects such as Mars, the Sun and its spots, and, of course, the Moon.[99] Some of these are done in haste, while others are highly finished. The drawings range in theme and purpose, technique and style, ranging in works done as a young man to those executed by an older man.

Indeed, many of Nasmyth's drawings reflect the formative impact of his father's teachings – a man who taught all eight of his children, along with a whole host of pupils in Edinburgh, how to draw and paint; the father being no other than Alexander Nasmyth, the distinguished Scottish landscape painter. It is thus little wonder that there is nothing more repeatedly emphasized in James Nasmyth's autobiography than the value of "the universal language of the pencil" for his life and vocation. For the "Graphic Language," in contrast to the tongue, can communicate scenes or ideas, take one back to places seen in distant travels, making it a "valuable art of graphic memoranda." But drawing is also a means of observation, one that "educates the eye," explains Nasmyth, "and the hand, and enables us to cultivate the faculty of definite observation." Producing forms on paper was a part of Nasmyth's means of communication; it was a language that stored ideas and memories; and it was a domesticator of hands, eyes and mind for the purposes of science.[100]

Elsewhere I have explored the productive connections between drawing and astronomical observation in much more detail.[101] Suffice it here to say that after many repeated drawings of the moon, Nasmyth gained a level of in-

98 James Nasmyth's "Scheme Book, c1838-1856" Institution of Mechanical Engineers, last accessed December 17, 2015, <http://www.imeche.org/flipbook/ils/ims_98/>.

99 The three albums I refer to are MS. 3241, MS.3242, and MS.3243, James Nasmyth Archives, National Library of Scotland. Also at the Science Museum of London, where many of Nasmyth's large drawings, photographs, and models of the moon are now to be found, even online.

100 Nasmyth, *James Nasmyth, Engineer*, 286, 56, 57.

101 Omar W. Nasim, *Observing by Hand: Sketching the Nebulae in the Nineteenth Century* (Chicago: University of Chicago Press, 2013).

timacy with the nooks and crannies of the lunar surface: "The Moon and I," he writes in a letter to Hill, "are still on the most intimate terms and he answers me very many authentic questions I put to him…"[102] The back and forth between the eyepiece and his sketches, night after night, not to mention the many detailed comparisons made between the sketches themselves, day after day, raised questions that were then put back to the moon in the next round of observations.[103] This continuous process of familiarization is a dialogue between the observer and the observed – one that is too often overshadowed by more popular snapshot versions of scientific observation. "These drawings," writes Nasmyth, "were again and again repeated, revised, and compared with the actual objects, the eye thus in advancing in correctness and power of appreciating minute details, while the hand was acquiring, by assiduous practice, the art of rendering correct representations of the objects in view."[104] It is not just the hands, in other words, that are at work in forming drawings but also the eyes – the two cooperate in order to acquaint one with the observed.

By drawing the surface features of the moon, Nasmyth gained a level of familiarity that was only to be enhanced when he "translated" them into plaster models. Nasmyth was by no means novice when it came to making models. As a boy in Edinburgh – that old city of lights and shadows, forged by volcanoes – when he was not busy drawing, he was constructing mechanical models in his father's workshop or in his own bedroom. Nasmyth displayed so much aptitude for constructing models that he was able to pay his own way through the university by selling them to lecturers. The combination of drawing and modeling went hand-in-hand; a combination that only increased the aspiring mechanic's chances at finding an apprenticeship as a mechanical engineer.[105] As for the knowledge gained through models, E. Ferguson reminds us that models "permit the observer to walk around it, to look down on and up at the object, and to receive tactile clues that help him make sense of the object … a three-dimensional model that supplies designers and builders with nonverbal, sensual, qualitative information – visual, tactile, muscular, and aural."[106] Nasmyth would have agreed, only to add:

---

102 Nasmyth to Hill, March 27, 1847, A9.42B, Royal Observatory of Edinburgh.

103 See also Nasmyth, "On the Physical Aspects of the Moon's Surface," 395-396.

104 Nasmyth and Carpenter, *The Moon*, viii-ix.

105 It was an exceptional combination for a young engineer to have in Britain at the time, see Smiles, *Industrial Biography*, 294-95.

106 Ferguson, *Engineering and the Mind's Eye*, 107.

> The truth is that the eyes and the fingers – *the bare fingers* – are the two principled inlets to sound practical instruction. They are the chief sources of trustworthy knowledge as to all the materials and operations which the engineer has to deal with... The nature and properties of the materials must come in through the fingers ends. Hence, I have no faith in young engineers who are addicted to wearing gloves. Gloves ... are perfect non-conductors of technical knowledge.[107]

The epistemology of the mechanic was not just galvanized for the purposes of educating students' of engineering, but as we have seen, it was activated in Nasmyth's approach to astronomical observations.[108] That "trustworthy knowledge" of astronomy might be transmitted through the eyes *and* bare hands by means of an intimate dialogue between the observer and observed unfolding through a series of glances, notes and sketches, and models – the eyes and hands, after all, are what grounds the work in tangibility and actuality, making the work, in Nasmyth's estimate, legitimate.[109]

However, there is another key component to this dialogue between observer and observed, and that is the mind. Among other things, its job is to make sense, order, and extrapolate from the many discrete impressions that are transmitted through eyes, hands, and ears, and combine them into meaningful and coherent wholes. It fills in the gaps, for instance, between observations in ways that accord with expectations created from observations and sketches previously made; in other words, it extends continuity from what has been observed into those moments and regions not accessible or not under the continuously watchful gaze. "There is an irresistible tendency in the mind," Nasmyth explains, "to pass beyond the actually *visible*, and to fill in with what

---

107 Nasmyth, *James Nasmyth, Engineer*, 95.

108 K. Hentschel is probably right to appeal to "Condillacian epistemology" when referring to Nasmyth's training as an engineer; see his *Visual Cultures in Science and Technology: A Comparative History* (Oxford: Oxford University Press, 2014), 131.

109 As we have already seen, geology was surely another science where the use of physical models was prevalent. Indeed, it has been argued that physical, geological models were used as forms of experiment – that they were even "technologies of the imagination"; see T. Brandstetter, "Time Machines: Model Experiments in Geology," *Centaurus* 53 (2011): 135-145; and N. Oreskes, "From Scaling to Simulation: Changing Meanings and Ambitions of Models in Geology," *Science without Laws: Model Systems, Cases, Exemplary Narratives*, eds. A.N.H. Creager, et. al. (Durham/London: Duke University Press, 2007), 93-124. For another take on the productive side of models, see Natasha Myers, *Rendering Life Molecular: Models, Modellers, and Excitable Matter* (Durham/London: Duke University Press, 2015).

it knows must exist those accessory features and phenomena that are only hidden from us by distance and by our peculiar point of view. Where the material eye is baffled, the clairvoyance of reason and analogy comes to its aid."[110] The mind's eye fills in the sensual gaps in order to construct its moonscape, filling it up with processes over time, and things in space.

When coherent, continuous wholes are built up using multiple, fragmented materials obtained from the outer world, an enriched and more fleshed-out world is seen with the mind's eye; as we have seen in the case of the translation from the confines of the bird's eye view to that of the moonscape.[111] What the inner eye acquires might be adjusted, refined, or tweaked, it is true, according to newly acquired outer impressions; but this is just the mind's dialogue with the rest, a dialogue that constrains and advances what is seen by means of the "constructive faculty" of the mind.[112] But what finally stands before it, built up from so many tangible, authentic parts, might find its way back into the material world as constructions that embody anew the mind's forms by means of the hands, and eyes. As such, the products are not free floating, flights of fancy, but are rather shaped and grounded, right from the start, by tangibility and concreteness, plausibility and coherence. These constructions may take on many guises including plaster models. In a passage from a late article on education, Nasmyth explains that "the faculty of observation [is] united to the powers of transmitting mental impressions in visible form through the agency of the fingers, whether by the pencil, the modelling-tool [sic], or the chisel," and that what results from this process is "entitled to be considered 'materialized thought.'"[113]

Just like the many drawings and models that contributed to actualizing the photographs, the latter too might be thought of as materialized thoughts, arising from the observational processes outlined above. As such, it is with Nasmyth's photographs that spectators ostensibly gain the most direct access to what stands before the inner eye.[114] And what stands out before the inner eye are forms, translated from drawings and models over into the mind and back

110 Nasmyth and Carpenter, *The Moon*, 158.

111 This connection is clearly made in Nasmyth and Carpenter, *The Moon*, 60.

112 Nasmyth, *James Nasmyth, Engineer*, 263.

113 James Nasmyth, "Hints on the Education of the Eye and Fingers," *James Murray's Magazine: A Home and Colonial Periodical for the General Reader*, May (1887): 312. A similar statement is made in Nasmyth, *James Nasmyth, Engineer*, 15. See also Nasmyth's article for the *Times* in defense of Sir Henry Cole's suggestion that the so-called three R's should be extended to include drawing and music, see "Mr James Nasmyth, C.E., on Freehand Drawing," *Journal of the Society of Arts* 28 (1880): 871-72.

114 Also see, Ferguson, *Engineering and the Mind's Eye*, 42.

again into photographs staged for this purpose. Forms are rematerialized in the photographs as interplay of light and shadow, an interplay that bewilders and defies, according to Nasmyth, "the power of brush or pencil to reproduce them."[115] The fact that Nasmyth selected photography for the purpose stems directly from his appreciation of it as the best medium for capturing and displaying shadows and hence forms. The shadows are the means to the forms we are most interested in as observers, because they "give information" of not just the moon and its history but also of the earth.[116] But the photographs are also materialized by means of the variety of reproduction methods selected for the quarto format of the first edition of *The Moon*, published in 1874.[117] This edition exemplifies photography's materiality best, because its 23 plates include six photogravures, four heliotypes, two lithographs and one chromolithograph; not to mention twelve mounted prints of photographs on eleven leaves that include ten autotypes and two woodburytypes. The second edition of *The Moon*, also published 1874 and in quarto format, drastically narrows down this materiality so that majority of prints are actually woodburytypes, a method of reproducing photographs that preserves a continuity of rich tones, without ink, and does so *in relief*, so that, in effect, the woodburtype reproductions are *cast-images*.[118]

Finally, the photographs take us to where the observer and the observed come together: "it is almost inevitable, we say, for such an observer to identify himself so far with the object of his scrutiny, as sometimes to become in thought a lunar being ... abstracted from terrestrial influences." When Nasmyth says that he is "invoking the conception of things that actually exist," it is the mechanic's mode of observation invoked, a process that ties tangibility and actuality together with conceptions and possibility; where the information obtained by gloveless hands and educated eyes is mentally extended further in ways that adhere with these ties.[119] Indeed, it is Nasmyth's multi-layered, multi-media, multi-sensual observational process that is supposed to guarantee the legitimacy of photography for science.

By linking up appearances and processes to drawing, modeling and staging, I have tried in this paper to put Nasmyth's photography back into touch with

115 Nasmyth and Carpenter, *The Moon*, 145.

116 Nasmyth and Carpenter, *The Moon*, 98, 147.

117 This point has been nicely shown in Robertson, "Science and Fiction."

118 In the third edition of *The Moon* (1885) too Nasmyth maintains his preference for the woodburtype. In this way we might, with Oliver W. Holmes, refer to these printed photographs as "morphotypes."

119 Nasmyth and Carpenter, *The Moon*, 157.

the hands, eyes and mind, without reducing them into groundless fantasy or science fiction. What binds them is the mechanic's mode of observation, where its resulting-products, though imaginatively inspired and constructively extended, are anchored in a variety of ways into the earth. From this perspective, we have met with a convergence of practices, techniques and methods from geology, ethnology, engineering and astronomy; a convergence that appealed to many sorts of audiences; and one that is made manifest by photography entrenched with drawings, paintings, sunlight, bodies, wood and gypsum – intermixed so as to give materialization to appropriate shadows and forms.

CHAPTER 6

Karl Friedrich Küstner's Moon Photographic Plates at Bonn Observatory

Michael Geffert

1 Introduction

Bonn observatory has a long tradition of photographic observation starting in 1900. The first plates were taken in Bonn itself, while after the Second World War an observatory for astronomical observations (Observatorium Hoher List) was built in the country about 85 km south of Bonn. When this observatory was closed in 2012, all photographic observations from Bonn observatory were transferred to the archives of the Argelander Institute in Bonn. These plates now belong to the "Sammlung Historischer Himmelsaufnahmen der Universität Bonn" (Bonn University collection of historical images of the heavens), where they are used for research, teaching, and public outreach purposes. The collection contains 15,000 photographic plates taken in Bonn (pre-1945), Göttingen (1930–1955) and at the Hoher List observatory (1954–1990). The observations were made on glass plates up to 30 × 30 cm in size. The collection also contains some ten historic telescopes and around 20 volumes of observations by Julius Schmidt, mainly from the 19th century.

Astronomers from Bonn conducted photographic observations mainly for spectroscopic, photometric and astrometric investigations. There were no scientific lunar observations. Only in the very first days of photographic astronomy in Bonn were plates of the moon taken by Karl Friedrich Küstner with the large refractor from Bonn. As far as we know today, Küstner did not use the plates for his scientific work. It is more likely that he was testing the quality of the telescope with observations of an extended object like the moon. Some of these plates were subsequently sent to Julius Franz in Wroclaw (Breslau).

Observations of other objects by Küstner were used by Bonn astronomers up to 2000, for example to determine proper motions of stars in the field of star clusters.[1] However, since the plates of the moon could not be used for scien-

1 Michael Geffert, Martin W. Kuemmel, and Hans Schmidt, "Internal motions in the Pleiades cluster," *Astron. Astrophys. Suppl.* 112 (1995): 229-234; Michael Odenkirchen, Peter Brosche, Michael Geffert, and Hans-Joachim Tucholke,"Globular cluster orbits based on Hipparcos proper motions," *New Astronomy* 2, 5 (1997): 477-499; Jörg Sanner, Martin Altmann, Jens

 | DOI 10.1163/9789004298873_008

tific projects, they were of no interest for astronomers in Bonn and were virtually forgotten.

It was only when a complete reinvestigation of all Küstner's old plates was undertaken that the attention of Bonn scientists was again drawn to his observations of the moon. In general, the "Sammlung Historischer Himmelsaufnahmen" was considered more an archive of material for science, education and public relations than a strictly scientific resource, and Küstner's observations of the moon have never before been analyzed. Hence this chapter represents a first attempt to explicate the background of the observations made in Bonn from 1900 to 1903 with respect to Küstner's work and to other research being conducted on the moon at that time.

2 The Beginnings of Astronomy at the University of Bonn

A year after the foundation of the university in Bonn in 1818, astronomy was established there as a new discipline. The first astronomer in Bonn, Karl Dietrich von Münchow (in Bonn from 1819 to 1835) came from Jena in order to establish an observatory for the university. However the university was unable to provide enough money either for instruments or for a suitable building. Only with the financial help of the King of Prussia, Friedrich Wilhelm IV, was von Münchow's successor, F.W.A. Argelander (in Bonn from 1837 to 1875), able to build Bonn University's first observatory in 1845. By chance Argelander had become the friend of the king when both were young. This friendship, which endured over many years, enabled the construction of the first observatory and the development of astronomy in Bonn.

Argelander and his collaborators attained considerable fame with the "Bonner Durchmusterung", the first large survey of more than 300,000 stars, as well as for his method for measuring the brightness variations of variable stars, and for the determination of the first distances of nearby stars. The main telescope of the first observatory in Bonn was the "heliometer", a refractor with a lens diameter of 16 cm and a focal length of 260 cm. The front lens of this telescope was split in order to enable measurements of small angular distances with high accuracy. All the observations at that time were done with the naked eye, and most of these observations are available today as numbers in catalogues and publications. Also during the era of Eduard Schönfeld (in Bonn from 1875 to 1891) visual observation remained the only astronomical technique available

---

Bunzendorf, and Michael Geffert, "Photometric and kinematic studies of open star clusters. II. NGC 1960 (M 36) and NGC 2194," *Astron. Astrophys.* 357 (2000): 471-483.

in Bonn. Schönfeld expanded the "Bonner Durchmusterung" in a southerly direction. Both "Durchmusterungen" contained stars within a declination of -20° to the pole which were brighter than V = 10.

3 The Schmidt-Dickert Moon Model in Bonn from 1854

Argelander's first collaborator, Julius Schmidt (in Bonn from 1846 to 1853) documented some of his observations with small sketches and drawings in his diaries. These logbooks were rediscovered in the archive of the Argelander Institute some years ago. Figure 6.1 shows one of the few lunar drawings made by Schmidt in 1846 from a visual observation with the heliometer in Bonn.

Throughout his life, Julius Schmidt was interested in the moon. In Bonn, he started his own lunar observations in 1846. Since his diaries are only now being examined in detail, the scope of his work on the moon during his time in Bonn is still unclear. However, in 1854 he motivated the curator of the National History Museum in Bonn, Thomas Dickert, to build the relief model of the moon[2] shown in Figure 1.8 in this volume. The model shows the surface of the moon as seen from the earth (not inverted). It was installed on a hemisphere of wood almost 3 m in diameter containing 116 spherical gypsum plates. The scale of the ups and downs on these gypsum plates was three times larger than that of the horizontal distances on the moon. Illuminating the model from different directions, one could effectively demonstrate the view of the moon during different phases. The model contained a great deal of information about the surface of the moon. As explained by Oliver C. Farrington in an article published in 1899 in the journal *Science*:

The surface details of relief shown upon the model are based upon the charts of Beer and Mädler[3], but many new localities were added from the observations of Dr. Schmidt himself. In all over 20,000 distinct localities are represented, modeled proportionally according to the relief which they present upon the moon.[4]

According to Farrington, the model was exhibited in the Field Columbian Museum in Chicago at the end of the nineteenth century. On this occasion Schmidt published an 8-page article entitled "Das Relief der sichtbaren Halbkugel des Mondes, angefertigt von Th. Dickert" (Relief of the visible hemi-

2 Johann F.J.Schmidt, *Das Relief der sichtbaren Halbkugel des Mondes* (Bonn: Carl Georgi, 1854).

3 Wilhelm Beer, W, Johann H. von Mädler, *Mappa selenographica totam lunare hemisphaeram visibilem complectens* (Berlin: Schropp, 1834).

4 Oliver C. Farrington, "The Schmidt-Dickert Moon Model," *Science* 9:210 (1899): 35-36.

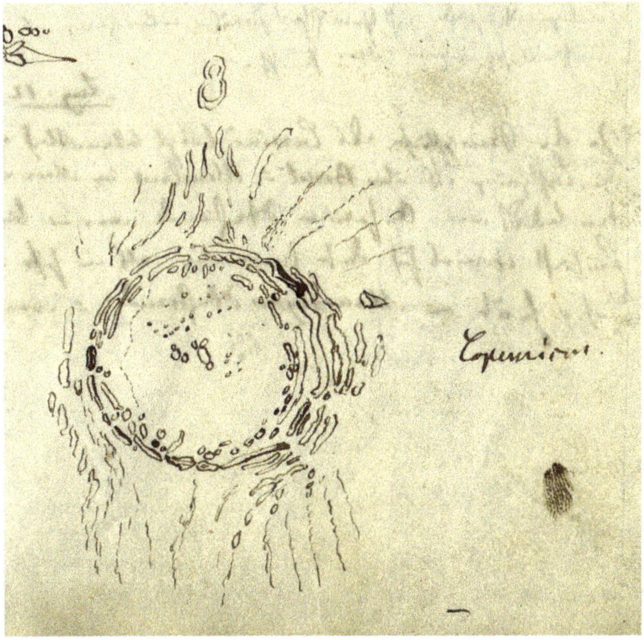

FIGURE 6.1
Drawing of the crater Copernicus by Julius Schmidt. The size of the drawing is 8 cm × 8 cm.

sphere of the moon prepared by Thomas Dickert), which includes a lunar map to explain the model.[5] Julius Schmidt later moved to Olomouc (now Czech Republic) and in 1858 became director of the observatory in Athens. There he published his famous charts of the moon with observations from 1840 to 1874.[6] These charts clearly belong to the most important visual observations of the moon from that period.

4 Karl Friedrich Küstner

Karl F. Küstner (Figure 6.2) was appointed director of Bonn observatory in 1891. Born in Görlitz / Eastern Germany, Küstner developed an early interest in astronomy and started studying astronomy at Strasbourg observatory.

In his PhD, which he wrote under the direction of August Winnecke, he analyzed (not his own) observations of the occultation of the Pleiades by the

5 Schmidt, *Das Relief der sichtbaren Halbkugel des Mondes*, 1854.

6 Johann F.J. Schmidt, *Charte der gebirge des mondes nach eigenen beobachtungen in den jahren 1840-1874 entworfen* (Berlin: Dietrich Reimer, 1878).

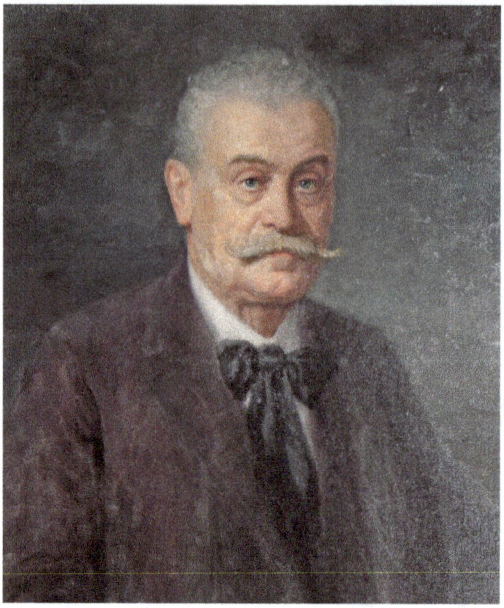

FIGURE 6.2
A portrait of K. F. Küstner (painting by W. Fassbender on the occasion of Küstners 75th birthday), COPYRIGHT: ARGELANDER-INSTITUT DER UNIVERSITÄT BONN

moon from 1839-1876. According to Hopmann, the final goal of this work was the accurate determination of the radius of the moon.[7]

During his time in Strasbourg Küstner proved himself a gifted observer, and he was invited to participate in the German expedition to Punta Arenas – a group of astronomers led by Arthur von Auwers – in order to observe the transit of Venus in 1882. At that time, transit observations of Venus were the key to determining the astronomical unit. Astronomers from different countries put major efforts into performing such observations with maximum accuracy.

Back in Germany, Küstner started working in Berlin with the meridian circle developed by Pistor and Martins. With this instrument he derived the positions of reference stars for the determination of precise data relating to comets and minor planets, and he also used the meridian circle to discover polar motion. During his time in Berlin Küstner became a leading visual astronomer.

In 1891 Küstner was appointed professor of astronomy and director of the observatory in Bonn (Figure 6.3). When he came to Bonn, the instruments in the institute seemed rather out-of-date. He immediately activated a Repsold meridian circle with an aperture of 16 cm: this telescope was already in Bonn but had not been used for years. After improving the instrument, Küstner used

7 Josef Hopmann, "Nekrolog Karl Friedrich Küstner," *Vierteljahresschrift Astron. Ges.* 72 (1937): 24-34.

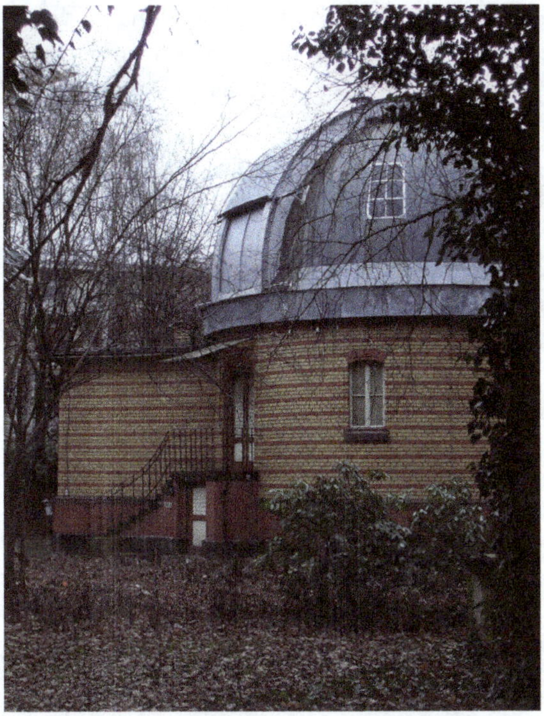

FIGURE 6.3
The dome of the double refractor in Bonn. COPYRIGHT: MICHAEL GEFFERT

it for many years to obtain observations of the highest quality. So much so that in 1910 he received the gold medal of the Royal Society for his work with meridian circles. Although he was extremely successful with the meridian circle, Küstner's main interest in Bonn was a new large telescope for photographic work. Around 1890 two developments had brought new horizons to astronomy. Photography grew in importance, because the limiting magnitude of observations had increased dramatically, and at the same time developments in spectroscopy enabled the determination of radial velocities and the classification of stars according to their spectra.

Küstner persuaded the university authorities to improve the observatory in Bonn by acquiring a new telescope, which was installed in 1899. After examining and improving the optical quality of this telescope, he started a series of spectroscopic observations in order to determine the radial velocities of stars. However, before taking these radial velocity measurements, he checked the spectroscopic method by the determination of a well-known quantity – the astronomical unit.

Based on the spectroscopic determinations of the radial velocity variations of Arcturus over one year, Küstner derived the velocity of the earth around the sun and hence the astronomical unit.[8] In two further publications he determined the radial velocity of 326 stars.[9] According to Hopmann (1937) these data were among the most accurate determinations of radial velocities at that time.[10]

However, in 1910 it became clear that the new and larger telescopes in the United States could measure the radial velocities of the stars faster, and Küstner stopped his work in this field and turned more intensively to astronomical photography. Beginning in 1914 he took direct images of more than 50 star clusters, proposing that a future generation of astronomers might use his measurements for the determination of proper motions. In fact around 1980 in Bonn under the direction of P. Brosche a project started to determine the proper motions of stars in star clusters. In this project astronomers re-measured the original plates of Küstner.

5 The Bonn Double Refractor

The history of the double refractor used by Küstner and details of the technical data are well documented[11] in annual reports (Figure 6.4).[12] Moreover, all Küstner's logbooks and diaries have been preserved, as well as most of his photographic observations.

The telescope arrived in Bonn in November 1899. It contains two tubes, one for visual and one for photographic observations. The visual tube has a focal length of f = 544 cm and a lens with diameter D = 36 cm, while the

---

8 Friedrich Küstner, "Eine spektrographische Bestimmung der Sonnenparallaxe," *Astronomische Nachrichten* 169 (1905): 241-264.

9 Friedrich Küstner, "Radial velocities of 99 stars of the second and third spectral classes observed at Bonn," *Astrophysical Journal* 27 (1908): 301-324; Frierich Küstner, "Radialgeschwindigkeiten von 227 Sternen des Spektraltypus F bis M beobachtet 1908 bis 1913 am Bonner 30 cm-Refraktor," *Astronomische Nachrichten* 198 (1914): 409-448.

10 Hopmann, "Nekrolog Karl Friedrich Küstner," 24-34.

11 Friedrich Küstner, "Der kugelförmige Sternhaufen Messier 56," *Veröff. Univ. Bonn* 14 (1920): 1-47; Martin Hoffmann, and Michael Geffert, "Photographische Astrometrie am Observatorium Hoher List", *Veroff. Univ. Bonn, 89 (1979): 1-12.*

12 Mainly in Friedrich Küstner, "Jahresbericht der Sternwarten für 1900: Bonn," *Vierteljahresschrift Astron. Ges.* 36 (1901): 83-86; Friedrich Küstner, "Jahresbericht der Sternwarten für 1901: Bonn," *Vierteljahresschrift Astron. Ges.* 37 (1902): 84-89; Friedrich Küstner, "Jahresbericht der Sternwarten für 1902: Bonn," *Vierteljahresschrift Astron. Ges.* 38 (1903): 86-88; F. Küstner, "Jahresbericht der Sternwarten für 1903: Bonn," *Vierteljahresschrift Astron. Ges.* 39 (1904): 63-65

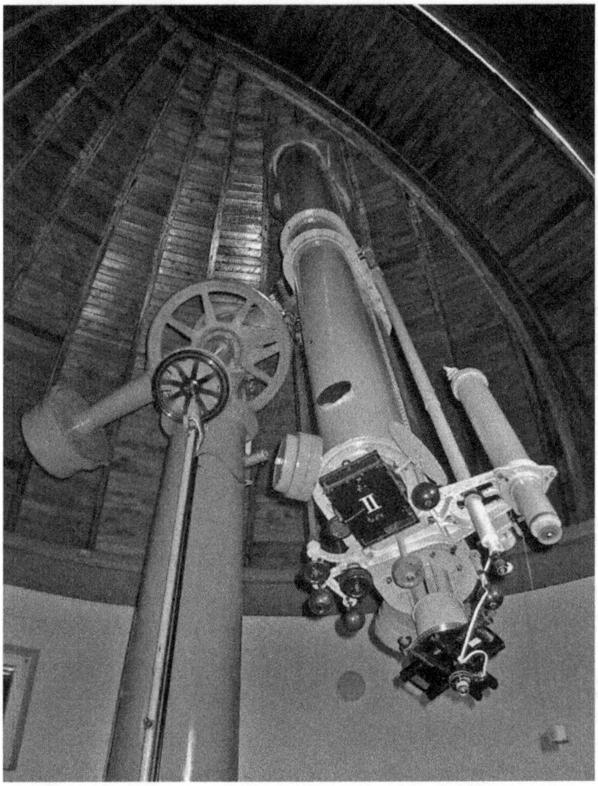

FIGURE 6.4
The double refractor of Bonn at Hoher List observatory. COPYRIGHT: MICHAEL GEFFERT

corresponding data for the photographic tube are f = 513 cm and D = 30 cm. The telescope's optics were by Steinheil of Munich, while the mounting was by Repsold. Küstner was the only observer to use that particular telescope. Only very few plates were taken by other Bonn astronomers.

Normally, the size of a field was 16 × 16 cm. Only in rare cases was a plate format of 9 × 12 cm used during the early years. Since a single plate covered a field of 1.6 × 1.6 degrees, the telescope was ideal for taking observations of star clusters. A limiting magnitude of B = 16.5 was achieved by Küstner. While earlier observations were made without filter, after 1978 a BG3 filter in combination with IIa-O, 103a-O and IIIa-J emulsion was used. The color range of the photographic tube corresponds roughly with the Johnson B band.

For his lunar observations Küstner used a Barlow lens manufactured by Steinheil. The Barlow lens enlarged the focal length of the system up to 12.5 m, corresponding to an image size of 11 to 12 cm for the moon on the photographic plates. This lens was only used for observations of the moon and of some double stars. All other plates were taken without the additional lens and had a scale of 40"/mm, corresponding to a focal length of about 500 cm.

During the first two years Küstner spent a lot of time comprehensively testing the telescope. He discovered that the visual tube could resolve double stars up to 0.5 arcsec and that even double stars with a separation of 0.3 arcsec could be detected by their elongated images. But he realized that the photographic tube images were affected by a slight distortion, which was corrected by dismounting and newly adjusting the objective lenses. Moreover, a correction technique was developed for the differential bending of the tubes with exposures longer than one hour. Based on his long experience, Küstner – as mentioned above – started his research by determining radial velocities and later moved to taking direct images of star clusters.

After the Second World War, astronomers in Bonn started to repeat Küstner's star cluster observations in order to determine proper motions. However, after several attempts this idea was soon given up. The Bonn observatory was located less than 1 km from the centre of the city, and since Küstner's time light pollution had increased so dramatically that deep sky observations were no longer possible with the refractor in its original place. Since astronomers at that time were convinced that proper motion work could only be done with plates from the same telescope, the refractor was moved to a new site in the Eifel hills some 85 km south of Bonn where atmospheric conditions were considerably better. The new "Hoher List" observatory had six telescopes: two were new, while four instruments (including the Bonn refractor) were brought from Bonn. In 1965 a new era of observation started. Proper motion work in Bonn began and a series of plates of star clusters was taken. Several scientific projects provided new information about the membership of stars in star clusters, and the motion of globular clusters in the Milky Way, as well as other results.

6 Küstner's Plates of the Moon – An Overview

Küstner observed the moon in Bonn between 1900 and 1904, but most of the plates still preserved in Bonn were taken in 1901 and 1902. The Bonn archives contain 14 plates from 1900, 36 from 1901, 31 from 1902, 3 from 1903 and 10 undated plates. Table 6.1 lists all of these. The moon plates are not subject to the numbering system, which Küstner used for all other direct images.

Küstner noted all observations made with the double refractor in his operation diaries (Figure 6.5). The plates were wrapped in sheets of paper with details written on them. However, the original sheets are available only for 25% of the plates of the moon, while the other plates were kept in new sheets. One reason is that Küstner – contrary to his normal procedure – kept several plates in one sheet. This treatment and the fact, that the plates do not appear in his

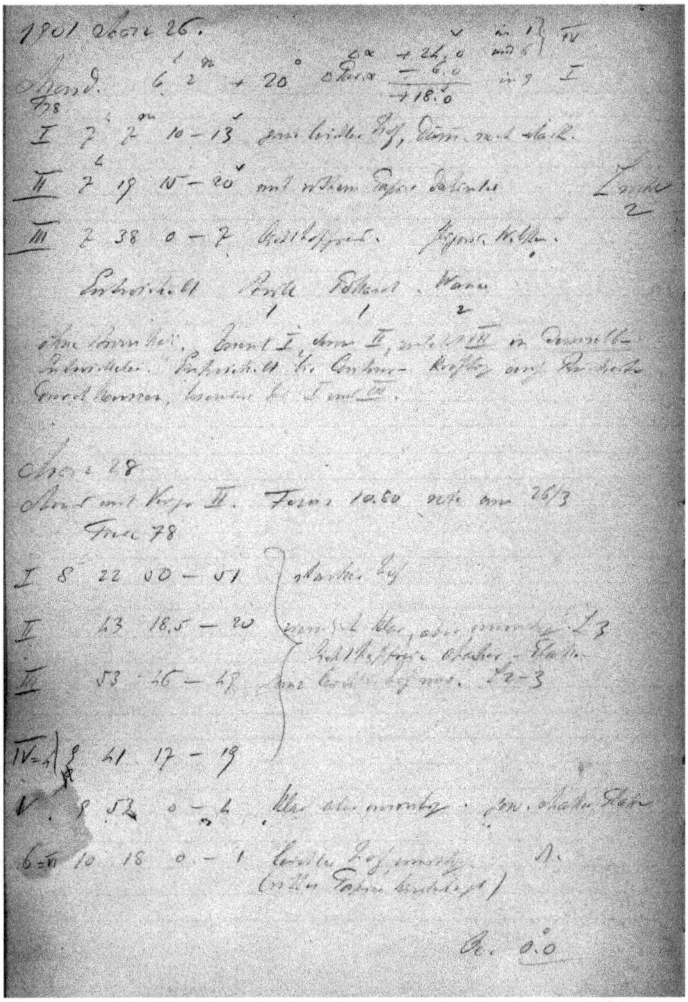

FIGURE 6.5 The observation diary for 26th and 28th of March 1901. COPYRIGHT: SAMMLUNG HISTORISCHER HIMMELSAUFNAHMEN DER UNIVERSITÄT BONN

normal counting system, underlines the assumption, that Küstner considered these plates only as test observations. On the other hand, most of the plates were signed by the astronomer either with his complete name or with the abbreviation "Kü". For most plates, the date as well as a single number is written directly on the plates. The meaning of the internal numbers is not well understood at the moment. Clearly the numbers are not a system applying only to the moon plates, because the numeration starts several times afresh between 1900 and 1903. The more than ten refractor logbooks are difficult to read today because they are in the long obsolete handwriting of the time, and analysis of their content is an ongoing project.

TABLE 6.1 A Summary list of Küstner's plates of the moon from 1900 to 1903

Year	Month	Day	Number of plates	Age of the moon [days]
1900	05	14	2	15
1900	05	19	2	20
1900	07	06	2	9
1900	07	14	2	17
1900	07	19	4	22
1900	08	02	2	6
1901	02	22	3	3
1901	02	26	3	7
1901	03	24	3	3
1901	03	26	3	5
1901	03	28	3	7
1901	03	29	4	8
1901	04	22	1	3
1901	04	23	1	4
1901	04	24	1	5
1901	04	25	1	6
1901	05	03	1	14
1901	05	20	1	2
1901	05	21	1	3
1901	05	22	3	4
1901	05	23	2	5
1901	05	24	1	6
1901	05	26	1	8
1901	05	27	1	9
1901	05	28	1	10
1901	05	29	1	11
1902	02	22	1	13
1902	02	26	1	17
1902	03	13	1	3
1902	03	14	1	4
1902	03	19	5	9
1902	03	20	2	10
1902	03	27	1	17
1902	04	14	7	5
1902	04	15	1	6
1902	04	18	3	9
1902	05	13	2	5
1902	05	17	3	9
1902	05	23	3	15
1903	07	17	2	22
1903	07	18	1	23

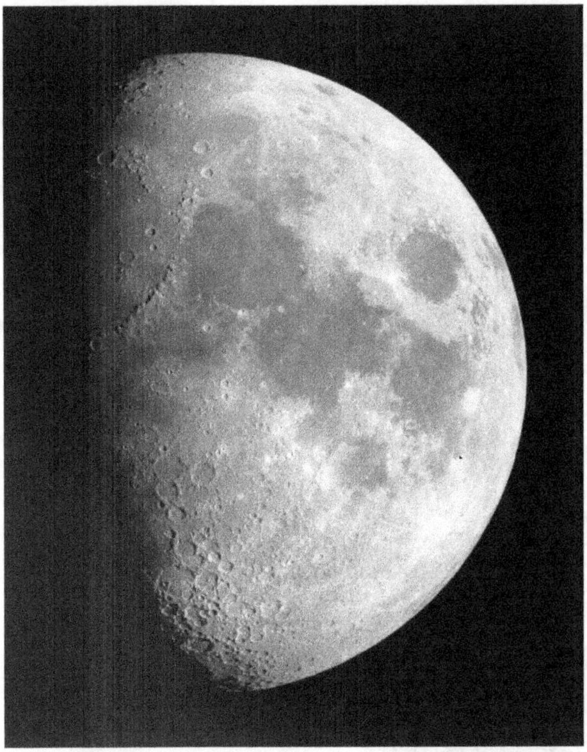

FIGURE 6.6
Observation of the Moon from 28th of March 1901.
COPYRIGHT: SAMMLUNG HISTORISCHER HIMMELSAUFNAHMEN DER UNIVERSITÄT BONN

It is interesting to note that the observations were mainly done during the waxing moon. Only 5 plates were observed at waning moon. The age of the moon in the 1901 series (around 5 days) may indicate that Küstner was trying to find the most suitable plates for Franz, who was at that time interested in the crater of the moon at its border (Figures 6.6, 6.7, and 6.8).

The SAO/NASA Astrophysics Data System does not list any publication by Küstner using his own lunar observations. Hopmann mentioned these observations only with the sentence "Die Mondaufnahmen erhielt Prof. Franz in Breslau…"[13] (The moon plates were sent to Prof. Franz in Breslau [Wroclaw]) – but given the large numbers of plates still in Bonn, while other plates marked "für Franz" (for Franz) are missing, this does not seem entirely correct. It is more likely that Küstner sent only a selection of his plates to Franz. In his annual report from Breslau, Franz mentioned having received 12 plates of the moon from the Bonn observatory.[14] Only in one publication did he use a plate

13 Hopmann, " Nekrolog Karl Friedrich Küstner," 24-34.

14 Julius Franz "Die Randlandschaften des Mondes," *Astronomische Mitteilungen der Koeniglichen Universitaets-Sternwarte zu Breslau* 2 (1913): 1-95

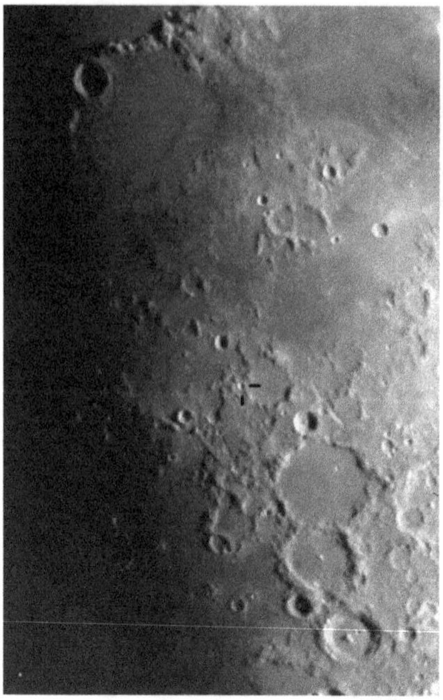

FIGURE 6.7
Detail of Figure 8. Crater Mösting A is indicated by the black lines. COPYRIGHT: SAMMLUNG HISTORISCHER HIMMELSAUFNAHMEN DER UNIVERSITÄT BONN

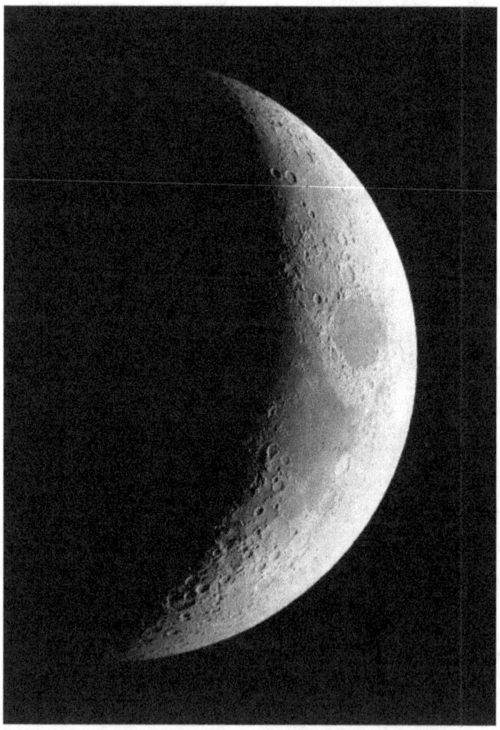

FIGURE 6.8
Plate of the Moon taken by Küstner in 24th of March 1901. Küstner took mainly plates of the waxing Moon. COPYRIGHT: SAMMLUNG HISTORISCHER HIMMELSAUFNAHMEN DER UNIVERSITÄT BONN

from Bonn, that of May 13, 1902.[15] Most probably the plates from Bonn remained in Breslau and were destroyed during the Second World War.

While Franz used the lunar observations e.g. to determine the positions of the craters of the moon, Küstner's interest in these observations was more technical. He wanted to test the image quality of his observations with the new telescope. Hence Küstner took series of up to 9 plates (April 14, 1902) within one night. Very often only part of such series has been preserved. Sometimes in the operation diary more plates are listed against one date than exist in the archive. Küstner varied exposure time, developer and emulsions. Emulsions with the designation "Matter" and "Schleussner" were used. Unfortunately all information on these emulsions, which were also used for the direct images of star clusters, has been lost. The exposure time varied from 15 seconds to less than 1 second. It is clear that exposure times imply a special quality of the night sky.

Our investigation of the Bonn moon plates is still in a preliminary phase. A complete view will only be possible when all the telescope logbooks have been studied and the complete plate archive in Bonn is properly organized and analyzed. Only in one annual report did Küstner give a summary of his lunar observations, and his remarks confirm his reputation as a severe critic of his own work:

> The photographs of the moon [...] when the air is still show the same details as those of Lick Observatory and even of Paris Observatory. However, for such an excellent quality one needs calm air, which one seldom finds in the neighborhood of the Rhine.[16]

7 The Work of Julius Franz

Julius Franz studied astronomy in Greifswald, Halle and Berlin and took his PhD 1872 in Berlin. In 1876 he became an astronomer in Königsberg. As a personal highlight, Franz led the 1882 expedition of German astronomers to Aiken for the measurement of the transit of Venus. However, in Neugebauer's (1913)

---

15 Franz, "Die Randlandschaften des Mondes", 1-95.

16 "Die Mondphotogramme ... dürften bei guter Luft denen der Licksternwarte und selbst der Pariser Sternwarte an Reichthum des Details nicht nachstehen. Allerdings ist dazu eine große Ruhe der Luft, wie wir sie gerade hier am Rhein nur selten haben, erforderlich."

opinion, the work of Franz on the moon was his most important topic: "Besonders geschätzt sind seine Forschungen über die Natur des Mondes".[17]

Franz's work on the moon started in Königsberg, where he published his first paper with new calculations of the physical librations of the moon.[18] The main problem for astronomers at that time was the determination of the mean position of the moon with high accuracy for celestial mechanics. Scientists mainly used the border of the moon for such measurements. However, the uneven border led to results that were less accurate than normal observations of stars. In 1889 Franz proposed the use of crater Mösting A[19] as a reference point in order to improve accuracy.[20] According to him this crater had a diameter of 6 arcsec and was the brightest point in that region. This made it an ideal object for measuring with a telescope and cross hair. However, for this procedure to be effective, the mean position of the moon depends on its libration. Therefore Franz also sought to improve the theory of libration.[21] In the following years he calculated the ephemerides for the observations of the crater Mösting A and analyzed the observational data. His results appeared in several papers in the Astronomische Nachrichten.[22] In his last paper from Königsberg Franz also determined the figure of the moon using five plates of the large Lick refractor (D = 91 cm, f = 1763 cm).[23]

In 1897 Franz became director of Breslau Observatory, where he continued his work on the moon by measuring the selenographic coordinates of craters[24] on plates from the Lick, Potsdam and Paris refractors. At a time when no spacecraft were available, the identification and measurement of the craters at the border of the moon was a difficult task for astronomers. After a long time spent

17 Paul V. Neugebauer, "Anzeige des Todes von Julius Franz," *Astronomische Nachrichten* 193 (1913): 443

18 Julius Franz, "Neue Berechung von Hartwig's Beobachtungen der physischen Libration des Mondes," *Astronomische Nachrichten* 116 (1886): 1-12

19 0.7°S, 5.9°W; diameter: 26 km; depth: 2.8 km: Eponym: Johan S. von Mösting.

20 Julius Franz, "Aufforderung zu Meridianbeobachtungen des Mondkraters Mösting A," *Astronomische Nachrichten* 122 (1889a): 237-240

21 Julius Franz, "Die Konstanten der physischen Libration des Mondes abgeleitet aus Schlüters Königsberger Heliometer-Beobachtungen," *Astronomische Beobachtungen in Königsberg* 38/1 (1889b): 1-47

22 e.g. Franz, 1886, 1889a.

23 Julius Franz, "Die Figur des Mondes," *Astronomische Beobachtungen in Königsberg* 38/5 (1899): 1-35

24 Julius Franz, "Ortsbestimmung von 150 Mondkratern," *Astronomische Mitteilungen der Koeniglichen Universitaets-Sternwarte zu Breslau* 1 (1901): I-VI, 1-51

analyzing data, Franz published a paper of measurements of the border of the moon.[25] In this publication he used at least one of the Küstner plates. He was also concerned with public outreach. His popular book on the moon, *Der Mond*, is a nice example of how a scientist could find a way to "translate" his scientific work into the language of laypeople.[26]

8 Conclusion

Nearly a hundred plates of the moon were exposed with the double refractor of Bonn between 1900 and 1903. The plates, which Küstner took of the moon, for were long forgotten in Bonn. Most probably Küstner observed the moon himself only for testing the optical quality of the telescope. Typical for his very comprehensive method for checking measurement errors, he took various series of observations with different exposure times and using different concentrations and types of developer. The varying phases of the moon and its sometimes low elevation in the sky, together with unfavorable weather conditions in Bonn, caused great problems for his observations. Nevertheless Küstner was satisfied with a few good observations demonstrating the excellent quality of the telescope. As a byproduct he sent several plates to Franz in Breslau for investigation of the border regions of the moon.

The double refractor was subsequently used in Bonn for further astronomic work (radial velocities, proper motions). Perhaps its most important period started after 1980 with the determination of proper motions with respect to the high quality first epoch observations of Küstner. The epoch of his observations varies mainly between 1916 and 1922, thus enabling proper motions of high quality. These data were used for the investigation of the motion of star clusters. Even if results from Gaia eventually supersede these data, the positions of the stars on Küstner's plates might be important for the determination of nonlinear motions of some stars. Few telescopes in the world have such a number of high quality early plates on which to base proper motion work.

For lunar observations, the use of a Barlow lens led to an increase of the focal length of the telescope from 500 to 1200 cm. With respect to the larger diameters and focal lengths of the corresponding telescopes in Paris (D = 82 cm; f = 1620 cm) and Lick (D = 91 cm; f = 1763 cm) the Bonn refractor seems to be too small to compete. This is clearly why Franz finally used only a few

25 Franz, "Die Randlandschaften des Mondes," *NOVA ACTA, Abh. Der Kaiserl. Leop.-Carol. Deutschen Akademie der Naturforscher*, Band IC, Nr.1 (1913): 1-95.

26 Julius Franz, *Der Mond* (Leipzig: Teubner Verlag, 1912).

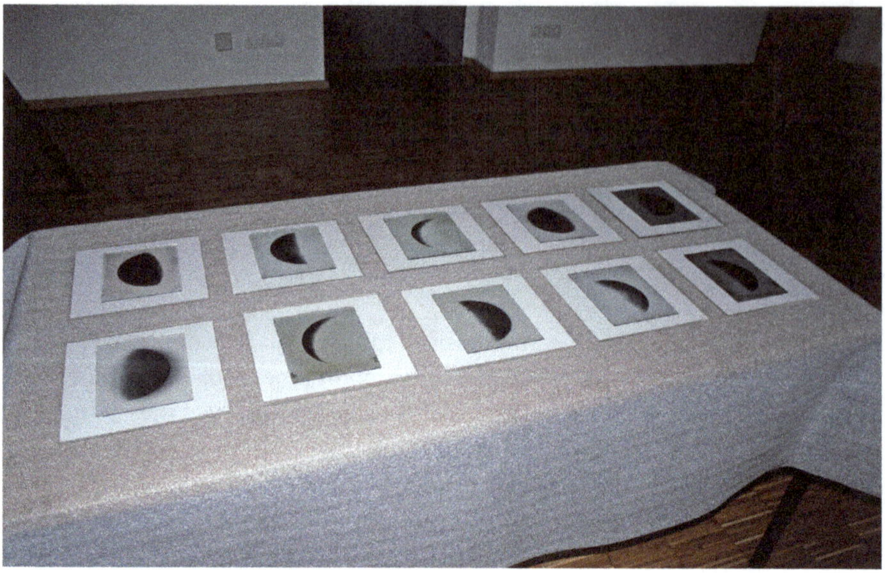

FIGURE 6.9 Exhibition of Küstners plates of the moon (2009) at Observatorium Hoher List.

observations from Bonn. For the determination of the positions of lunar craters the plates from Bonn were of only minor importance.

9 Epilogue

As the former German chancellor Helmut Schmidt once proposed, all scientists paid by the community for their work have a duty to share their scientific results with the community. They are obliged to find a way to communicate scientific progress in their disciplines. For two reasons, astronomers seem to have a special responsibility. Sometimes special events like comets or eclipses excite the interest of many people. On the other hand, astronomy is related to fundamental questions of human life like the birth and end of planets and stars – or even of the universe. Moreover, it has been shown that astronomical observations attract the attention of many people.

When Küstner's plates were rediscovered in Bonn it was clear that his observations were only of historical interest. However, historical objects of astronomy may be highly relevant to the public understanding of science. They demonstrate how astronomical research has developed with the instruments and methods of the particular time. People may understand astronomy better by following the historical traces of astronomical research. On the other hand, historical objects like instruments and plates appear also as art objects of their own epoch and may be presented as such in exhibitions.

With Küstner's moon plates we have taken this path, presenting them as art objects. Figure 6.9 shows some of the plates as exhibited at Hoher List Observatory. This exhibition contained modern paintings and historical instruments, and similar projects uniting art and astronomy have used the Küstner plates, along with other historical material, in recent years.

CHAPTER 7

The Digital Sky of Hamburg Observatory

Bringing Astro-photographic Plates from the 20th into the 21st Century

Detlef Groote

1 Introduction

Soon after their invention photographic plates were also used for celestial objects, and for a century they were *the* detector and storage medium for astronomical and astrophysical research. During the last quarter of the 20th century, with growing CCD pixel size and sensitivity, they were slowly replaced by the CCD technique. Finally, in the last quarter of the last century, they were used mainly for surveys with large Schmidt-telescopes and upcoming digitization. Worldwide about 7 million photographic plates still exist[1] but are unusable for modern research techniques as long as they are not available as digital images.

Twentieth century digitization processes used measuring machines like the PDS-machine, a state-of-the-art scanning device developed around 1980 consisting of a granite basis, a quickly moving table as plate holder for the x-direction and a granite arm moving slowly stepwise in the y-direction. The weight of the smaller 10 × 10 inch version used at Hamburger Sternwarte was about one ton, the scan time for plates of size 24 × 24 (all format sizes in [cm] if not explicitly given) was about 10 hours, and therefore too slow, and the storage media too expensive and too short-lived to think of a complete digitization of all photographic plates. Nowadays, the situation is quite different. Terabyte RAID systems[2] have become affordable, scanner technology has improved in quality and velocity, and the Internet offers the possibility of worldwide use of astronomical images. In other words, the general framework available offers the possibility to realize such projects. Nevertheless, the costs and time effort required to scan some 10,000 photographic plates are still huge. But it must be kept in mind that hardly any of these plates have been analyzed by modern computer techniques, which can analyze all the objects on all the plates

1 Rene Hudec, "Astronomical Photographic Plate Archives," in *Astroplate 2014 proceedings*, ed. Linda Mišková et al. (Prague: Institute of Chemical Technology, 2014): 53, 127

2 RAID: Redundant Array of Independent Disks, is a data storage virtualization technology that combines multiple physical disk drive components into a single logical unit for the purposes of data redundancy, performance improvement, or both.

 | DOI 10.1163/9789004298873_009

objectively and exactly, in a comparable manner, and without suffering from fatigue.

There are several scan projects under way worldwide, but only with essential funding can real progress be expected. One of the largest projects is the DASCH[3] project at Harvard, which comprises 500,000 plates, of which 170,000 had already been scanned by May 2017. New and ongoing projects (European and worldwide) can also be found at the AstroPlate website[4], where international meetings are organized every two years. The project presented here is unique inasmuch as all plate images are immediately available online with OpenAccess and the appropriate hand-written material is also connected with the plate scans. A newly developed search mask offers a SQL-language free search so that all meta-data such as e.g. emulsion, filter, telescope mode, and observer can be easily used to extract plates of interest. Additionally, a PhotoViewer offers the possibility to look through photographic databases taken on expeditions, around the construction phase of the observatory, or from telescopes, instruments or persons.

2 The Observatory and Its Historical Background

Founded in 1833, the Hamburger Sternwarte was moved to Bergedorf in 1909.[5] The new observatory was already equipped with cameras attached to the telescopes and with objective prisms to serve the upcoming interest in astrophysical research.[6] The oldest photographic plates taken by observatory staff are from a solar eclipse expedition to Algeria in 1905. From the start of observations at the observatory in 1911 until the end of the century more than 40,000 photographic plates and films were exposed and preserved, some of them taken at other observatories worldwide. Only about 10% of the youngest plates were already digitized in the last century explicitly for use in astronomical research.

3 Jonathan Grindlay, Sumin Tang, Edward Los, and Mathieu Servillat, "Opening the 100-Year Window for Time Domain Astronomy," *New Horizons in Time-Domain Astronomy: Proceedings IAU Symposium* 285 (2012): 29-34

4 <http://www.astroplate.cz/>.

5 Richard Schorr, *AN* 180 (1909): 293; Gudrun Wolfschmidt, ed. *Sonne, Mond und Sterne Meilensteine der Astronomiegeschichte* (Hamburg: tredition (Nuncius Hamburgensis, Band 29), 2013).

6 Gudrun Wolfschmidt, ed. *Kometen, Sterne, Galaxien – Astronomie in der Hamburger* (Hamburg: tredition (Nuncius Hamburgensis; Band 24), 2014).

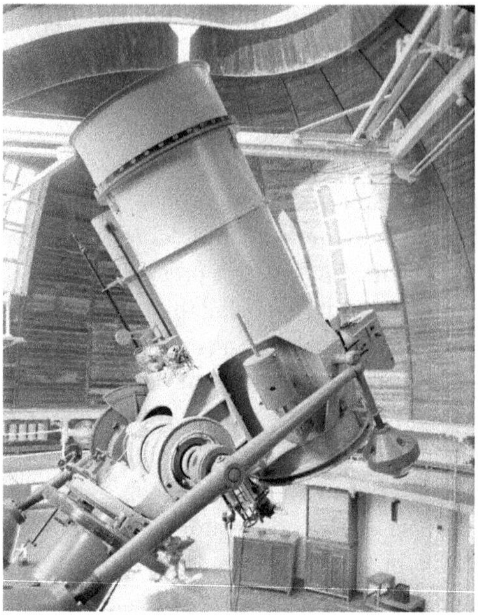

FIGURE 7.1
The 1m-telescope in 1912. COURTESY OF HAMBURGER STERNWARTE.

3 The Telescopes and Their Purpose

The use of the telescopes and their equipment changed with time and new research interests. The most important telescopes for photographic work were initially the 1m telescope and the Lippert Astrographs.

The former (see figure 7.1), was at its launch the fourth largest telescope in the world. A reflector telescope with a parabolic mirror, it was designed to look deep into the sky and take pictures of galaxies like the Andromeda nebula (see figure 7.10 later in this chapter). But larger telescopes were soon available at other observatories. Due to its large comatic aberration[7], the telescope was later often used with a diaphragm to improve image quality. After World War II it was used primarily with a spectrograph to expose spectra of single stars.

The original Lippert Astrographs consisted of three telescopes on a single mounting so that three plates could be exposed simultaneously using different plate emulsions or filters, or fitting one telescope with an objective prism.

Two astrographs, the 4 element Petzval P and the Triplet K had focal lengths of 1.5 meters, while the third (Triplet L) was built as a 'Carte du Ciel' astrograph with 3.4 m focal length and 33 cm aperture. In 1957 and 1958 these telescopes

7 The size of stellar images increases with distance from the plate center. In the plate corners the images are deformed and look a bit like small comets (see Figure 10).

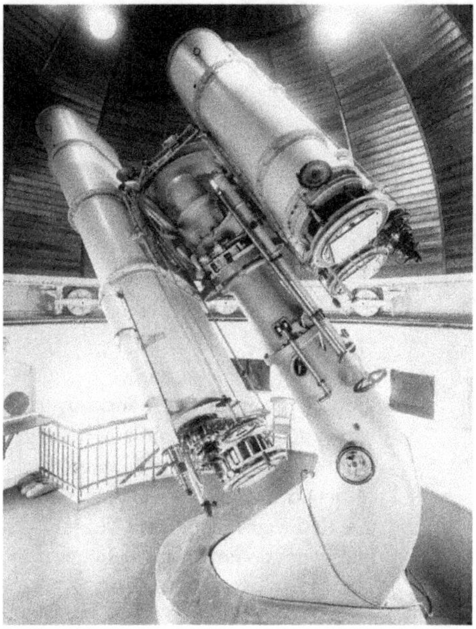

FIGURE 7.2
The Lippert-astrographs (1912). COURTESY OF HAMBURGER STERNWARTE.

were almost completely removed and replaced by a single telescope using the 60 cm parabolic mirror from the 'Doppel-Reflektor-Teleskop' (double-reflector telescope). This telescope had two reflector telescopes on one mounting (see figure 7.3, the Schmidt-Telescope is the longer one on the right-hand side); it was constructed by Bernhard Schmidt, on behalf of Richard Schorr, to prove the superior quality of a 'Schmidt-telescope' (spherical mirror plus correction plate) compared to a similar telescope with parabolic mirror only.[8]

But the differences in image quality for the chosen telescope dimensions were smaller than expected. Additionally, the Schmidt-telescope was wind-sensitive and its correction plate often became damped. Therefore, the parabolic mirror with good optical quality was mainly used instead. Only 470 plates were taken with this Schmidt telescope.

4 The Plates and Their Content

Large numbers of plates were produced with the main telescopes of the observatory such as the Lippert Astrographs (9233 plates by 1972), the 1m Spiegelteleskop (5825 plates by 1947 plus 2301 spectral plates by 1971), and the large

8 Richard Schorr, "Anzeige des Todes von Bernhard Schmidt," *AN* 258 (1936): 45

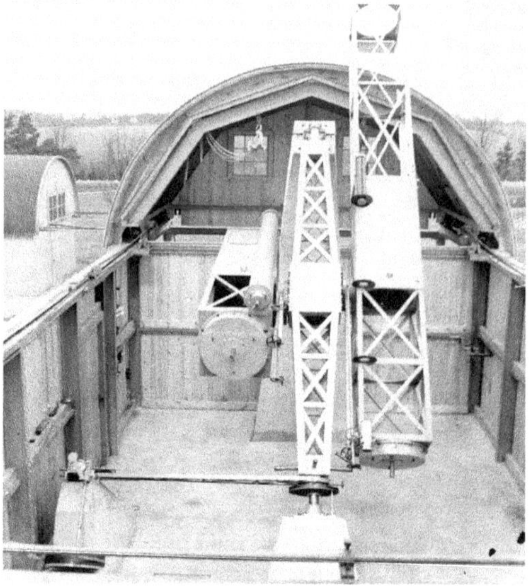

FIGURE 7.3
The double-reflector telescope: on the left side the 60 cm parabolic mirror with 3 m focal length, and on the right hand side the comparable Schmidt camera, which is twice as long (6 m) due to the correction lens. These telescopes do not exist any longer.
COURTESY OF HAMBURGER STERNWARTE.

Schmidt telescope (Großer Schmidtspiegel, 5771 plates by 1975) which commenced observations in 1955.[9] This telescope was later moved to the Calar Alto Observatory in southern Spain, where observations started in 1980. Nearly 3292 plates were exposed with this instrument by the end of the extragalactic survey (HQS, see below), which was also the end of observations in April 1998.

According to need, plate sizes range from the smallest at 6.5 × 9 to 30 × 30 for the largest plates. Exceptions are the solar eclipse plates with sizes up to 70 × 80 or the moon photographs of Bernhard Schmidt at 30 × 40. Films were only used for the original Schmidt telescope (with a diameter of 9 cm, see below). For such a telescope the focus lies on a sphere, not on a plane, and glass can only be bent very slightly. The larger Schmidt telescope (Großer Schmidtspiegel) use thin glass plates (only 1 mm thick and 24 × 24 in size) with a special plate-holder which allows the plates to be pressed into the required form.

The advantage of photographic plates as a storage medium is their high density of information, much of which has now kept for up to a century. On the other hand, they are a non-digital medium that suffers from aging and

9 Jochen Schramm, "Der Große Hamburger Schmidtspiegel". In *Kometen, Sterne Galaxien - Astronomie in der Hamburger Sternwarte, Zum 100-jährigen Jubiläum der Hamburger Sternwarte in Bergedorf*, edited by Gudrun Wolfschmidt. Hamburg: tredition (Nuncius Hamburgensis; Band 24), 2014.

bad storage conditions such as high temperatures, humidity, and dust, and that scratches when frequently used. Moreover, after a century some of the photographic plates show damage such as darkening and silvering, or dissolution and rupture of the emulsion. But compared to modern storage media like CDROMs or DVDs, photographic plates still seem remarkably long lasting, as those digital media will be readable for only about a decade, and it seems unlikely that the hardware required for reading and writing them will last much longer.

For these reasons, a pilot project was started in 2010 to test the feasibility of a complete digitization of the Hamburger Sternwarte plate archives, using an Epson EXPRESSION 10000XL flatbed scanner – a type used in many similar projects worldwide. Additionally, all handwritten material was to be digitized and all data to be made freely available (Open Access) via the Internet. First (optimistic) estimates of the time necessary for such a project suggested 20-30 years, but that seemed reducible by involving more people and scanners.

Turning now to the plates and their content: though the sky seems at first sight largely unchangeable, if we look at the details we see many differences. Each photograph of the sky presents a unique view of the objects at the time of exposure. Tiny differences in position, if measured from year to year, may accumulate to large values on a time scale of a hundred years. Many objects such as stars and quasars are variable in brightness, or show sudden short jumps in brightness like novae, super novae or gamma ray bursts. The images of all such events lie possibly frozen on photographic plates and may still be hidden in the immensity of star counts of more than a hundred thousand on a single exposure.

Photography of the moon – the second brightest celestial object, visible both night and day – was mostly done for sentimental reasons, its beauty being visible even with the naked eye. So far, some 170 plates with exposures of the moon have been found in the archives, although they were not used for moon research. Nevertheless, for the Hamburger Sternwarte and its development, the moon photography performed by Bernhard Schmidt (1879–1935), an ingenious optical instrument engineer with high technical skills, soon became important. In 1917 he offered 5-6 plates (see figure 7.4) taken at Mittweida, where he lived and worked at the time (see figure 7.6), to Richard Schorr, the first and longtime director of the Hamburger Sternwarte.[10]

These plates were exposed at different moon phases and were taken with a self-polished mirror of 31 m focal length and 30 cm aperture fitted to a horizon-

10 Otto Heckmann, "Richard Schorr," *MittAG* 3 (1951): 3.

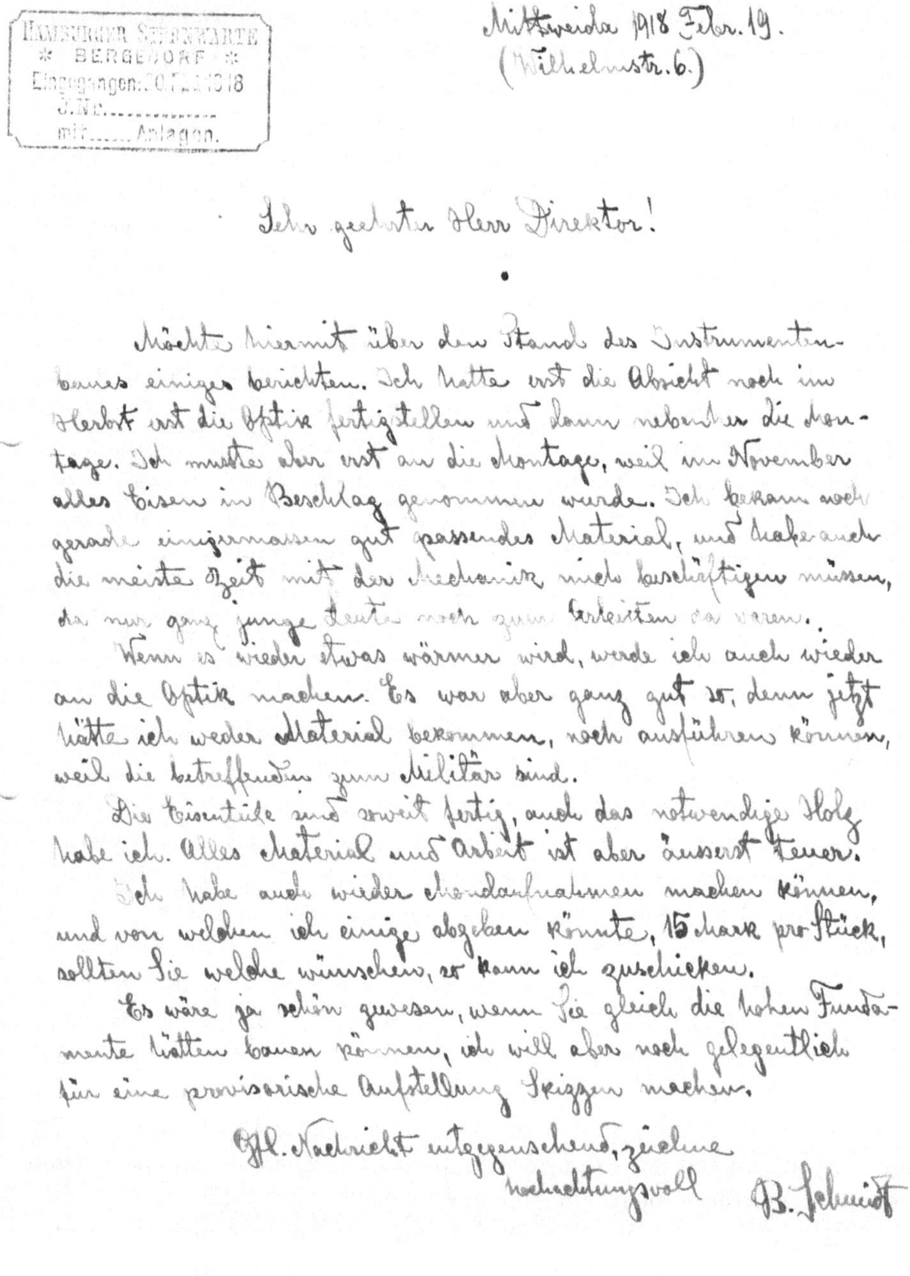

HAMBURGER STERNWARTE
* BERGEDORF *
Eingegangen: 20.FEB.1918
J.Nr. ....
mit .... Anlagen.

Mittweida 1918 Febr. 19.
(Wilhelmstr. 6.)

Sehr geehrter Herr Direktor!

Möchte hiermit über den Stand des Instrumentenbaues einiges berichten. Ich hatte erst die Absicht noch im Herbst erst die Optik fertigzustellen und dann nebenher die Montage. Ich musste aber erst an die Montage, weil im November alles Eisen in Beschlag genommen wurde. Ich bekam noch gerade einigermassen gut passendes Material, und habe auch die meiste Zeit mit der Mechanik mich beschäftigen müssen, da nur ganz junge Leute noch zum Arbeiten da waren.

Wenn es wieder etwas wärmer wird, werde ich auch wieder an die Optik machen. Es war aber ganz gut so, denn jetzt hätte ich weder Material bekommen, noch ausführen können, weil die betreffenden zum Militär sind.

Die Eisenteile sind soweit fertig, auch das notwendige Holz habe ich. Alles Material und Arbeit ist aber äusserst teuer.

Ich habe auch wieder Mondaufnahmen machen können, und von welchen ich einige abgeben könnte, 15 Mark pro Stück, sollten Sie welche wünschen, so kann ich zuschicken.

Es wäre ja schön gewesen, wenn Sie gleich die hohen Fundamente hätten bauen können, ich will aber noch gelegentlich für eine provisorische Aufstellung Skizzen machen.

Gefl. Nachricht entgegensehend, zeichne
hochachtungsvoll B. Schmidt

FIGURE 7.4a Correspondence between Bernhard Schmidt and Richard Schorr concerning the purchase of 6 large photographic plates of the moon. Schorr bought the plates, recognized his skills, and took him therefore to Hamburg (For the answer to Bernhard Schmidt, see Figure 7.4b. on the facing page).
COURTESY OF HAMBURGER STERNWARTE.

Hamburger Sternwarte

Direktor: Professor Dr. R. Schorr.

J.-Nr. ....

Fernsprecher:
1) Sternwarte in Bergedorf: Bgdf. 160.
2) Zeit-Zentrale in Hamburg, Holstenwall: Hbg. 1. 2000.
3) Telefonisches Zeitsignal: Hbg. 4. 10000.

Bergedorf, 1918 Februar 21.

Sehr geehrter Herr Schmidt !

Aus Ihrem Schreiben vom 19.d.M. ersehe ich, dass Sie Ihre Absicht, im vorigen Jahre die Spiegel möglichst noch fertigzustellen, nicht ausgeführt und sich statt dessen der Ausführung der Montierung gewidmet haben. Bei der jetzigen Sachlage scheint dies allerdings zweckmässig gewesen zu sein, denn die Ausführung aller Bauten und Eisenkonstruktionen ist ja jetzt immer mehr erschwert worden. Sobald es aber der Zustand der Witterung erlaubt, nehme ich an, dass Sie sich der Herstellung der Optik widmen werden.

Von Ihren Mondaufnahmen würde ich ganz gern 4–6 Stück zu je 15 Mark für unsere Sternwarte kaufen und zwar möglichst Aufnahmen, die sich auf die ganze Lunation verteilen.

Da jedoch, wie ich aus dem Zustande der bisher übersandten Pakete entnehme, die von Ihnen, als feindlichen Untertan, abgesandten Packete dort von Seiten der Zensur geöffnet und geprüft werden, scheint mir Gefahr vorzuliegen, dass die dünnen Glasplatten dabei zerbrechen. Ich möchte dah daher empfehlen, dass Sie, wenn e möglich ist, das Paket mit den Glasplatten bei Ihrer dortigen Prüfungsstelle offen zur Durchsicht abgeben und dann in deren Gegenwort selbst verschliessen, damit kein Schaden eintritt

Hochachtungsvoll

Direktor der Sternwarte.

FIGURE 7.4b

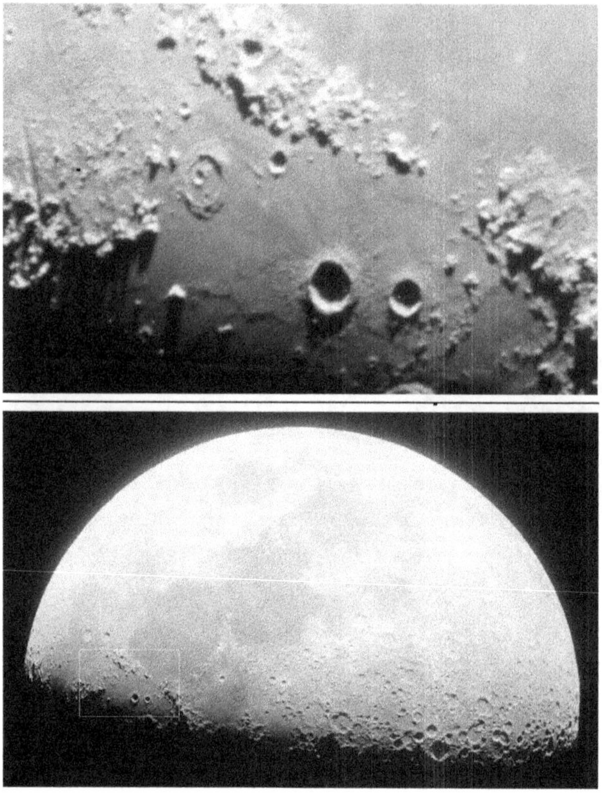

FIGURE 7.5
Moon photography done by Bernhard Schmidt in 1917 with a 31m horizontal-mirror-telescope and 30 cm aperture. The exposure time is about 10 sec. The excerpt exhibits many details. We can see four craters that lie in the eastern Mare Umbrium. From bottom to top: Autolycus (coordinates: 30.7°N, 1.5°E; diameter: 39 km; depth: 3.4 km); Aristillus (coordinates: 33.9°N, 1.2°E; diameter: 55 km, depth: 3.6 km); Theaetetus (coordinates: 37.0°N, 6.0°E; diameter: 25 km; depth: 2.8 km); Cassini (coordinates: 40.2°N, 4.6°E; diameter: 57 km; depth: 1.2 km). COURTESY OF HAMBURGER STERNWARTE.

tal-mirror telescope.[11] The plate size is 30 × 40 and the moon diameter on the plates is about 27.5 cm (see figure 7.5).

The plates are of excellent quality for the age, taking into account the circumstances (see figure 7.6) and length of exposure. Schorr bought them for 15 marks each. Recognizing his abilities, he employed Schmidt, who later (in 1932) invented a wide-angle telescope with a spherical mirror and a correction lens.[12] Its original name, *Schmidt'sches komafreies Spiegelteleskop*, can be found on the notebook cover used by the observers during the night. This telescope produced point-like images all over the plate, even in its corners.

11 Such systems are used for long focal lengths and use a primary, turnable flat mirror (coelostat), which reflects the light to a fixed mirror.

12 Roger Ceragioli and Walter Stephani, "Optiken für die Hamburger Sternwarte. Bernhard Schmidt und Richard Schorr – Zeugnisse und Ergebnisse einer Zusammenarbeit," in *Kometen, Sterne, Galaxien – Astronomie in der Hamburger Sternwarte*, ed. Gudrun Wolfschmidt (Hamburg: tredition (Nuncius Hamburgensis; Band 24), 2014).

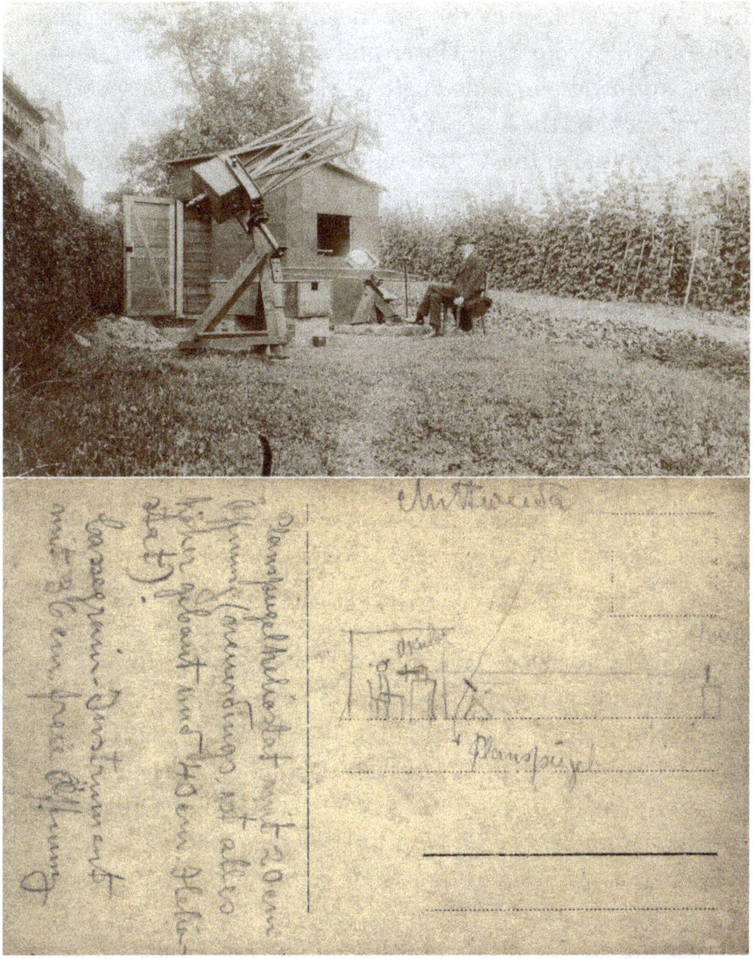

FIGURE 7.6 Bernhard Schmidt with his horizontal-mirror-telescope in Mittweida, presumably before 1916, a comparable one to that used for his moon photographs, later. The coelostat can be seen just in front of his hut at the right-hand lower corner. The mirror must have been then in direction to the reader approximately 30 meters away and the camera inside the hut. On the lower part a sketch of the telescope outline can be found, drawn by himself. COURTESY OF HAMBURGER STERNWARTE.

Today, telescopes of this type are commonly known as Schmidt telescopes, although at the time of his early death in 1935 his invention had not yet been accepted by the astronomic community, for the simple reason that nobody believed such a complex lens feasible – a skepticism encouraged by Schmidt's own secrecy about it. Only 10 years after his death the importance of his invention was recognized by the American observational astronomer Nicholas Ulrich Mayall[13] from Lick Observatory. Later, large Schmidt telescopes were used worldwide for wide-angle all-sky surveys and especially at Calar Alto Observatory[14] and the European Southern Observatory[15] (ESO). Two large surveys were initiated at these observatories namely the Hamburg Quasar Survey[16] (HQS) and the Hamburg ESO Survey[17] (HES). These surveys were the last ones done with photographic plates, but they were already designated for digitization using a PDS-machine. Early last century, the discovery of new objects was the dominant aim. After exploring our solar system for new planets and comets, interest shifted to variable stars or novae, and soon to galaxies, in the context of exploration of the universe and its origins. At the same time, observational techniques improved with increasing telescope sizes, and an increasing knowledge of theoretical astrophysics led to a better understanding of what had been newly detected by the observers.

Large surveys such as the *Bergedorfer Spektraldurchmusterung*, a survey of about 130,000 stars, used an objective prism in front of the telescope that opens the dot-like image of each star into an elongated spectrum. Such spectra display not only the energy distribution of the stars, but also exhibit spectral lines[18]. Such a spectrum is like a fingerprint that can be used to determine physical properties of the object such as temperature, gravitational acceleration, and chemical composition. Spectra can be used for classification and ordering of objects according to their physical parameters. Parallel to such research, catalogs with accurate positions of the stars were urgently needed, and projects of the *Astronomische Gesellschaft*[19] such as the AGK2 and AGK3 catalogs were launched, resulting in thousands of plates being exactly measured.

Besides their use for astronomical purposes, photographic plates were also used for documentary photography, and hundreds of plates in the archives

---

13 Nichollas U. Mayall, "Bernhard Schmidt and his coma free reflector," *PASP* 58 (1946): 282.

14 <http://www.caha.es/telescopes>.

15 <https://www.eso.org/public/germany/teles-instr/lasilla/1mschmidt/>.

16 Hans J. Hagen, Detlef Groote, Dieter Engels, and Dieter Reimers, "The Hamburg Quasar Survey. I. Schmidt observations and plate digitization," *A&AS* 111 (1995): 195.

17 Lutz Wisotzki, T. Köhler, Detlef Groote, and Dieter Reimers, "The Hamburg/ESO survey for bright QSOs I. Survey design and candidate selection procedure," *A&AS*, 115 (1996): 227.

18 Spectral lines are produced by the absorption or emission of light added mostly onto a continuum of radiation emerging from the stars. Each chemical element has large numbers of possibly visible lines in the spectrum at distinct wavelengths.

19 <http://www.astronomische-gesellschaft.de/de>.

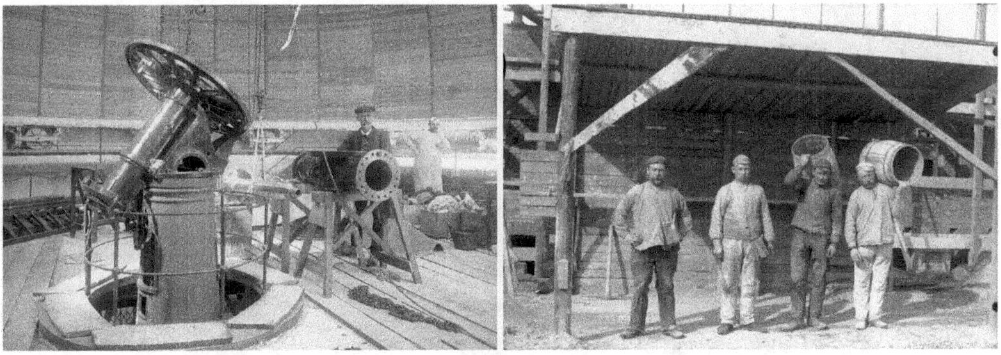

FIGURE 7.7 Construction work at the 'Großer-Refraktor' of Hamburger Sternwarte (left-hand image). The mounting is already built-on but the 9 m long tube is still not mounted. The large lifting platform (filling the whole dome) is at its highest position. Workers for the construction of the buildings (right-hand image). COURTESY OF HAMBURGER STERNWARTE.

document the construction of the Hamburger Sternwarte, its telescopes and the people who worked on it (see figure 7.7).

Today these plates are virtually useless if they are not digitally cataloged and their images and handwritten material made available to modern computer techniques.

5 The Scanning Process and Data Storage

Compared to the state-of-the-art PDS machines of the 1980s, modern flatbed scanners have the advantage of speed. A complete scan of a 24 × 24 plate with the PDS machine took about 10 hours in 1984. The modern flatbed scanners used in our project can do a comparable scan in 7 minutes. But you can't have everything. The density range of such a scanner is somewhat smaller, though this may be negligible for use with most old plates and emulsions. A more serious disadvantage is that while positional accuracy along the CCD is comparable to a PDS machine, the CCD wobbles in the direction of scan. Fortunately, this inaccuracy can be overcome with modern analysis techniques. In figure 7.8, the deviation in position of ~100,000 stars is shown before and after correction.[20]

20 Taavi Tuvikene (Estonia) 2015 (private communication).

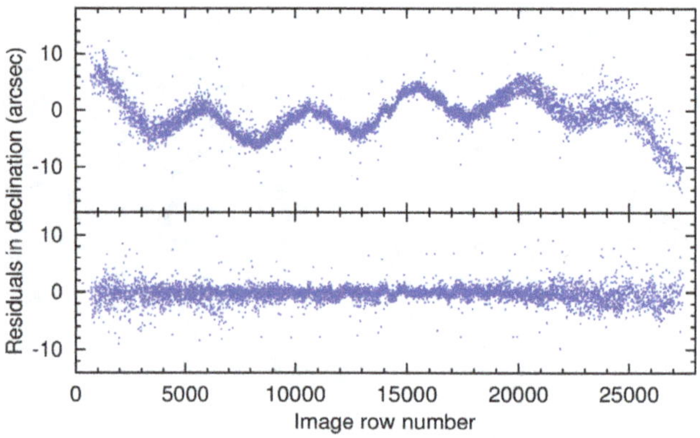

FIGURE 7.8
Astrometric residuals in the plate LA00784 as a function of image row number (Y-coordinate) after initial solution (top) and after refined solution in sub-fields (bottom). The wobbles, introduced by a flat-bed scanner, disappear after refinement. FIGURE CREATED BY TAAVI TUVIKENE TARTU/ESTONIA, A CO-WORKER IN THE APPLAUSE PROJECT.

This correction profits from the availability of online catalogs with precise celestial coordinates and well-known magnitudes.[21] From these, exact object positions on the plates and precise celestial coordinates can be derived for billions of objects. Furthermore, it is possible today to calibrate the measured brightness of most objects by constructing local characteristic curves that were not available at the time when the plates were exposed.

The scanning process includes careful cleaning of the plates from dirt on the glass and dust on the emulsion side. As a first step the plate envelopes are scanned in low resolution (150 dpi). Before cleaning the plates – but only when there are marks on the glass side – a low-resolution color scan (300 dpi) is made; this is necessary to retain informative marks pointing to objects of interest on the plates. Then the back of the plates is cleaned with moistened microfiber cloths. The high-resolution scan is done at 2400 dpi, which is the equivalent of 10.8 micron pixel on the plate. For most plates the low-resolution scan will be calculated automatically from the high-resolution scans.

Parallel to plate scanning, handwritten material – e.g. logbooks, where information on each plate (plate number, object name, coordinates, emulsion type or other important details) is neatly written in one line – must be scanned in low resolution (150 dpi). This information is comparable to meta-data files. Even more important are the observer notes written next to the observation, which may contain additional information on weather conditions or technical problems.

All scanned images are saved to the local hard-disk of the scan-PC. During the night, processes will start automatically to convert the scan data (16 bit

21 Magnitude is an old logarithmic unit of measurement for the apparent brightness of stars.

PNG-files) into the FITS[22] format generally used in astronomy, and to transfer the data to a local RAID6-system where it is kept for further processing. A back-up copy will be sent to a tape robot at Hamburg University, 20 km away. Another copy will be kept at the Leibniz Institute for Astrophysics in Potsdam to be used in the context of the APPLAUSE project (see below).

6 The Web Pages and Archive Search

The immense work of scanning all the plates only makes sense if this data is made available to the worldwide astronomical community. After the images are copied to the RAID system, web pages are created automatically. On December 1st 2011, the 100th anniversary of the first plate exposed at the newly built Hamburg observatory, trilingual web pages (German, English, Spanish) were made public[23] in a ceremony at the observatory. At that time only 3700 of the oldest plates had been scanned. Since then, all scanned plates have been published on the day after their scanning. At the time of publishing this book more than 40,000 plates will be available online.

But as well as scanning the plates, plate envelopes, and books, it is necessary to record the meta-data for each plate. This meta-data comprises the plate number – generally 2 capital letters for the telescope followed by a 5 digit number or characters plus number, and sometimes by an additional character if two or more plates have the same plate number – the center coordinates, the object or field name (if available), and the date of observation. Such data can usually be found on the plate envelope. Additionally, there is other information like the plate size, sky conditions, multi-exposure information, time stamps of beginning and end of observation (important for variable stars and phase calculations), the emulsion type and filter type if used, special modes of the telescope such as objective prisms and/or gratings, and the observer's name and other information concerning the observation. Such data is often found in the observer notes.

But to make all such data usable, search functions are needed. An important question is e.g., how to find all the plates on which a certain object might be visible. To achieve this, a SQL database has been created detailing not the data files themselves, but simply the path to locate them. As the project aims are not only confined to future research but include conservation of information and presentation to the general public – e.g. amateur astronomers, historians

22 Flexible Image Transport system developed by NASA.

23 <http://plate-archive@hs.uni-hamburg.de>.

or other interested people – a search mask has been developed that uses SQL syntax-free input and allows searches of all available meta-data. However, it is still possible to use the SQL syntax for professional searches (see figure 7.9).As an example of a search for visually attractive objects, one could enter in the object field 'Nebel' (nebula) and in the exposure field 120 minutes. The result would be a list of long exposures (> 120 min) including objects like the Orion and Andromeda nebulae (Figure 7.10).

Three exceptional plates were exposed simultaneously using the Lippert astrographs with an exposure time of 1558 minutes each, i.e. about 26 hours (Figures 7.11a and 7.11b).

This entry is not erroneous. In those days astronomers thought that a longer exposure time would always result in a deeper image. But looking at the faintest objects on the plate today, it becomes clear that the same depth (density) was already reached after about 3 hours of exposure. Nevertheless, this observation, exposed in 4 consecutive nights with more than 6 hours of uninterrupted guiding is a masterpiece of observational technique. Any fault in guiding, any technical insufficiency would have resulted in a corrupt image.

To use the database for research it should be possible to search for plates with similar emulsions or filters. Via three drop-down menus on the search mask, all the emulsions and filters used, and the mode of the telescope (objective prism, grating, etc.), can be activated for or excluded from the search. Astronomers familiar with virtual observatories (VOs) and SQL language can make SQL queries. All queries result in a list of objects that can be downloaded as a standard CSV[24] file.

In the top row of the search mask one or more telescopes may be chosen, though searches over the whole database are also possible. The plate number entry in the first column helps trace a specific plate, while the coordinate search (box or circle) helps find a known object on all plates, or on a specific plate. The second column contains dates (useful if you are looking for a plate exposed on your birthday, for example) and date ranges. The third column contains minimum and maximum exposure times and interim ranges. Plate size, multi-exposures, and/or objective prism plates can be selected or deselected by using a '+' or a '-' sign. It is also possible to search for objects, which comprise a certain string or substring in their information field.

All selections may be reset in the final column. Table size can be limited by entering the number of lines to be displayed or skipped. At the top of the output table the SQL search command currently in use will be displayed. Most of the additional information is handwritten, often in Sütterlin, the old German

24 *Comma Separated Values.*

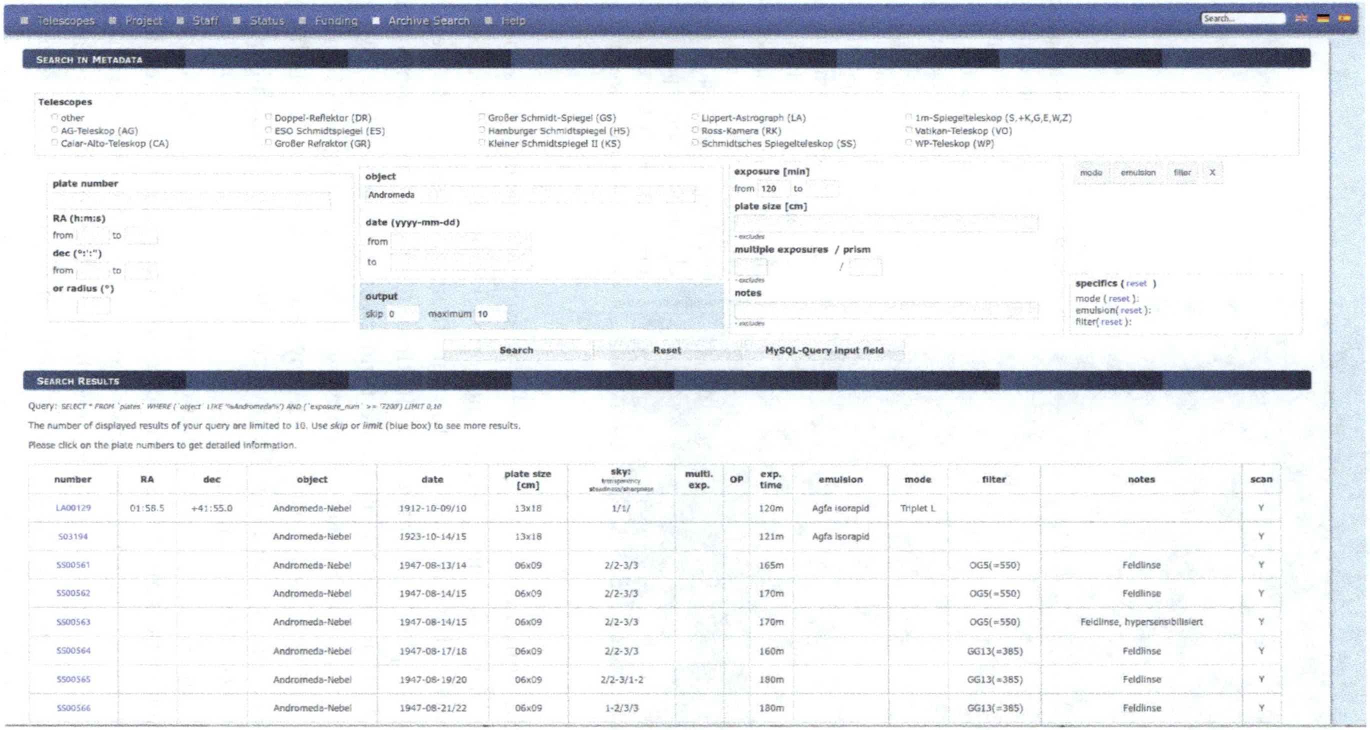

FIGURE 7.9 Search mask and some results for the two entries object = Andromeda and exposure time = 120 minutes. The plate number in the resulting table can be clicked and will open the wep-page (see figure 7.10) for the chosen plate. The other search entries can be found below. LAYOUT OF THE WEB PAGE WAS DEVELOPED BY CHRISTIAN OFFEN AND DETLEF GROOTE.

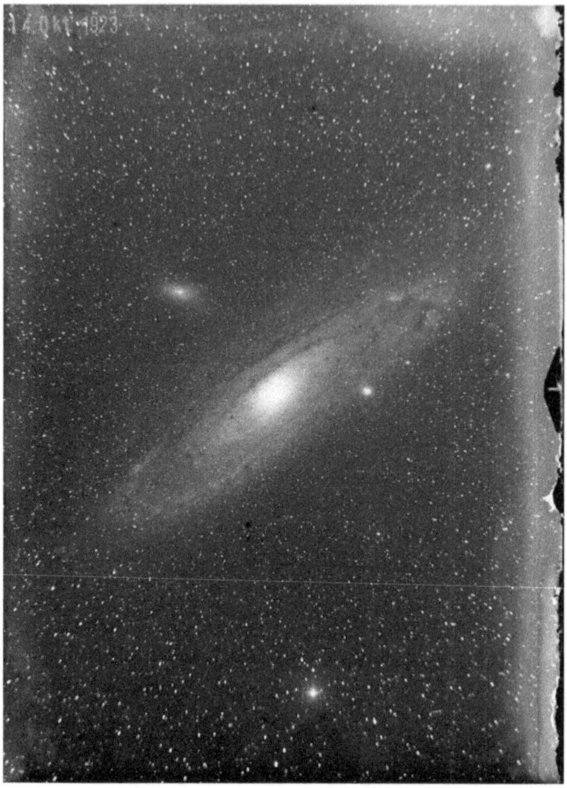

FIGURE 7.10
Inverted image of plate So3194, a two hours exposure of the Andromeda nebulae from Oct 14, 1923 exposed with the 1m-Spiegelteleskop. Large stellar images in the plate corners are due to comatic aberration of the parabolic mirror. Aging problems can be seen at the right-hand rim of the plate. The darkening of the original image due to the glue of the paper envelope and the already beginning separating of the emulsion. COURTESY OF HAMBURGER STERNWARTE.

script, and therefore not usable for OCR[25]-software. All this data must be entered manually, a time-consuming and not error-free process; to date, some 3 million characters have been entered in the database. Special software was written involving plausibility checks on the time stamps and comparing calculated exposure time with the exposure time given on the envelope. Where these times differ, the values are compared with the original data and corrected manually, in rare cases by best guess. In any case, as all material is published together, any user may draw her or his own conclusions after a careful inspection of all the available information.

A widely format-free input interpretation is integrated to ease usability even for astronomic laypersons. The search for coordinates can be entered in all common formats (e.g. 17:30:12 or 17:30.2 or 17.033). For the δ-coordinate a minus sign can be used in front (with or without blank) while the plus sign can be omitted. Before sending a complete new query, or if the result was zero, it is

25 *Optical Character Recognition.*

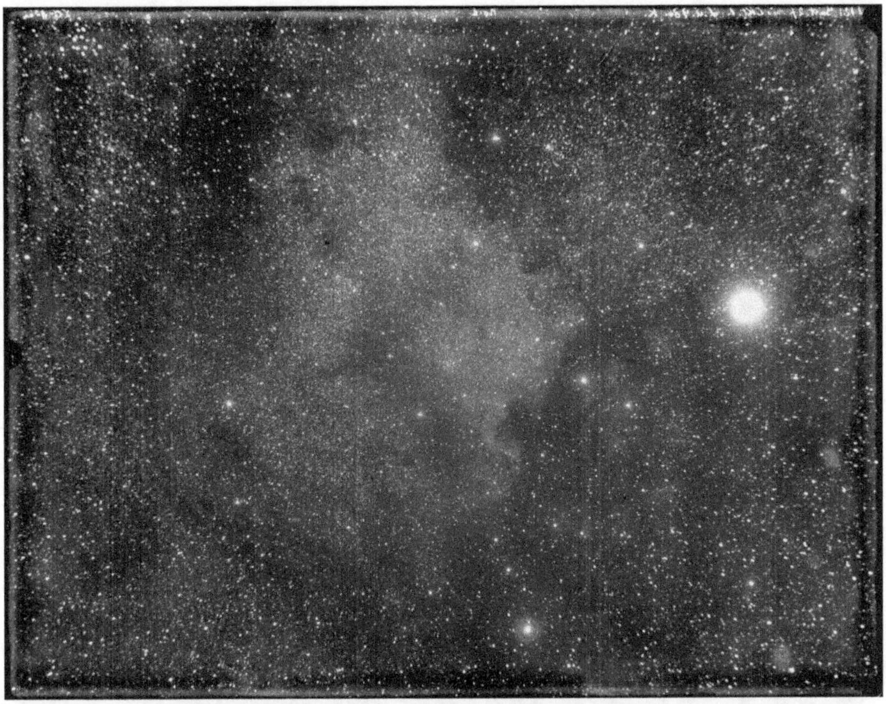

FIGURE 7.11a Plate LA00986 (Lippert-Astrograph) of the so called North-America-Nebulae. COURTESY OF HAMBURGER STERNWARTE.

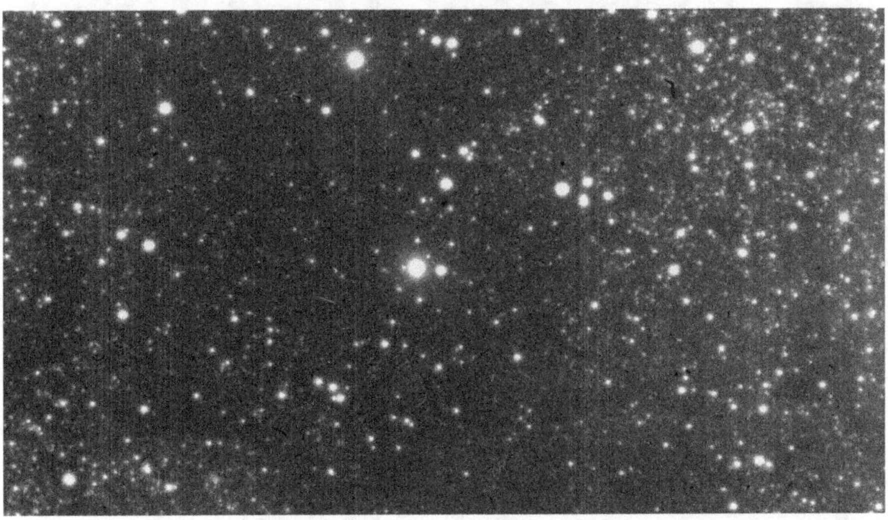

FIGURE 7.11b Cutout near the center of Fig. 11a. displays the good image quality even at 26 h exposure time. COURTESY OF HAMBURGER STERNWARTE.

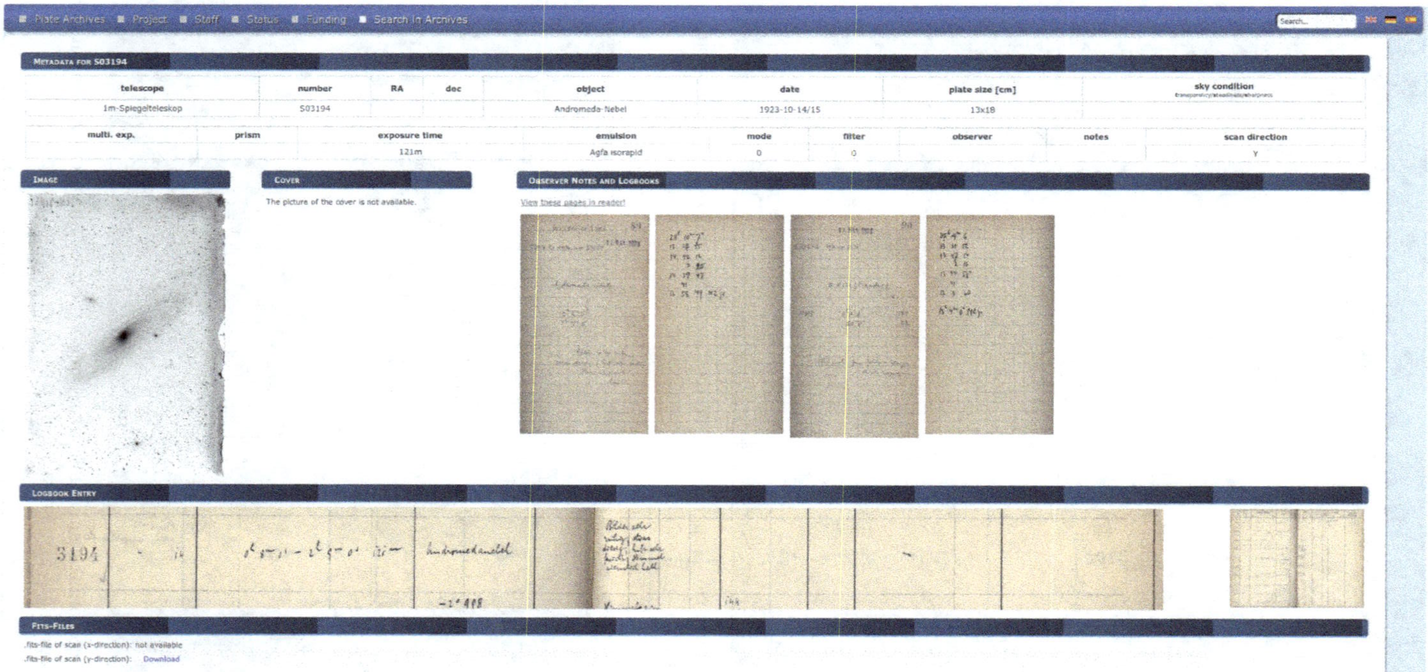

FIGURE 7.12 Web-page for a single plate. At the top the metadata are displayed. In the middle part a low-resolution image can be seen which enlarges at mouse-over automatically, though by clicking it will be opened in an extra window where it can be enlarged even more. The same holds for the plate envelope as well as for the pages of the observer notes. In the bottom row the logbook pages can be found (right-hand small image) and the logbook entry for that plate. Layout of the web page was developed by Christian Offen (at that time Hamburger Universität) and Detlef Groote.

advisable to reset the input mask. In the list of any database query each plate can be inspected by clicking onto the plate number. The opening web page includes all relevant information concerning that plate (Figure 7.12).

7 Project Status and Scientific Possibilities

When the pilot scheme was started in 2010 at the Hamburger Sternwarte with a first scanner bought from charitable contributions, it seemed impossible to finish such a project in a reasonable time. But downscaling was possible, given the number of people helping in the project and the number of additional scanners. All the factors that had made such a project impossible (scanner velocity, available disk-space and hardware costs) were now acceptable. We are, in fact, living in the age of digitization, where almost everything that has been published before on paper can be, and is, digitized.

The rapid growth of affordable storage space is most important for any such project. Luckily, in 2011 the German Research Foundation (DFG) opened funding for Scientific Library Services and Information Systems, to which three research groups applied successfully with a joint project led by the three German observatories: Potsdam, Bamberg and Hamburg. A three year project was started with the goal of scanning 80,000 plates and publishing them in the German Virtual Observatory (GAVO).[26] These institutes launched an archive-project called APPLAUSE (Archives of Photographic PLates for Astronomical USE).[27] In the context of this project a first analysis of the plates was started, with the aim of identifying all the objects on each plate, and determining their accurate celestial coordinates, and brightness.

Such data can be used by researchers all over the world to extract information about special objects by using published tables. Since June 2012 at the Hamburger Sternwarte, two assistants have been scanning plates and books, aided at times by two employees from the observatory and up to four honorary helpers. Actually, more than 41,000 plates have been scanned to date (most of them twice in high resolution), plus the same number of plate envelopes. Additionally, so far 175 observer notebooks and logbooks have been scanned, leading to an approximate total of more than 190,000 image files. The disk space so far needed amounts to roughly 43 TB.

---

26 <http://www.g-vo.org/>.

27 <https://www.plate-archive.org/applause/>.

8 First Research Done with Historical Plates

There are two main reasons for using historical photographic plates in modern research. The first is the actual time when the plate was exposed which results in a unique measurement, and the second is the large time difference if images are compared with today's images as tiny changes in only one year will accumulate and become finally measurable. In the first place there are singular events e.g. novae. These are stars in a near binary system, where one component is a white dwarf (WD) already, while the other star is evolving towards the red giant phase and already fills its so called 'Roche lobe'[28] so that matter from that star spirals into a disk moving down to the surface of the white dwarf. Due to the high gravitation of the WD it may happen that temperatures reach extreme levels, initiating a thermonuclear explosion in which large parts of the accreted matter are blown away, forming an expanding shell. Such a shell will be illuminated by the central star and will be seen as an emission line spectrum.

Some of the oldest Lippert Astrographs plates, depicting Nova Gem 1912, were exposed especially for this object, but without knowing anything about it other than its sudden appearance as a new star, which was appropriately called a "nova" (Figure 7.13).

To compare such an unknown nova spectrum with other stellar spectra, a bright star with two additional short exposures was placed above and below it. First, a cool star was used as a comparison but its spectrum did not resemble that of the nova. So a hot stellar spectrum was tried, with the result that hydrogen absorption lines of the hot star (bright lines in a dark spectrum) coincided perfectly with many dark emission lines in the nova spectrum. It became clear that the radiation was emitted from a hot hydrogen-rich gas. At that time such spectra could only be described in words, as no digitization process was available. Today, we can even calculate theoretical models for such shell spectra, and a comparison of scanned spectra with calculated models provides information on such physical parameters of the shell as temperature, density, and expansion velocity. Thus it is now deemed possible to analyze an event that took place more than a hundred years ago, and to compare it to known objects of similar type observed much later and at better (= higher) resolution. An example is shown in figure 7.14, which represents the scan of the spectrum of Nova Her 1934 from plate LA04542 compared to a calculated model spectrum.

---

28 Gravitationally bound region in a binary stellar system.

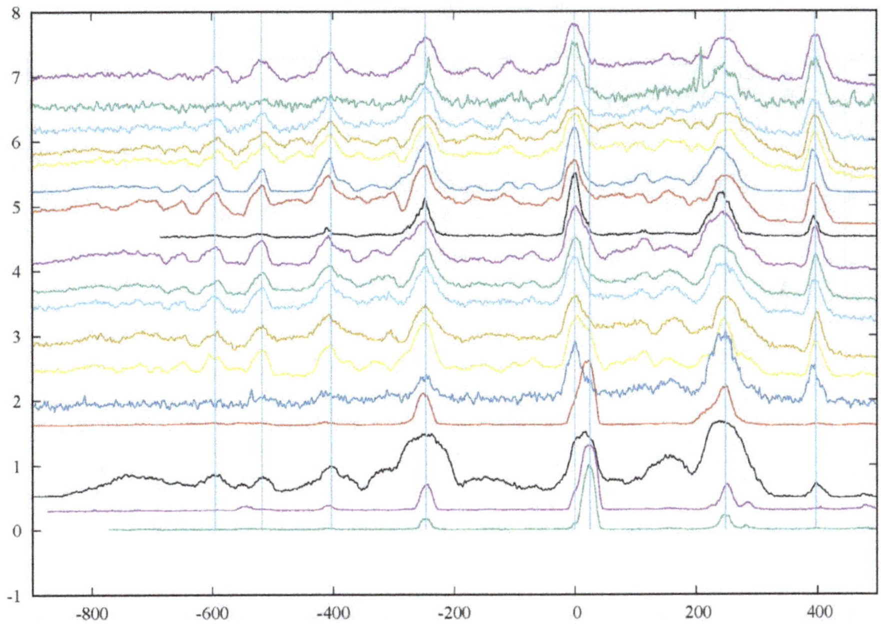

FIGURE 7.13 18 spectra of Nova Gem 1912. The non-linear x-axis is given in pixels with 0 equal to 434 nm. The spectra are background subtracted, scaled (arbitrarily), and shifted. They are ordered by time from top to bottom. The plates were taken during aproximately one year with non- equidistant time steps. Strong changes of the expanding shell can be seen clearly, especially at later phases. FIGURE BY DETLEF GROOTE.

But only a qualitative comparison can be given here, as for a quantitative comparison a previous calibration of the photographic plate will be essential.[29]

A model temperature of 15,000 K was assumed, which is normal for such objects, and the expansion velocity came out to 1500 km/sec also normal for such *slow* novae. The spectrum has additional emission lines, e.g. a line feature at 467.0 nm, which belongs to the element iron.

A second attempt to use the database at Hamburger Sternwarte produced a bachelor's thesis that was later published in a scientific journal (Wertz et al., 2016,)[30]. When determining the historical magnitudes of Quasars, a brightness

29 The theoretical spectrum was kindly provided by Ivan de Gennaro Aquino, 2015 (private communication).

30 Myriam Wertz, Dieter Horns, Detlef Groote, Taavi Tuvikene, Stefan Czesla, and Jürgen Schmitt, "Hamburger Sternwarte plate archives: Historic long-term variability study of

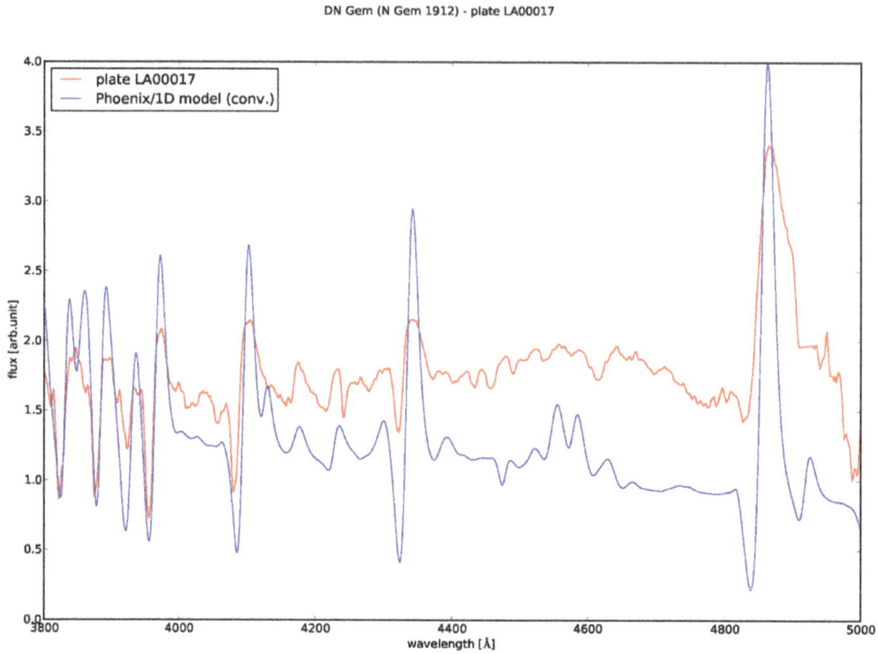

FIGURE 7.14 Observed nova spectrum (red line) compared to a roughly determined theoretical model spectrum (blue line). Here, the wavelengt-scale is given in Ångström (= 0.1 nm). FIGURE BY IVAN DE GENNARO AQUINO (AT THAT TIME HAMBURGER STERNWARTE), DETLEF GROOTE PROVIDED THE OBSERVED SPECTRUM.

drop was found for one of the probed objects, namely the blazar[31] 1ES 1215 + 303. This implies a weakening of 80% of stellar light during a comparatively short period of less than 300 days.

These are only some examples illustrating the usefulness of such a database containing billions of objects which, using *Open Access*, can be elaborated easily worldwide by anyone who is interested in astronomical research. Today, amateur astronomers can use equipment that is as powerful as the instrumentation of professional astronomers a hundred years ago, and with the help of readily available computer software they can even contribute to science by comparing their observations with historical data.

The oldest and largest plates available were taken on a solar eclipse expedition to Algeria (see figure 7.15).

---

active galaxies based on digitized photographic plates," *Astron. Nachr./AN* (2016): 1-6. DOI:10.1002/asna.201613201.

31 Very compact quasar (quasi stellar object) of type BL Lac.

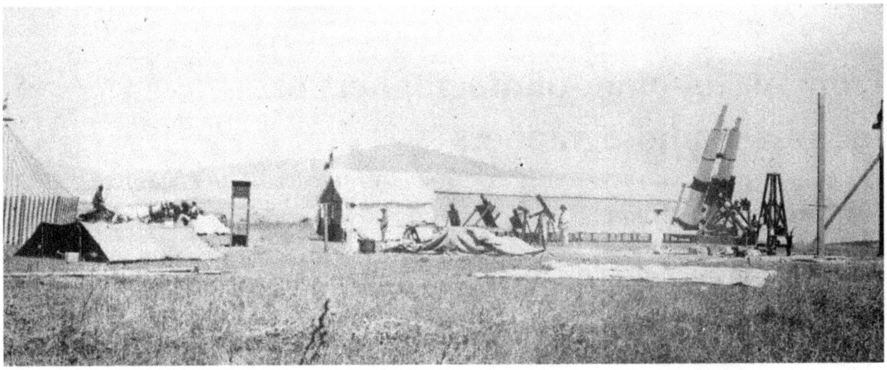

FIGURE 7.15 Picture taken at the observing site in Souk Ahras in Algeria(1905). COURTESY OF HAMBURGER STERNWARTE.

The main purpose of that expedition was to photograph the inner corona of the sun. A large telescope was procured (see the photograph) with a special lens fabricated by Carl Zeiss. The photographs were made by Schorr accompanied by K. Graff, who did the drawings for the plastic overlay and the final lithograph published in the work. For 3¾ minutes total eclipse time immense logistic efforts and costs were necessary. The main result of that successful expedition was 5 plates 70 × 80 in size with exposure times between 4 and 70 sec, exposed with the 20 m horizontal mirror telescope.[32]Other expeditions[33] – like that to Central-Asia (Dshisak/Turkestan) in 1907 – were less successful: a blizzard destroyed the already adjusted equipment. Comparably unsuccessful was the expedition to the Crimea in 1914 which ended dramatically with the imprisonment of the observers on the day of their arrival due to the outbreak of World War I (the last member of the group only returned after a year in captivity).

The scientific results from such expeditions were small compared to the publicity that they achieved at the time. Nevertheless, the photographs of the sun and the photographs documenting the expeditions, as well as e.g. the moon photographs of Bernhard Schmidt plus the documentation are of great historical value and should therefore also be made available by digitization and publication.

32 Ricahard Schorr, "Die Hamburgische Sonnenfinsternis-Expedition nach Souk-Ahras (Algerien)," in *Jahrbuch der Hamburgischen Wissenschaftlichen Anstalten XXII 1904* (Hamburger Sternwarte 1905)

33 Jochen Schramm, "Reisen zur verfinsterten Sonne," *Sterne Weltraum, Special* 4 (*Sonne*) (1999): 28.

CHAPTER 8

From Astronomer-photographers to Astronaut-photographers

Close-up Detailed Observations Performed by Spacecraft and Manned Exploration of the Moon

Pedro Ré and Carmen Pérez González

Photographing from spacecraft changed the concept of a human being behind a camera, or more precisely, a human being behind a telescope coupled with a camera: *how* was the camera actually operated from the spacecraft?* What was the role of the astronauts in the photographic act? Beyond that, a new kind of lunar photograph and photographer was born: photographs taken directly on lunar soil, by pioneer astronaut-photographers who were lucky enough to be selected to step onto the lunar surface. We will focus especially on the central role played by the Hasselblad camera as a space camera in this fundamental chapter of the history of lunar exploration.

In this chapter we explore how the concept of lunar photography changed or was developed when the space era started in 1957 with the first unmanned space mission, *Sputnik II,* which took the first animal into space, a dog named *Laika,* who died in orbit.[1] We will explore the main milestones in the history of lunar photography as accomplished by the different unmanned and manned missions launched by the USA, the USSR, Japan, China and India. In the course of these missions, the number of lunar photographs increased remarkably, as did the tasks achieved by them.

* Thanks are due to the unknown peer-reviewer of the book, who made useful and constructive comments on this chapter. We are grateful to Elsa Modin (Hasselblad Foundation in Göteborg) for facilitating some images and documents for this chapter.

1 For a table with the space missions that sent animals into space, see: "Vorbereitungsflüge für die bemannte Weltraumfahrt", in Michael Maegraith, ed., *Mondlandung. Dokumentation der Weltraumfahrt USA und UdSSR* (Stuttgart: Besler Verlag, 3rd edition, September 1969): 200.

 | DOI 10.1163/9789004298873_010

1 Unmanned Spacecraft Photographic Missions 1957–1967

The last lunar photographic atlas before the space age was published in 1960[2] by the Dutch-American astronomer Gerard Peter Kuiper. Photographs obtained at five different observatories (Mount Wilson, Lick, Pic du Midi, MacDonald and Yerkes) were used in this atlas.[3] The original atlas was joined by four supplements published in the following years.[4]

The Consolidated Lunar Atlas includes some of the best lunar photographs obtained with Earth-based telescopes until 1967. It consisted of loose-leaf high resolution photographic prints reproduced in limited numbers and distributed to the members of the space community to support the Apollo missions to the moon (Figure 8.1). The images were taken in two groups: group 1 (non-full-moon images) were obtained with the 61-inch NASA Telescope at Catalina Observatory; group 2 (full-moon images) were taken with the 61-inch U.S. Astrometric Telescope at U.S. Naval Observatory.[5] *The Consolidated Lunar Atlas* images have medium resolution (about 1 km), consistent image scale, large field of view and multi-phase coverage for every nearside region of the moon. Only a few intact copies of the *Consolidated Lunar Atlas* have survived.

After that publication, and during the final four decades of the 20th century, robotic lunar probes and the Apollo missions taught us more about the moon than had been learnt during the previous four centuries of lunar research. At the dawn of the space-age in the late 1950s the study of our satellite became a priority for both the United States and the Soviet Union.[6]

---

2 Kuiper began collecting images as early as 1955.

3 Gerard P. Kuiper, David W. G Arthur, E. Moore, J.W. Tapscott, and Ewen A. Whitaker, *Photographic Lunar Atlas Based on Photographs Taken at the Mount Wilson, Lick, Pic du Midi, McDonald and Yerkes Observatories* (Chicago: Chicago University Press, 1960).

4 David W.G. Arthur and Ewen A. Whitaker, *Orthographic Atlas of the Moon-Supplement Number One to the Photographic Lunar Atlas* (Tucson: University of Arizona Press, 1960); Ewen A. Whitaker, Gerard P. Kuiper, William K. Hartmann and L.H. Spradley, *Rectified Lunar Atlas – Supplement Number Two to the Photographic Lunar Atlas*. (Tucson: University of Arizona, 1963); Gerard P. Kuiper, R.G. Strom, Ewen A. Whitaker, J.W. Fountain and S.M. Larson, *Consolidated lunar atlas. Supplement no. 3 & 4 to the USAF Photographic lunar atlas*. Air Force Cambridge Research Laboratory (Office of Aerospace Research) Lunar and Planetary Laboratory (Tucson: University of Arizona, 1967).

5 A digital version of the Consolidated Lunar Atlas is available at: <http://www.lpi.usra.edu/resources/cla/>.

6 For a good overview of successful lunar photography missions, see Raymond L. Heacock's article *Lunar Photography: techniques and results* is an excellent source: Raymond L. Heacock, "Lunar Photography: techniques and results," *Space Science Reviews* 8 (1968): 214-257. Especially interesting is TABLE I published in this article, which gives as a graphic

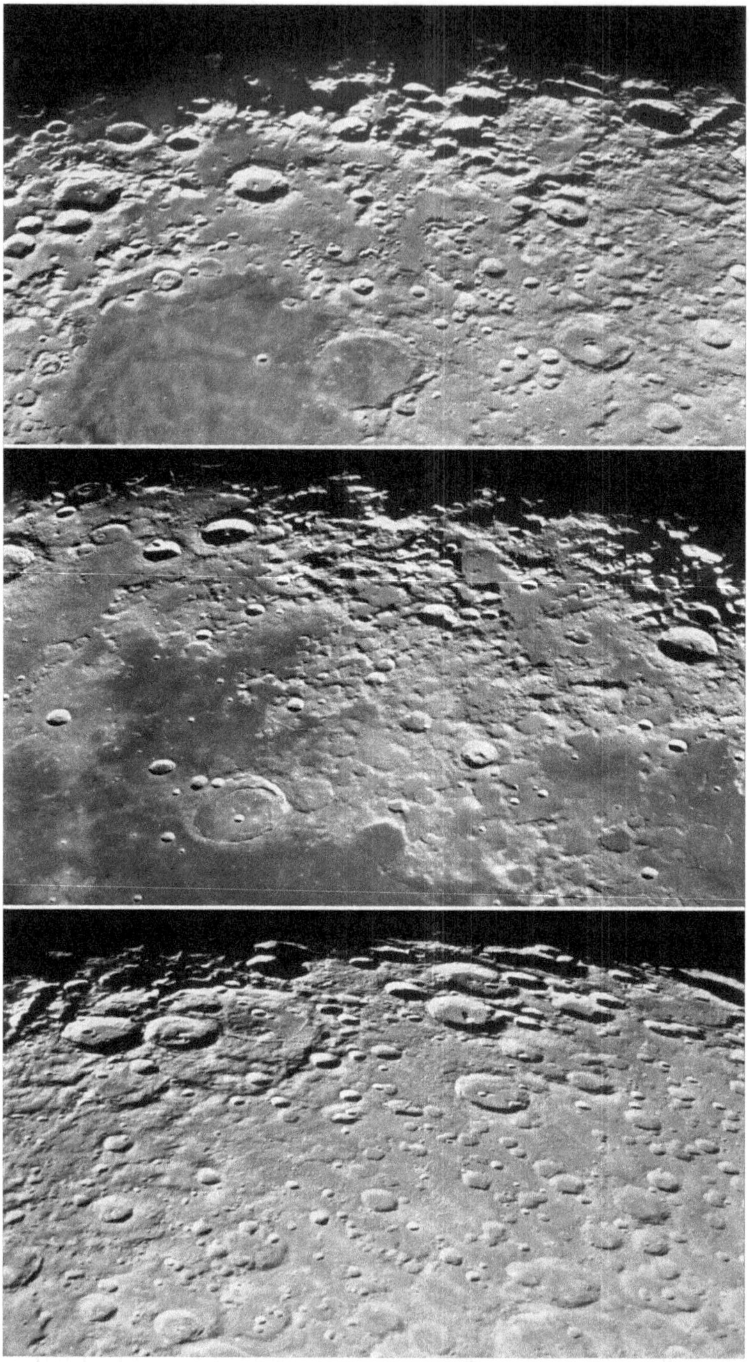

FIGURE 8.1 Plates F6, B5 and G6 from the Consolidated Lunar Atlas obtained on 13 October, 1965. These plates have a 0.2" resolution (400 m). LUNAR AND PLANETARY INSTITUTE

As pointed out by Merton E. Davies and Bruce C. Murray, there are two ways to exploit the extraordinary resolution and information capacity of film for space photography:

> First, one can expose the film in space and then return it undeveloped to Earth, using either manned or unmanned transportation systems. Return to Earth, however, is expensive and complex. In the second method, film can be developed and scanned by an automatic system aboard a spacecraft; as the scanning signal is transmitted to Earth, high quality pictures can then be reconstructed simultaneously on the ground. This latter approach, termed "film readout" has been used by both the U.S. and Soviets.[7]

1.1 *Soviet Lunar Probes*

The Soviet program was conducted through a series of two dozen probes launched between January 1959 and August 1976.

Luna 1 was launched in 1959 (2 January). Nine months later (12 September) Luna 2 became the first spacecraft to impact the lunar surface east of Mare Serenitatis near the Aristillus[8], Archimedes[9], and Autolycus[10] craters. The

---

resumé of all USA and USSR missions, the photo system type used, and the number of photographs obtained: 215. A very good source of information resuming the previous missions was also published in German right after the first successful Moon landing in July 1969: *Mondlandung. Dokumentation der Weltraumfahrt USA und UdSSR* (first edition published 20th July 1969, 70,000 print run). The second edition, published only 10 days later (71,000-106,000 print run), was followed by a third edition, published two months later (107,000-135,000 print run): *Mondlandung. Dokumentation der Weltraumfahrt USA und UdSSR* (Stuttgart: Besler Verlag, 1st edition, 20th July 1969; 2nd edition, 30th July 1969; and 3rd edition, 25th September 1969. Especially interesting are the tables published at the end of the book: *Übersicht aller bemannten Raumflüge der USA und UdSSR; Tabellarische Übersicht: Astronauten und ihre Raumflüge; Tabellarische Übersicht sämtlicher bemannter Raumflüge; Vorbereitungsflüge für die bemannte Weltraumfahrt:* 196-200.

7 Merton E. Davies and Bruce C. Murray, *The view from space. Photographic exploration of the planets* (New York and London: Columbia University Press, 1971), 35. The authors explain that although the format and resolution of the film readout pictures might be comparable to that of returned film, automatic processing in the spacecraft must be substituted for laboratory processing and there is always degradation of the image in the scan mechanization. Certainly, they explain further, one of the greatest advantages of film return is in photogrammetry for the production of topographic maps (46).

8 Aristillus (Lat. 33.6 °N, Long. 4.5 °E)

9 Archimedes (Lat. 31.6 °N, Long 1.5 °W)

10 Autolytus (Lat. 30.9 °N, Long 2.2 °E)

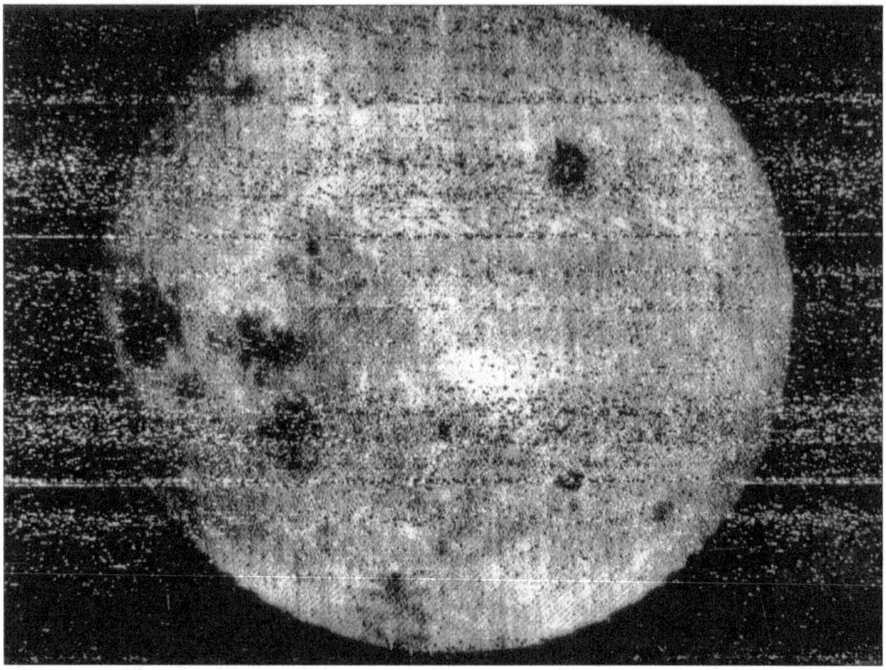

FIGURE 8.2 Far side of the Moon photographed by the soviet Luna 3 probe (7 October 1959). NASA

first photographs of the *far side* of the moon were obtained by the Soviet Luna 3[11] probe in October, 1959. These first images showed a very different topography. Only two dark regions were found which were named Mare Moscoviense (Sea of Moscow) and Tsiolkovsky (large dark-floored plain) (Figure 8.2).

After a series of failures, Luna 9 was the first spacecraft to land on the moon on 3 February 1966. Luna 9 preceded the U.S. Surveyor 1 soft lander by four months. This successful landing proved that the moon surface could support the weight of the probe and was not covered by a thick layer of dust as some models had predicted. The first panoramic photographs of the moon surface were also obtained by Luna 9 (Figure 8.3).

Luna 10 became the moon's first artificial satellite in April 1966. During a two-month mission the spacecraft obtained important scientific data. The moon's magnetic field was found to be weak and no radiation belts were detected. Luna 13 transmitted high-resolution panoramic images of the moon surface after soft landing in Oceanus Procellarum on 24 December 1966. Each panoramic view transmission required more than 100 min. The probe also tested the lunar soil and found it to be similar in density to terrestrial soil. In

11 The third soviet space probe to be sent to the neighborhood of the moon.

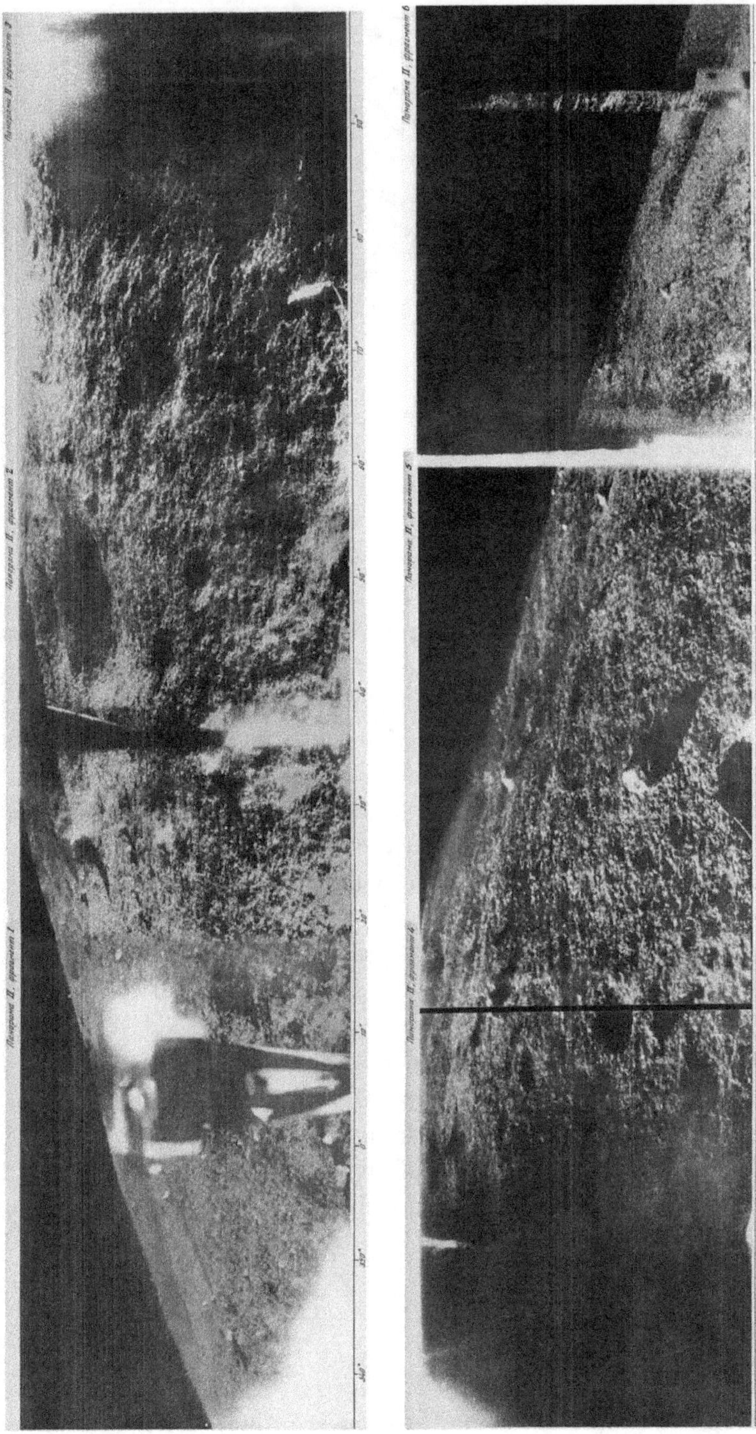

FIGURE 8.3 Luna 9 panoramic view of Oceanus Procellarum (3 February 1966). NASA

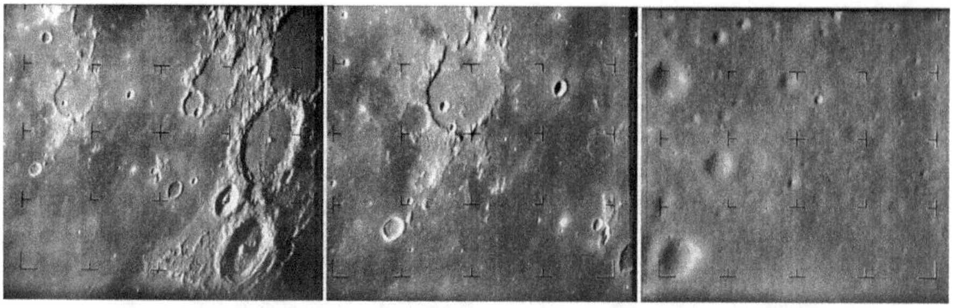

FIGURE 8.4 Ranger 7 images before impacting the lunar surface (31 July 1964). NASA

September 1970, Luna 16 landed in Mare Fecunditatis and became the first unmanned probe to return a sample to Earth. A long extensible arm drilled holes into the surface, sampling about 100 g of lunar soil that were deposited in the spacecraft's upper stage. This upper stage was returned to Earth a day later. This was the first successful robotic sample-return mission.

Luna 17 transported a lunar rover (Lunokhod 1) to the moon in November 1970. This eight-wheeled vehicle was controlled from Earth via a basic TV system. Four high-resolution TV cameras investigated lunar topography. During a 322-day mission Lunokhod 1 travelled 10.5 km, taking 20,000 photographs and 200 panoramic photos. In January 1973, Luna 21 landed on the moon surface and deployed Lunokhod 2. This lunar rover was a more advanced version of its predecessor. In its lifetime of four months, Lunokhod 2 travelled more than 37 km, returning valuable lunar data to Earth. The main objectives of the Luna 21 mission were to image the lunar surface, determine the feasibility of conducting astronomical observations from the moon, perform laser ranging experiments, observe solar X-rays and study the properties of lunar surface materials.

1.2 *United States Lunar Probes*

Robotic lunar exploration was conducted by the United States with three different probe programs: i) Ranger; ii) Surveyor and iii) Lunar Orbiter.

Ranger spacecraft were designed to transmit high-resolution images of the lunar surface during the last minutes of flight up to impact.[12] For 17 minutes before impact (31 July 1964), Ranger 7 transmitted to Earth several pictures of the area 100 km south-west of Fra Mauro. The final image before impact was taken from a height of 1.5 km showing features as small as 1 meter across (Figure 8.4).

12 Cargil R. Hall, and Paul Dickison, *Lunar Impact: A History of Project Ranger.* NASA SP-4210 (Washington, D.C.: NASA, 1977): <http://history.nasa.gov/SP-4210/pages/TOC.htm>.

Ranger 8 obtained high-resolution images of Mare Tranquilitatis, impacting the lunar surface on 20 February 1965. The impact area was located about 60 km north-east of where Apollo 11 would land four and a half year later. The final Ranger mission returned images of the crater Alphonsus[13] on 14 March 1965. Ranger 9's mission was a huge success.[14] Its final image was taken from 530 meters and showed features as small as 250 mm.

From 1966 to 1968 several Surveyor probes landed on the moon. Five of the seven Surveyors conducted successful missions. The main objectives of the Surveyor program were to support the coming crewed Apollo landings.

Surveyor 1, the first US soft landing success, explored the flat plains of Oceanus Procellarum in June 1966. The next successful Surveyor[15] landed in Mare Cognitum in April 1967. This mission found that lunar soil had the proprieties of fine sand, capable of supporting the weight of astronauts and their equipment. On 19 November 1969 the Apollo 12 Lunar Module landed within about 180 m of the Surveyor 3 spacecraft. Surveyor 4 failed minutes before its planned soft landing. Surveyor 5 landed successfully in Mare Tranquillitatis and returned thousands of television images.[16] The final Surveyor[17] landed on a mountainous region close to the crater Tycho[18] on 10 January 1968 (Figure 8.5).

In 1964 the Lunar Orbiter program was initiated. All five Lunar Orbiter missions were successful and 99% of the moon was mapped with unprecedented accuracy.[19]

The main purpose of the program was to choose the landing sites for the later Surveyor and Apollo missions. Specific objectives were:

1. *Photography.* To obtain detailed lunar topographic and geological information of various lunar-terrain types and to increase man's scientific understanding of the moon.
2. *Selenodesy.* To provide precision trajectory information that would improve definition of the lunar gravitational field.

13 Alphonsus: 13.4°S, 2.8°W

14 Ranger's 9 images were broadcast live on national US television.

15 Surveyor 3.

16 The instrumentation for Surveyor 5 was similar to that of the previous Surveyor probes and included survey television cameras and numerous engineering sensors.

17 Surveyor 7.

18 Tycho: 43.31°S, 11.36°W.

19 99% of the moon surface was mapped from photographs taken with a resolution of 60 m, a tenfold increase in resolution as far as earth-based photographs are concerned.

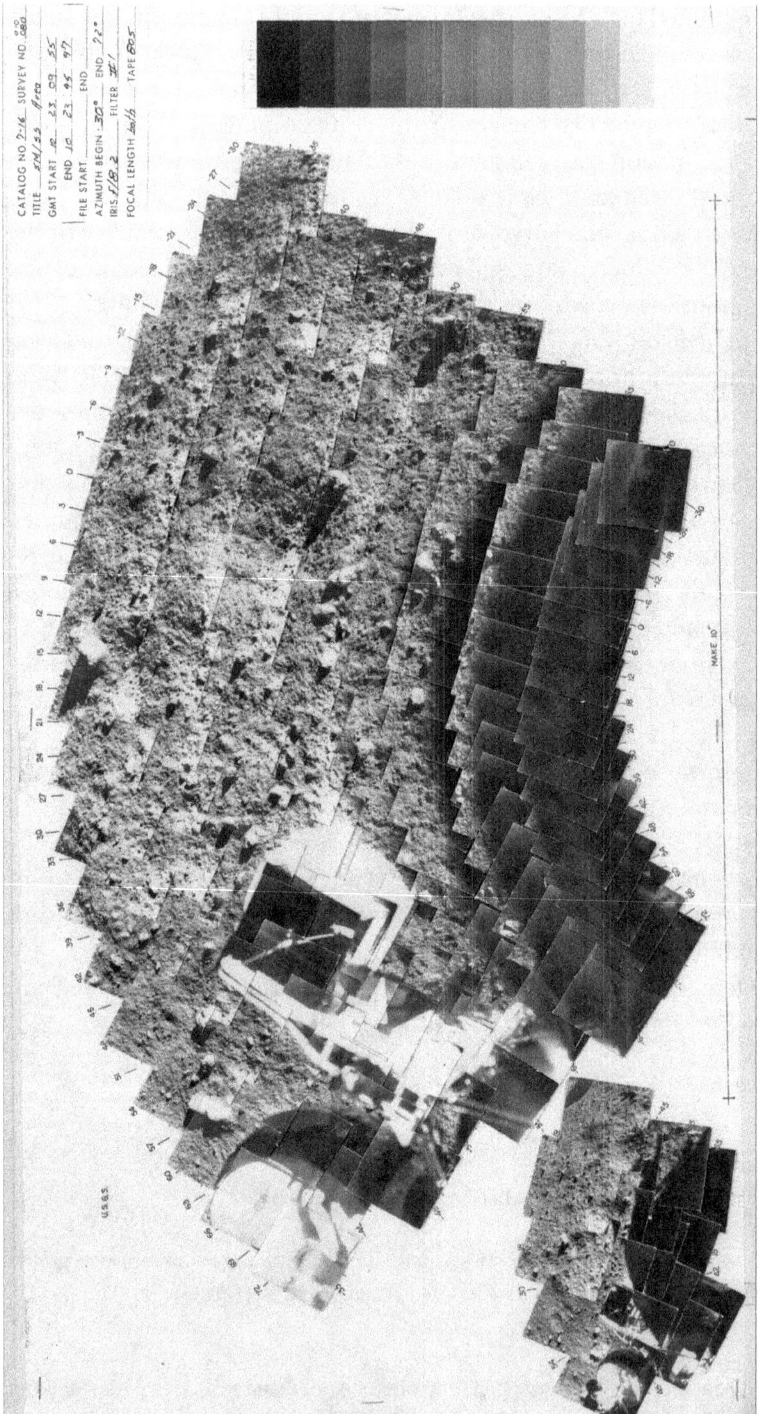

FIGURE 8.5 Mosaic images taken by Surveyor 7 near the crater Tycho. NASA

3. *Analysis of the moon's environment.* To provide measurements of the micrometeoroid and radiation flux in the lunar environment for spacecraft performance analysis.

The first three Lunar Orbiters were mainly dedicated to imaging prospective Apollo landing sites. *Lunar Orbiter 4* was dedicated to providing complete imaging coverage of the lunar nearside and *Lunar Orbiter 5* provided high-resolution imaging of selected sites. A nearly complete coverage of the far side of the moon was obtained over the course of the five missions, with 1654 photographs taken by the five *Lunar Orbiters.*[20] *Each Orbiter had two 70 mm film cameras that were used simultaneously and had different resolutions, field, and views. Two lenses were used: 610 mm (high-resolution) and 80 mm (low-resolution).*[21] *The Lunar Orbiter Photographic Atlas of the Moon (LOPAM) was published in 1971.*[22] *The LOPAM Atlas is* considered the definitive reference manual for global photographic coverage of the moon.[23] The digital version of the LOPAM Atlas was published in 2004.[24] It includes the 675 plates of the original version with many improvements (Figure 8.6).[25]

2 Apollo Missions and Hasselblads 1968-1972

Apollo 8 astronauts were the first humans to witness the Earth rising over the moon's horizon. The "Earthrise" photograph is considered by many to be one

20 Of the 1654 Lunar Orbiter photographs, 840 are areas photographed on the basis of Apollo mission requirements and obtained mainly during missions I, II and III. The remaining 814 photographs were taken primarily during missions IV and V and include 703 of the near side of the moon, 105 of the far side of the moon and 6 of the earth. The first series were taken from low altitude, providing detailed coverage of 22 areas located on the equatorial belt of the near side of the moon; the second series provided a broad coverage of the entire moon and covered 88 areas on the near side.

21 Each camera had a fixed aperture of f/5.6 with controllable shutter speeds of 0.001, 0.02, or 0.04 seconds. *The 70 mm film was developed on board and transmitted to Earth.*

22 David E. Hughes, and J. Kendrick Bowker, *Lunar Orbiter Photographic Atlas of the Moon.* NASA SP-206. (Washington, D.C.: NASA, 1970). <http://www.lpi.usra.edu/resources/lunar_orbiter/book/lopam.pdf>.

23 The images contained in the atlas are excellent for studying lunar morphology, because they were obtained at low to moderate Sun angles.

24 Digital Lunar Orbiter Photographic Atlas of the Moon <http://www.lpi.usra.edu/resources/lunar_orbiter/>.

25 Multiple search capabilities were added to the database to expedite locating images and features of interest. Surface feature information was updated and improved.

the most important photographs of the twentieth century (Figure 8.7).[26] Nevertheless, the main photographic objectives of the Apollo 8 mission were to: (i) obtain vertical and oblique overlapping photographs during at least two revolutions; (ii) photograph specific "targets of opportunity"[27] and (iii) photograph potential landing sites.

Between 1969 and 1972, twelve US Apollo astronauts walked on the moon. Six of the missions (Apollo 11, 12, 14, 15, 16, and 17) achieved this goal. Apollo 7 and 9 were Earth-orbiting missions.[28] Apollo astronauts, particularly command module pilots, contributed significantly to our understanding of the moon's surface features and geology. Additional photographs obtained from lunar orbit during the Apollo missions were very interesting from a geological point of view.

We know that there are 13 Hasselblad cameras on the Moon, taken there by different missions of the Apollo Program: the next image is of great interest, as it shows 8 of the Hasselblad cameras that were left on the Moon in Frau Mauro (Figure 8.8) – 6 of the 8 cameras that appear in the photograph must have been left by Apollo 12 and 14 astronauts, as both missions landed very close to the Fra Mauro area.[29] The third Apollo mission to land close to Fra Mauro was Apollo 16; this may have left the other two cameras there to complete the set of the photograph.[30]

But why and how did the Hasselblad camera become the space camera? Ultimately responsible for this was the astronaut Walter Marty Schirra (1923–2007), who knew Hasselblad well. In 1962 he took a standard Hasselblad 500 C on board the Mercury capsule Sigma 7, which orbited Earth 6 times. The camera had been modified so that it could be used under exceptional conditions:

26 This epoch-making photograph was obtained on 24 December 1968 by William A. Anders, Apollo 8 Lunar Module Pilot. The "Earthrise" photograph is believed to have inspired worldwide concern for Earth and its environment. As Robert Godwin explains in the introduction of the Apollo 11 NASA Mission Reports, "within two years of the return of that startling picture of Earth by Apollo 8, we saw the formation of the United States Environmental Protection Agency, Greenpeace, and the inspiration of the Earth Day": *Apollo 11 – The NASA Mission Reports,* volume 2: 3. For the story of the first photographs of Earth from space and their impact, see: Robert Poole, *Earthrise. How Man First Saw the Earth* (New Haven and London: Yale University Press, 2008).

27 Because of their military training, the phrase "targets of opportunity" was coined by astronauts to designate areas to be photographed.

28 Edgar M. Cortright, *Apollo expeditions to the Moon* (Scientific and Technical Information Office National Aeronautics and Space Administration (Washington, DC.: NASA, SP-350, 1975): <http://www.hq.nasa.gov/office/pao/History/SP-350/toc.html>.

29 Apollo 12 landed in Oceanus Procellarum and Apollo 14 landed in Fra Mauro.

30 Apollo 16 landed near the crater Descartes.

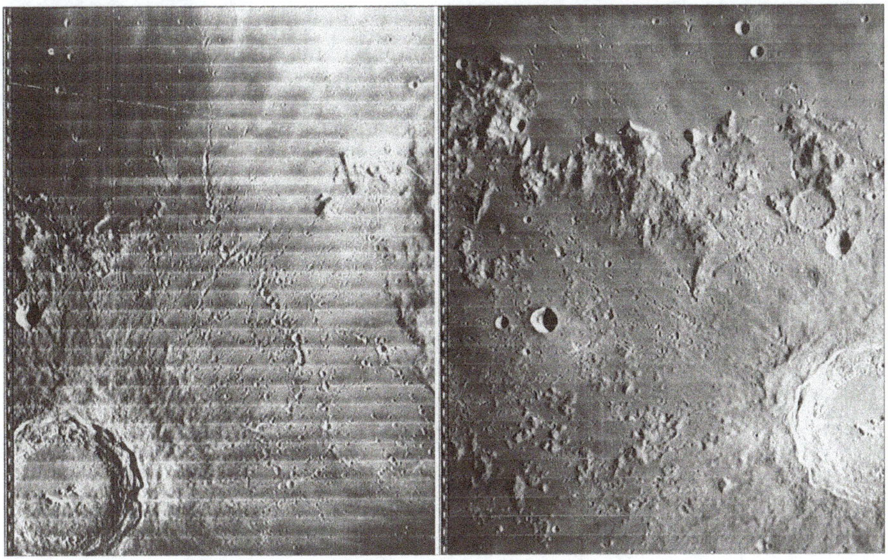

FIGURE 8.6 LOPAM *photos* IV-121-H2 and IV-126-H2, Copernicus crater. NASA

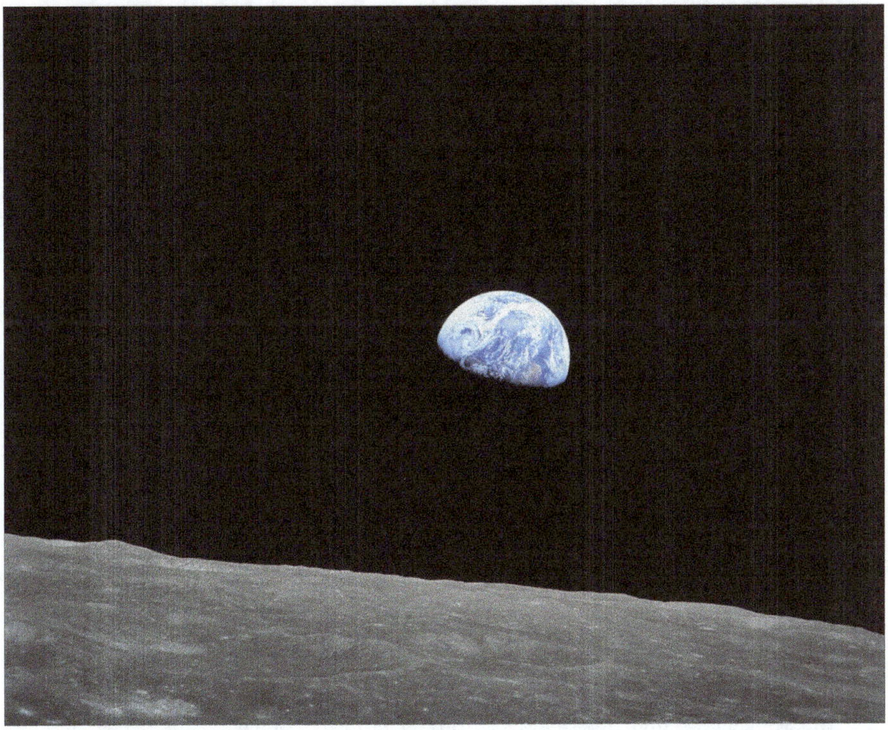

FIGURE 8.7 "Earthrise" Photographed in 24 December 1968 by Apollo 8 astronaut William A. Anders. NASA

the mirror and auxiliary shutter were removed, and optical viewfinder was fitted to the accessory shoe, the leather covering was removed and the entire camera was given an anodized black finish.[31] An 80 mm f. 2.8 Zeiss Planar lens was used. The images that he brought back from the mission were of excellent quality and NASA decided to make Hasselblad their camera of choice for future space missions.

An interesting chapter in the history of space cameras was the 21 minutes spent floating weightlessly in space by the American astronaut Edward Higgins White (1930–1967)[32] in June 1965, as they were of special photo-technical interest: a camera was used for the first time in a complete vacuum and produced excellent pictures admired worldwide.[33] In June 1966, the Hasselblad Super Wide with a Zeiss Biogon 4.5/38 mm made its debut on the Gemini 9 flight. Later, astronaut Michael Collins reported in dismay from a photography mission outside of the Gemini 10 capsule – "Houston, I lost my EVA Hasselblad inadvertently, I am sorry to say." The headlines of Swedish newspapers announced humorously: "Hasselblad Camera, Sweden's first satellite."[34]

In the November 1972 Hasselblad Press Release Report we can read:

> Hasselblad holds a unique record in the relatively brief history of space research. On October 3, 1962, the first Hasselblad camera was taken into space. Since that time it has been a constant companion of astronauts on every subsequent space flight. 1972 will thus mark Hasselblad's 10th anniversary in space. The past decade has witnessed astounding success for the Hasselblad system, the product of years of intensive effort.
>
> The basic principles of the Hasselblad – high precision, a large film and interchangeability of lenses and magazines – were just what the client was looking for. And the client in this case was NASA (National Aeronautics and Space Administration), USA.[35]

NASA's technicians knew how the Hasselblad cameras worked on earth. But the standard Hasselblad camera was not suitable for working on the Moon. NASA and the Hasselblad Company exchanged thousands of letters, cables and telex messages, and specialists traveled back and forth from Houston to Göte-

---

31 Nogéus, Gunilla, "Hasselblad Space Cameras – A Brief History," *Hasselblad Magazine* 1970: 3, 28.

32 Edward Higgins White died during prelaunch testing for the first manned Apollo mission at Cape Canaveral.

33 Nogéus , "Hasselblad Space Cameras," 28.

34 Hasselblad Press Info: *Hasselblad – 30 Years on the moon!* (Göteborg, July 2, 1999): 1.

35 Hasselblad Press Release: *Hasselblad 10 years in space. A unique photographic anniversary 1962-1972* (Göteborg, November 1972): 1.

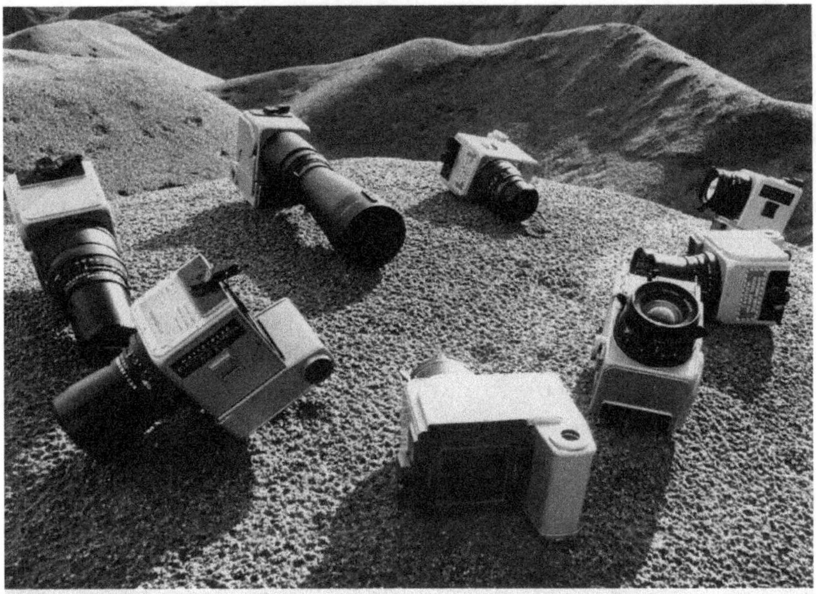

7. By the "shores" of the Ocean of Storms or the Sea of Tranquillity, in the area around the massive Fra Mauro crater or in the environs of Hadley Rille, a total of 8 Hasselblad 500EL Data Camera bodies with Reseau plates and various top-flight Zeiss lenses can be found among a great deal of other scientific equipment. A lunar map will help you to locate the sites. But no film magazines were left behind. They and their precious film contents were all brought back to earth by the astronauts.

NEWS RELEASE
FROM
VICTOR HASSELBLAD AKTIEBOLAG
GÖTEBORG C SWEDEN

FIGURE 8.8 Hasselblad Cameras in the Moon – News release 6, Victor Hasselblad Aktiebolag, HASSELBLAD FOUNDATION ARCHIVE, GÖTEBORG, SWEDEN

borg on a joint effort to produce a Hasselblad camera suitable for work under the extreme physical and climatic conditions of the Moon. Finally the first slightly modified Hasselblad 500C was produced, and made its debut in space on the 3 October 1962 in the hands of the astronaut Schirra, as mentioned above.[36]

The effective NASA-Hasselblad teamwork continued over the decades, and a special department was established at the Hasselblad factory in Göteborg to carry out further development work on the space cameras. The first of these two cameras was for use inside the Lunar Module cabin and was called the Intravehicular Camera (IVC). The second one was for use outside the Lunar Module cabin and was called the Extravehicular Camera (EVC).

The two Hasselblad cameras on board Apollo 11 were built around the Hasselblad 500EL body, featuring a largely automated process due in part to an electric motor. The astronauts needed only to set distance, aperture, and shutter speed.

The first astronaut-photographer to take a photograph on and of the lunar surface was Neil Armstrong (1930–2012). Armstrong, Edwin Buzz Aldrin (1930) and Michael Collins (1930), the three astronauts that flew in the Apollo XI to the Moon, carried 3 Hasselblad 500 EL cameras, one of them with an 80 mm Zeiss Planar, and another one with a 250 mm Zeiss Sonar lens. As soon as Armstrong stepped onto lunar soil, he started taking photographs with the camera he was assigned to use – a 70 mm Hasselblad with two film magazines (Figure 8.9) – after having secured the camera on a bracket mounted on his chest:

> Looking up at the LM, I'm standing directly in the shadow now looking up at Buzz in the window [Aldrin was operating the sixteen millimetre camera]. I can see everything quite clearly. The light is sufficiently bright, backlighted into the front of the LM, that everything is very clearly visible […].
>
> Armstrong: I'll step out and take some of my first pictures here.
>
> MCC: Roger, Neil, we're reading you loud and clear. We see you getting some pictures and the contingency sample..... Neil, this is Houston, did you copy about the contingency sample? Over.

36 Hasselblad Press Release (Göteborg, November 1972): 1.

> Armstrong: Rog, I'm going to get that just as soon as I finish these pictures.[37]

Obviously Mission Control were more concerned about the contingency sample, while Armstrong himself must have been completely fascinated by the amazing images that he was enjoying for the first time in the history of humanity. Mission Control pushes Armstrong again to take the samples:

> MCC: OK, going to get the contingency sample now, Neil?

> Armstrong: Right.

During the 2 hours and 20 minutes that Armstrong and Aldrin walked on the Moon, they took many photographs, some of them becoming icons for the history of photography. One of the very first ones must have been the photograph of a footprint (Figure 8.10), which marked the fulfillment of an age-old dream. Aldrin said on his return to earth:

Those footprints belong to each one of you. A symbol of the true human spirit.[38]

One of the most astounding photographs taken by Armstrong is a portrait of Aldrin (Figure 8.11). All is depicted on the photograph, which provides the viewer a tremendous narrative of what was happening in that very moment in the Moon, that night of the 20th July 1969: Aldrin walks across a small crater near one of Eagle's foil-wrapped landing probes. The visor reflects his own long black shadow, the solar wind collector, the white figure of Armstrong (the photographer), and the lunar module. We can see that Aldrin has his Hasselblad already located properly on his chest.

In the next image, we see the solar wind collector that was reflected on Aldrin's visor (Figure 8.12). The solar wind composition experiment was designed to measure the abundance and isotopic compositions of the noble gases in the solar wind (He3, He4, Ne20, Ne21, Ne22, Ar36, and Ar38). As explained in the *Apollo 11. NASA Mission Reports,*

> The experiment consisted of a specially prepared aluminum foil with an effective area of 0.4 square meter. When exposed to the solar wind at the

37 Peter Ryan, *The Invasion of the Moon 1969. The Story of Apollo 11* (Harmondsworth: Penguin, 1969), 122.

38 National Geographic, *First Explorers on the Moon. The Incredible Story of Apollo 11 in Five Parts* 136/6, December 1969, 772.

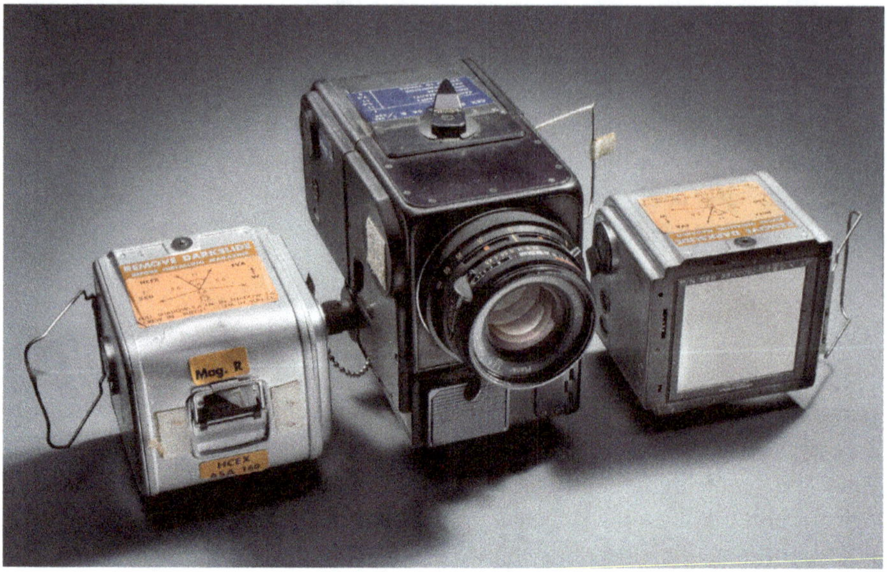

FIGURE 8.9 Eric Long, Neil Armstrong used this 70 mm Hasselblad camera with two films magazines during his Apollo 11 moonwalk, NATIONAL AIR AND SPACE MUSEUM, SMITHSONIAN INSTITUTION, INV. NR. 2009-4813

lunar surface, solar wind particles which arrived with velocities of a few hundred kilometers per second would penetrate the foil to a depth of several millionths of a centimeter and become firmly trapped. Particle measurements would be accomplished by heating the returned foil in an ultra high vacuum system. The evolving atoms would then be analyzed in statistically operated mass spectrometers, and the absolute and isotopic quantities of the particles determined.

The experiment was deployed approximately 6 meters from the lunar module. The staff of the experiment penetrated 13.5 centimeters into the surface.[39]

On the pages devoted to announcements in the National Geographic special issue devoted to the Moon landing, there is one full page of Hasselblad cameras, *What did we learn from building the moon camera?,* which shows Armstrong with a Moon Hasselblad camera and an unknown photographer also with a Hasselblad, and the text compares both kinds of Hasselblad cameras:

39 Robert Godwin, *Apollo 11. The NASA Mission Reports,* Volume Three (Burlington: Apogee Books Space Series, 1999),143.

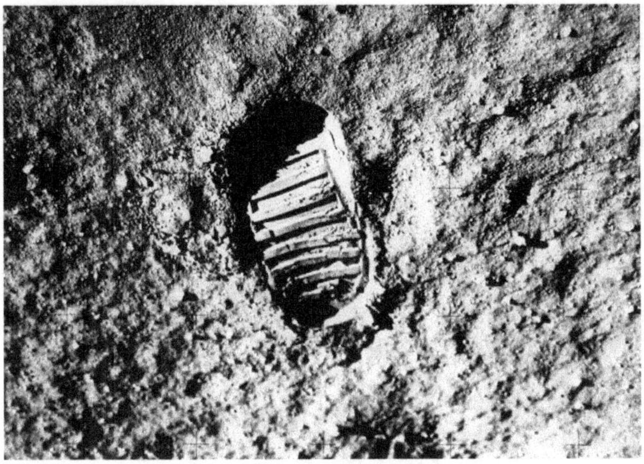

FIGURE 8.10
Footprint in the Moon, HASSELBLAD FOUNDATION, GÖTEBORG

We learned what a good camera we already had on earth. Because the moon Hasselblad is basically our electrically-driven Hasselblad 500 EL.

We also learned that NASA's photographic needs were much the same as the needs of serious photographers anywhere.

And they give a series of these needs: high resolution photographs, great shooting capacity, simplicity of operation, fail-safe reliability. But they also point out the obvious differences between moon and earth Hasselblads:

For one thing, the moon Hasselblad has wings on the diaphragm and shutter speed rings so they can be operated with bulky gloves on.

It has an oversized shutter release button for the same reason.

It has a longer handle on the film magazine slide for the same reason again.

And it has to keep the film from floating off into space during weightlessness.

The earth Hasselblad doesn't have any of these things because it does not need them.

On the other hand the earth Hasselblad has things the moon Hasselblad does not have. Or need. Like interchangeable film transport mechanisms and interchangeable viewers.[40]

40 National Geographic 1969, advertisement at the front.

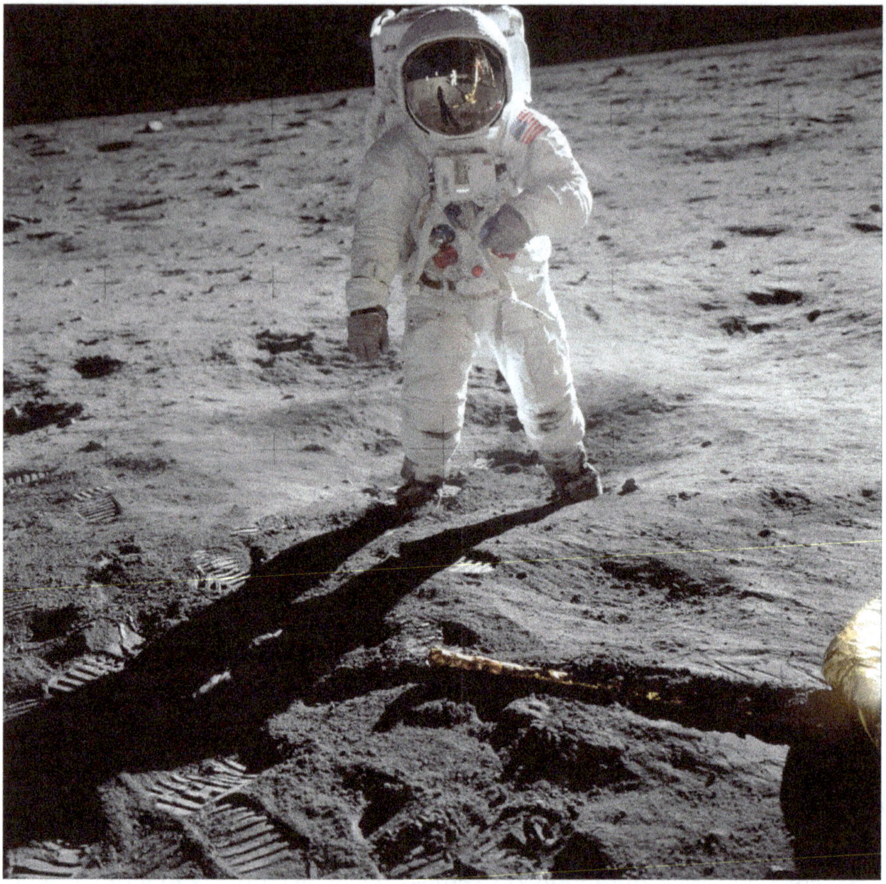

FIGURE 8.11 Iconic photograph of astronaut Buzz Aldrin taken by Apollo 11 mission commander Neil Armstrong (20 July 1969). NASA

These differences in the needs translated into some technical adjustments that the moon cameras must have. The camera used for Apollo 11 and the following six missions to the moon were Hasselblad Electric Data cameras, essentially a modified Hasselblad EL. The camera had to be specially lubricated, as the normal lubricant of the shutter would boil in a vacuum. Camera operation had to be assured for extreme temperatures. All possible dangerous materials used in the camera (glue, plastic, paint, etc) had to be approved by NASA specialists. All electrical equipment had to be carefully sealed to avoid any risk of causing the oxygen saturated atmosphere in the capsule to ignite. And, of course, as

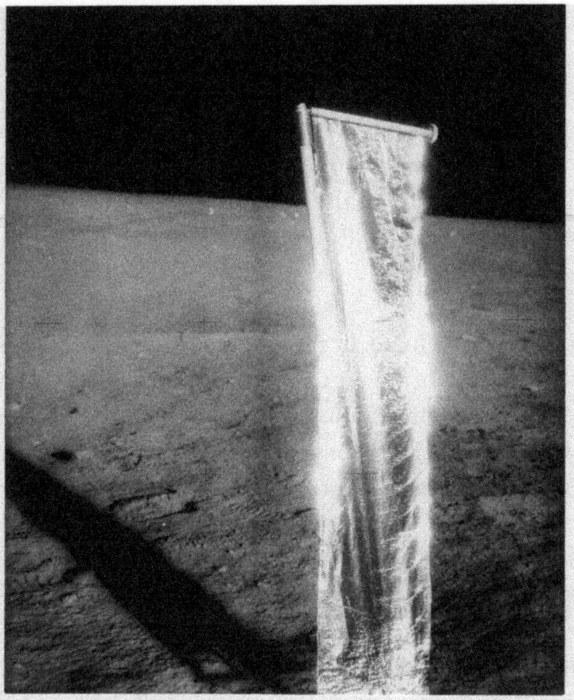

6.

A solar wind composition experiment was carried out near the lunar module. Before the return flight the aluminium curtain was taken down and packed together with the lunar gravel. Particles which had collected on it were analyzed later in the laboratories on earth.

Hasselblad 500EL Data Camera with the Reseau plate, Zeiss Biogon 5. 6/60 mm, Synchro- Compur shutter.

NEWS RELEASE
FROM
VICTOR HASSELBLAD AKTIEBOLAG
GÖTEBORG C SWEDEN

FIGURE 8.12
Solar Wind – News release 6, Victor Hasselblad Aktiebolag, HASSELBLAD FOUNDATION ARCHIVE, GÖTEBORG, SWEDEN

explained in the *National Geographic* advertisement, the camera exterior had to be modified to allow operation with spacesuit gloves.

Apollo 11 was the first manned spacecraft landing on the moon. The *Eagle* Lunar Module landed in Mar Tranquilitatis on 20 July 1969. Apollo 11 astronauts were the first to observe and document scientific phenomena on the lunar surface.[41] Several Hasselblad cameras were used during *extravehicular activities* (EVA).[42]

In addition to documenting their *extravehicular activities*, the astronauts performed two other photographic tasks: (i) close-up stereoscopic photography and (ii) panoramic photography. The panoramic photographs were taken at five stations around the lunar module. In the first volume of the Apollo 11 NASA Mission Report, we can read in detail about the photographic tasks planned for the Apollo 11 astronauts:

> Still and motion pictures will be made of most spacecraft maneuvers as well as of the lunar surface and of crew activities in the Apollo 11 cabin. During lunar surface activities after lunar module touchdown and the two hour 40 minute EVA, emphasis will be on photographic documentation of crew mobility, lunar surface features and lunar material sample collection.
>
> Camera equipment carried on Apollo 11 consists of one 70 mm Hasselblad electric camera stowed aboard the command module, two Hasselblad 70 mm lunar surface superwide angle cameras stowed aboard the LM and a 35 mm stereo close-up camera in the LM MESA.
>
> The 2.3 pound Hasselblad superwide angle camera in the LM is fitted with a 38 mm f/4-5 Zeiss Biogon lens with a focusing range from 12 inches to infinity. Shutter speeds range from time exposure and one second to 1/500 second. The angular field of view with the 38 mm lens is 71 degrees vertical and horizontal on the square-format film frame.
>
> The command module Hasselblad electric camera is normally fitted with an 80 mm f/2.8 Zeiss Planar lens, but bayonet-mount 60 mm and a 250 mm lens may be substituted for special tasks. The 80 mm lens has a focusing range from three feet to infinity and has a field of view of 38 degrees vertical and horizontal.
>
> Stowed with the Hasselblads are such associated items as a spotmeter, ringsight, polarizing filter, and film magazines. Both versions of the Hasselblad accept the same type film magazine.[43]

---

41 Apollo 11 Lunar Photography. Nasa NSSDC Report 70-06 (1970).

42 One 70-mm Hasselblad electric camera, two 70-mm lunar surface superwide-angle cameras, one Hasselblad El data camera. Other cameras were two 16-mm Maurer data acquisition cameras, one 35-mm surface close-up stereoscopic camera and a television camera.

43 Robert Godwin, *Apollo 11. The NASA Mission Reports,* Volume One (Burlington: Apogee Books Space Series, 1999), 48.

As also written in that report, Armstrong and Aldrin will use the Hasselblad lunar surface camera extensively during their surface EVA to document each of their major tasks. Additionally, as reported further:

> they will make a 360-degree overlapping panorama sequence of still photos of the lunar horizon, photograph surface features in the immediate area, make close-ups of geological samples and the area from which they were collected and record on film the appearance and condition of the lunar module after landing.[44]

During the mission, as reported in the third volume of the aforementioned report, all nine of the 70 mm and all 13 of the 16 mm film magazines carried onboard the spacecraft were exposed. Approximately 90 percent of the photographic objectives were accomplished including about 85 percent of the requested lunar photography and about 46 percent of the targets of opportunity.[45] The report pinpoints clearly the lunar surface photographic objectives of the mission:

1. Long distance coverage from the command module.
2. Lunar mapping photography from orbit.
3. Landed lunar module location.
4. Sequence photography during descent, lunar stay, and ascent.
5. Still photographs through the lunar module window.
6. Still photographs on the lunar surface.
7. Close-up stereo photography.[46]

The photographic results of the mission are also reported in detail:

> Lunar photography from the command module consisted mainly of specified targets of opportunity together with a short strip of vertical still photography from about 170 to 120 degrees east longitude. Most of the other 70 mm command module photography of the surface consisted of features selected by the crew.
>
> The 16 mm sequence camera photography was generally excellent. The descent film was used to determine the location of the landed lunar module. One sequence of 16 mm coverage taken from the lunar module

44 Godwin, *Apollo 11*, 48.

45 Godwin, *Apollo 11*, Volume Three, 143.

46 Godwin, *Apollo 11*, Volume Three, 143.

> window shows the lunar surface change from a light to a very dark color wherever the crew walked.
>
> The quantity and quality of still photographs taken through the lunar module window and on the lunar surface were very good. On some sequences, to insure good photography the crew varied the exposures one stop in either direction from the exposure indicated. The still photography on the surface indicated that the landing site location determined by use of the 16 mm descent film is correct.
>
> The close-up stereo photography provides good quality imagery of 17 areas, each 3 by 3 inches. These areas included various rocks, some ground surface cracks, and some rock, which appears to have been partially melted or splattered with molten glass.[47]

As reported in the **Apollo 12** NASA Mission Report, during the mission, all but two of the total of twenty-five 70 mm and 16 mm film magazines carried on board were returned exposed. A partially exposed 70 mm magazine had jammed and was inadvertently left on the lunar surface, and one 16 mm magazine was not used. Approximately 53 percent of the suggested targets of opportunity from lunar orbit were photographed.[48]

The photographic objectives are also defined in the mission's report, which informs that the lunar surface photographs included:

1. Long-distance photography from the command module during *translunar* and *transearth* coast for documentation purposes
2. Surface photography from lunar orbit, including multispectral strip photography and selected targets of opportunity for selenographic purposes and for use in planning and training for future missions
3. Photography of the lunar surface during descent and ascent
4. 4. Sextant photography of the Lansberg area from orbit
5. Photography of the lunar module and experiment equipment f. Photography of the crew performing various lunar surface tasks
6. Photography of the surface environment
7. Panoramic and stereo photographs of samples, sample areas , selenogolic features , and the traverse regions for documented scientific study
8. Photography of selected portions of the Surveyor III spacecraft and surrounding surface.[49]

---

47 Godwin, *Apollo 11*, Volume Three, 144.

48 Godwin, *Apollo 12*, 3-41.

49 Godwin, *Apollo 12*, 3-41 and 42.

The reports explains at length the objectives and results of photographs of candidate exploration sites: to obtain sufficient photographic data on candidate lunar landing sites for future missions, the following coverage of lunar surface areas Lalande, Fra Mauro, and Descartes was planned:

a. a. 70-mm stereoscopic photography of the ground track from terminator to terminator during two passes over the three sites, with concurrent 16-mm sextant sequence photography during the first pass
b. b. Landmark tracking of a s eries of four landmarks bracketing the three sites included in the stereoscopic photography, and performed during two subsequent , successive orbits
c. c. 70-mm high resolution photographs using a 500-mm lens, and additional high resolution oblique photography.[50]

As regards photographing candidate landing sites, the report explains:

> Because of a crew error in site identification, the first high-resolution photographs were taken of the Herschel area instead of Lalande. However, a substitute target to the south of Lalande, assigned in realtime, was subsequently photographed. A first attempt to obtain high-resolution photographs of Fra Mauro and Descartes was unsuccessful because of a camera malfunction. However, on a second attempt, photographs were obtained of Fra Mauro and an area slightly east of the Descartes target area, and high-resolution oblique photography was also accomplished.[51]

Among its photographic objectives **Apollo 14** had dim-light photography: low-brightness astronomical light sources were photographed using the 16 mm data acquisition camera with the 18 mm lens. The sources included zodiacal light, galactic light, the lunar libration region (L_4) and the dark side of the earth.[52]

An additional objective was command and service module orbital science photography:

> This photography consisted of general coverage to provide a basis for site selection for further photography, interpretation of lunar surface features

---

50 Godwin, *Apollo 12*, 12-1.

51 Godwin, *Apollo 12*, 12-2.

52 Godwin, Robert, *Apollo 14. The NASA Mission Reports* (Burlington: Apogee Books Space Series, 2001), 4-5.

> and their evolution, and identification of specific areas and features for study. The Apollo lunar missions have in the past obtained photographs of these areas, as targets-of-opportunity or in support of specific objectives. The Apollo 13 S-IVB impact area was given highest priority in orbital science photography. The target was successfully acquired on revolution 34 using the Hasselblad camera with the 500-mm lens, and the crew optical alignment sight to compensate for the spacecraft's motion. Second priority was given to the lunar module landing target, which was obtained with the lunar topographic camera on revolution 14. However, the camera malfunctioned and subsequent photography with this camera was deleted. A total of eight photographic targets was planned for hand-held photography using color film; three were to be taken with the 500-mm lens (a total of 35 lunar degrees), and five with the 250-mm lens (a total of 130 lunar degrees). The 500-mm targets were successfully acquired. Three of the five 250-mm targets were deleted in real-time for operational reasons (60 lunar degrees), and two were successfully acquired (70 lunar degrees). A total of 65 percent of off-track photography has been successfully acquired. The earthshine target was successfully acquired using both the Hasselblad data camera with the 80-mm lens and the 16-mm data acquisition camera with the 18-mm lens.[53]

As for the task of taking photographs of a candidate exploration site, the report informs us that, high-resolution photographs were required to search for safe landing sites. The report informs us further that,

> They also provide data for crew training and onboard navigational data. The photographs on this mission were to be taken with the lunar topographic camera on revolution 4 (low orbit), and 27 and 28 (high orbits). During revolution 4, malfunction of the lunar topographic camera was noted by the Command Module Pilot. On revolutions 27, 28, and 30, the 70-mm Hasselbald camera with the 500-mm lens (lunar topographic camera backup system) was used to obtain the required photography. About 40 frames were obtained of the Descartes region on each revolution using the crew optical alignment sight to compensate for image motion. The three targets were successfully acquired. To support the photography, a stereo strip was taken with the Hasselblad data camera with

---

53 Godwin, *Apollo 14*, 4-6.

> the 80-mm lens from terminator-to-terminator including the crew optical alignment sight maneuver for camera calibration.[54]

Apollo 15, 16 and 17 used a Lunar Rover Vehicle, also known as moon buggy for greater surface mobility (Figure 8.13). The Lunar Rover was first used on 31 July 1971.[55] Previous Apollo astronaut teams were restricted to short walking distances around the landing site. The use of a Lunar Rover Vehicle expanded considerably the range of the exploration area.[56] Lunar Rovers had a maximum speed of 13 km/s and a range of 9.5 km.

The **Apollo 15** NASA Mission Report informs us extensively of the photographic objectives, activities and achievements during the mission:

> A total of 1152 photographs were taken on the lunar surface with the 60-mm and 500-mm focal-length cameras. At least 360-degree 60-mm panorama was taken at every station except stations 3 and 4.[57]

As the report states, Apollo 15 was the first mission using a 500-mm focal length lens mounted on the 70-mm Hasselblad electric data camera hand-held by a crewman. Good photography was obtained of distant photographic targets such as the Apeninne Front and across and inside of the Hadley Rille.

The **Apollo 16** NASA Report informs us of the photographic equipment and the photographic activity during the mission:

> Still and motion pictures will be made of most spacecraft maneuvers and crew lunar surface activities. During lunar surface operations, emphasis will be on documenting placement of lunar surface experiments, documenting lunar samples, and on recording in their natural state the lunar surface features.
>
> Command module lunar orbit photographic tasks and experiments include high-resolution photography to support future landing missions, photography of surface features of special scientific interest and

---

54 Godwin, *Apollo 14*, 4-7.

55 Godwin, Robert, *Apollo 15. The NASA Mission Reports* (Burlington: Apogee Books Space Series, 2001).

56 The astronauts remained however within walking distance of the lunar module, taking into account that the rover could break at any point. The LRV could be operated by either astronaut.

57 Godwin, *Apollo 15*, 31.

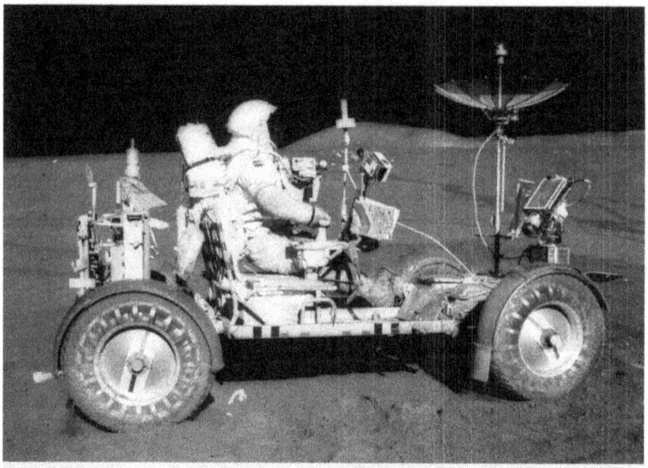

4.

Just as during the successful Apollo XV mission in January-February 1971, a lunar rover will also be used for the transport of various equipment, this time in the Descartes area. The car - and the Hasselblad cameras minus magazines - will be left behind on the moon. This will leave two cars and seven Hasselblad cameras on our satellite. And a Hasselblad SWC dropped in space is probably still in orbit around the earth.

NEWS RELEASE
FROM
VICTOR HASSELBLAD AKTIEBOLAG
GÖTEBORG C SWEDEN

FIGURE 8.13
News release from Victor Hasselblad Aktiebolag, HASSELBLAD FOUNDATION, GÖTEBORG, SWEDEN

astronomical phenomena such as solar corona, Gegenschein, zodiacal light, libration points, and galactic poles.

Camera equipment stowed in the Apollo 16 command module consists of one 70 mm Hasselblad electric camera, a 16 mm Mauer motion picture camera, and a 35 mm Nikon F single-lens reflex camera. The command module Hasselblad electric camera is normally fitted with a 80 mm f/2.8 Zeiss planar lens, but a bayonet-mount 250 mm lens can be fitted for long-distance Earth-Moon photos. A 105 mm f/4.3 Zeiss UV Sonnar is provided for the ultraviolet photography experiment.

The 35 mm Nikon F is fitted with a 55 mm f/1.2 lens for the Gegenschein and dim-light photographic experiment.

The Mauer 16 mm motion picture camera in the command module has lenses of 10, 18 and 75 mm focal length available. Accessories include

> a right-angle mirror, a power cable and a sextant adapter, which allows the camera to film through the navigation sextant optical system.
>
> Cameras stowed in the lunar module are two 70 mm Hasselblad data cameras fitted with 60 mm Zeiss Metric lenses, an electric Hasselblad with 500 mm lens and two 16 mm Mauer motion pictures cameras with 10 mm lenses. One of the Hasselblads and one of the motion picture cameras are stowed in the modular equipment stowage assembly (MESA) in the LM descent stage.
>
> The LM Hasselblads have crew chest mounts that fit dovetail brackets on the crewman's remote control unit, thereby leaving both hands free. One of the LM motion picture cameras will be mounted in the right-hand window to record descent, landing, ascent and rendezvous. The 16 mm camera stowed in the MESA will be carried aboard the lunar roving vehicle to record portions of the three EVAs. Descriptions of the 24-inch panoramic camera and the 3-inch mapping/stellar camera are in the orbital science section of this press kit.[58]

And the final Apollo mission, **Apollo 17**, maintained the same line of using photography for an increasing number of tasks. The photographic achievements of Apollo 17 were to provide precisely oriented mapping camera photographs and high-resolution panoramic camera photographs of the lunar surface, to support a wide variety of scientific and operational experiments, and to document operational tasks on the lunar surface and in flight. As explained by M.C. McEwen and Uel S. Clanton from the NASA Lyndon B. Johnson Space Center:

> Lunar surface photographs are primarily of three types: (1) operational photographs, to document the condition performance, orientation, or setting of equipment and the effectiveness of procedures; (2) scientific documentation photographs, to record samples in their undisturbed condition as well as their location, orientation, and detailed setting or to record features or materials that were not collected; and (3) panoramic views to provide for the accurate location of traverse stations and to provide the capability to reconstruct the geologic setting of the landing site.
>
> Orbital photographic tasks, other than mapping camera operation, were both for operational and scientific purposes. Orbital science pho-

58 Robert Godwin, *Apollo 16. The NASA Mission Reports*, Volume 1 (Burlington: Apogee Boos Space Series, 2002), 69.

> tography planned before the mission included (1) areas of geologic interest; (2) near terminator areas, where details of relief are enhanced; (3) areas not covered by photography from other missions; (4) areas in earthshine; (5) low-light-level astronomical phenomena, such as the solar corona and zodiacal light. In addition, film was allotted for crew-option photographs to be taken on the basis of real-time observations.[59]

Three more Apollo missions were planned to take place after Apollo 17. However, in 1970 NASA Administrator Thomas Paine announced budget cuts and Apollo 18, 19 and 20 were cancelled. After six Moon landings and $24 billion spent, the Apollo program came to an end.

As explained in the November 1974 Hasselblad Press Release, one mishap will go down in the history of photography: during one space walk, astronaut Michael Collins happened to drop his Hasselblad SWC (Super Wide C) and attached magazine in space. He "dived" for it several times, but it wandered out of reach into space, providing Hasselblad with its own earth satellite.[60]

In Table 8.1 we have summarized the history of the Moon Hasselblad Camera: all the missions that used Hasselblads, who the photographers were, and the results of their photographic activity. Over 15,000 photographs had been taken by Hasselblad cameras in space by 1970.[61]

The next table (Table 8.2), published in the Hasselblad Foundation's report "Hasselblad 10 years in space", summarizes the activity of the Hasselblad space cameras during the first decade of their use:

59 M.C. McEwen, and Uel S. Clanton, "Photographic Summary," in *Apollo 17. Preliminary Science Report* (Washington D.C., Lyndon B. Johnson Space Center, 1973), 4-1 to 4-31. In the same volume there are a few chapters on the photographic activity and results of the mission: James W. Head, and Douglas Lloyd, "Appendix: Near-Terminator and Earthshine Photography," 4-33 to 4-39; Sherman S.C. Wu; Francis J. Schafer; Gary M. Nakata; and Raymond Jordan, "Repeatability of elevation measurements – Apollo photography," 33-35 to 33-40; and chapter 34 devoted fully to photography: R.D. Mercer; L. Dunkelman, R.E. Evans, "Part A: Zodiacal Light photography"; and R.M. MacQueen, C.L. Ross, and R.E. Evans, "Part B – Solar Corona Photography".

60 Hasselblad Press Release, November 1974: 3. The camera plunged to its destruction about three years later, experts say.

61 Nogéus 1970: 28. Table 1 has been elaborated mostly based on information written in Nogéus' article.

TABLE 8.1 History of the Moon Hasselblad Camera

Mission	Date	Astronauts	Hasselblad Cameras	Nr. of photographs	Time on the Moon
Apollo VII	Oct. '68	Schirra – Eisele – Cunningham	standard 500 C Hasselblad		
Apollo VIII	Dec. '68	Borman – Lovell – Anders	2 × 500 EL/70's Hasselblad	1,100 photographs – Iconic photograph of the Earth with the Moon surface in the foreground	
Apollo IX	Mar '69	James McDivitt – David Scott – Russell Schweick-art	2 × 500C's Hasselblad – SWC wide angle camera – 4 × 500EL/70's	Very satisfactory results that provided photo-interpreters with a great many interesting facts	
Apollo X	May '69	Stafford – Young – Cernan	1 × 500EL/70		
Apollo XI	Jul '69	Neil Armstrong: commander – Michael Collins: command module pilot – Edwin Aldrin: lunar module pilot	3 × 500EL's Hassel-blad: one with 80 mm Zeiss Planar – one with 250 mm Zeiss sonar	First images taken in lunar soil	2 hours and 20 minutes
Apollo XII	Nov '69	Charles "Pete" Conrad: commander – Richard Gordon: command module pilot – Alan Bean: lunar module pilot	2 × 500EL Data: Camera with Reseau Plate, af.5.6./60 mm Zeiss Biogon lens and a 70 mm magazine; 4 × 500EL/70's incronised on a cluster	Terrain photography (with the two 500EL Data Cameras; Multispectral photography of the Moon surface from his "parking orbit"	2 EVAS: 5 hours and 30 minutes, and 1 mile and a half
Apollo XIII	Apr '70	James Lovell – Fred Haise – John Swigert		unsuccessful mission	
Apollo XIV	Jan '71	Alan Shepard: commander – Stuart Roosa: command module pilot – Edgar Mitchell: lunar module pilot	70 mm Hasselblad EL Camera; 3 × 70 mm Hasselblad Data Camera		2 EVAS: 4 hours and 49 minutes
Apollo XV	Jul '71	David Scott: commander – Alfred Worden: command module pilot – James Irwin: lunar module pilot	3 × 70 mm Hasselblad Cameras (2 × 60 mm focal lenses; 1 × 500 mm focal lens); 1 × 70 mm Hasselblad EL Camera	2654 photographs, out of them more than 2/3 were taken from the surface	3 EVAS: 18 hours and 37 minutes; they traveled 17.5 km with the first car on the Moon

TABLE 8.1 History of the Moon Hasselblad Camera (cont.)

Mission	Date	Astronauts	Hasselblad Cameras	Nr. of photographs	Time on the Moon
Apollo XVI	Apr '72	John Young: commander – Thomas "Ken" Mattingly: command module pilot – Charles Duke, Jr.: lunar module pilot	3 × 70 mm Hasselblad Cameras (2 × 60 mm focal lenses; 1 × 500 mm focal lens); 1 × 70 mm Hasselblad EL Camera		2 EVAS: 20 hours and 14 minutes
Apollo XVII	Dec '72	Eugene Cernan: commander – Ronald Evans: command module pilot – Harrison "Jack" Schmitt: lunar module pilot	3 × 70 mm Hasselblad Cameras (2 × 60 mm focal lenses; 1 × 500 mm focal lens); 1 × 70 mm Hasselblad EL Camera		3 EVAS: 22 hours and 4 minutes

TABLE 8.2 Hasselblad – a decade in space

Space statistics	
First Hasselblad in space	October, 3 1962
Number of space flights with Hasselblad cameras	22
Number of astronauts in these flights	28
Number of Hasselblads used	49
Number of different Hasselblad models	4
Number of film magazines used	244
Number of space and lunar photographs taken with Hasselblad (approx.)	30.000
Number of orbits of the earth	952
Number of orbits of the moon	288
Hasselblad hours logged in space	8,583 hours, 39 minutes
Number of moon landings	5
Hasselblad hours logged on the moon	224 hours, 44 minutes
Number of miles in space with Hasselblad equipment approx.	34.7 million
Number of camera bodies with attached lenses left behind on the moon	8
Number of Hasselblad cameras lost in space	1
Number of Apollo flights to the moon	1
Next NASA flight to the moon scheduled for	1986 (?)

According to the previous table, the number of photographs taken with Hasselblad cameras was over 30,000. Several of them became iconic photographs and shaped the visual memory of a whole generation. But what do viewers of the moon landing of Apollo 11 remember of that fascinating moment that happened during the night of the 20-21 July 1969? Over 53 million households tuned in to watch the Apollo 11 Mission on TV, and over 530 million viewers worldwide watched the moon landing.[62]

In a paper entitled "Organizing images of futures-past: Remembering the Apollo Moon Landings", the authors discuss the way that NASA constituted both its own past and future significance through the remediation and premediation of key images of the Apollo space program.[63] The authors made an experiment within the framework of an exhibition at the National Space Centre in the UK, where visitors could give feedback on the topic exhibition by writing "memory cards" of the moon landings. Over 420 cards were compiled over six months, from February to August 2010. The authors of the aforementioned article analyzed the cards in terms of the social and cultural practices of remembering. Interesting for us would be to know what the situation was regarding visual memories of the moon landing beyond Europe and America, and particularly, in Asia. To the best of our knowledge, no study on this has been made to date.

The memory cards that the exhibition's visitors could use started with the sentence: *My strongest memory of the moon landing is....* One of the cards exemplifies nicely what a tremendous impact those first images of Neil Armstrong descending to reach the lunar soil must have had on the millions of TV viewers: this is CARD no. 54:

> *My strongest memory of the moon landing is ...*
>
> ... Getting up in the middle of the night to watch Armstrong and Aldrin walk on the moon; and my father taking photos of the TV screen as Armstrong climbed down the ladder. (Card 54)[64]

---

62 *Apollo Mission Overview,* NASA: <https://www.nasa.gov/mission_pages/apollo/missions/apollo11.html> (retrieved: 21.06.2017.)

63 L.Goodings, S.D. Brown, and M. Parker, "Organizing images of futures-past: Remembering the Apollo Moon Landings," in *International Journal of Management Concepts and Philosophy* 7/3-4 (2013): 268-288.

64 Goodings "Organizing images," 277.

One might wonder how other people around the world, who did not have access to TV, experienced and visualized the event. For instance, many people followed the landing on the radio, which means they had to use their imagination to reconstruct the event, as illustrates the memories of a then 7 year old boy, Reza, imagining the moon landing under the starry Iranian sky, in Darreh Zereshk near Yazd, while listening to the radio:

> We were all gathered in the open verandah elevated from the garden overlooking the pool of water that would be irrigated by the village *qanat*, where dad would be the only person to dare swimming in the green chilly waters every morning. That night the grown ups sat around playing backgammon, drinking tea, reading, doing the *mosha'ere* (a game where you test each other on the amount of poetry you know by heart). Older cousins were happy to be allowed to stay up late; they were bopping in and out. My sister had passed out. I was struggling to stay awake. We did not have TV. The radio was on. The only source of light was an oil lamp. The wind catchers caught the wind and blew the breeze over us. It must have been close to midnight when Armstrong stepped on the moon. Dad was the only "science dude" in the crowd; very excited indeed.[65]

3 Clementine and Lunar Prospector

Lunar exploration was resumed after nearly two decades with the *Clementine* probe. The objective of the mission was to test sensors and spacecraft components under extended exposure to the space environment and to make scientific observations of the moon and the near-Earth asteroid 1620 Geographos. The asteroid observations were not made due to a malfunction in the spacecraft. *Clementine* – also known as the *Deep Space Program Science Experiment* – was a joint US Department of Defence/NASA mission. The Clementine probe studied the moon's topography, structure, composition and mineralogy. A comprehensive survey of the moon was carried out by this spacecraft from February to May 1994. More than 2 million images of the moon were obtained from an eccentric moon orbit.[66]

65 E-mail communication: 23.06.2017

66 For detailed information on this mission, see NASA's website: <https://nssdc.gsfc.nasa.gov/nmc/masterCatalog.do?sc = 1994-004A>. See also: S. Nozette, *et al.*" The Clementine mission to the Moon: Scientific overview," *Science* 266/5192 (1994): 1835-1839; and T.C. Sorensen, and Paul D. Spudis, "The Clementine mission – A 10-year perspective," *J. Earth Syst. Sci.* 114/6 (2005): 645-668.

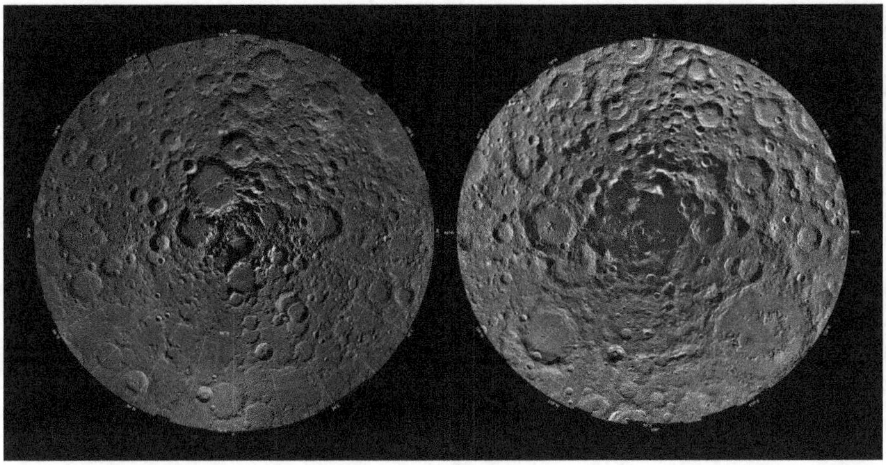

FIGURE 8.14 Clementine mosaic images of north (left) and south (right) Polar Regions of the Moon. NASA

The *Clementine* probe carried four different imaging systems, obtaining images at visual, infrared, and ultraviolet wavelengths.[67] During each five-hour lunar orbit up to 5000 images were transmitted. *Clementine* provided our first complete coverage of the lunar surface, including the poles. It also found evidence of the presence of water ice in the moon's surface soils (Figure 8.14).[68]

Lunar Prospector, launched on January 6, 1998, was the last US moon probe of the 20th century.[69] *The Lunar Prospector probe was* designed for a low polar orbit investigation of the moon, including mapping of surface composition and possible deposits of polar ice, measurements of magnetic and gravity fields, and study of lunar outgassing events. It complemented the Clementine data and improved the understanding of the origin, evolution and current state of the moon.

The spacecraft carried six different experiments:[70] gamma ray spectrometer, neutron spectrometer, magnetometer, electron reflectometer, alpha particle spectrometer, and a Doppler gravity experiment. The mission was a huge

67 The spacecraft also had two star tracker cameras, used mainly for altitude determination but also as wide-field cameras for various scientific and operational purposes.

68 The polar regions of the moon are of special interest because of the postulated occurrence of ice in permanently shadowed areas.

69 Lunar Prospector Mission Handbook <http://geo.pds.nasa.gov/missions/lunarp/schandbk.pdf>.

70 No cameras were carried by the Lunar Prospector spacecraft.

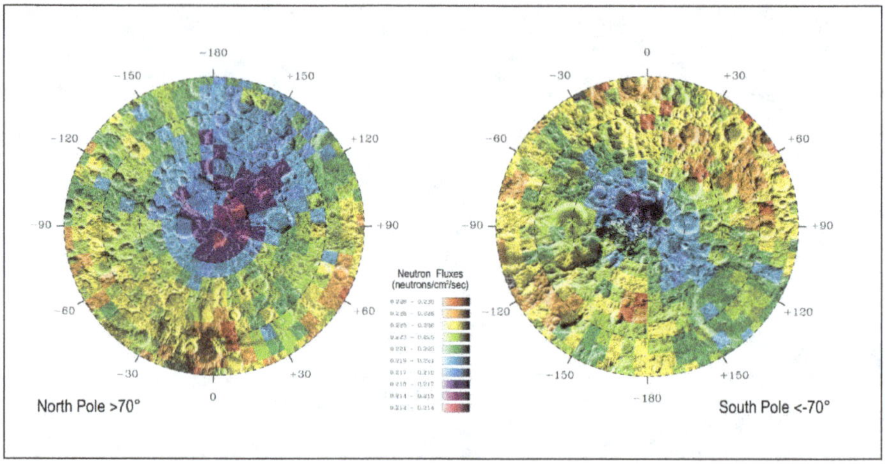

FIGURE 8.15 Lunar prospector evidence for water ice at the north (left) and south (right) lunar poles. NASA

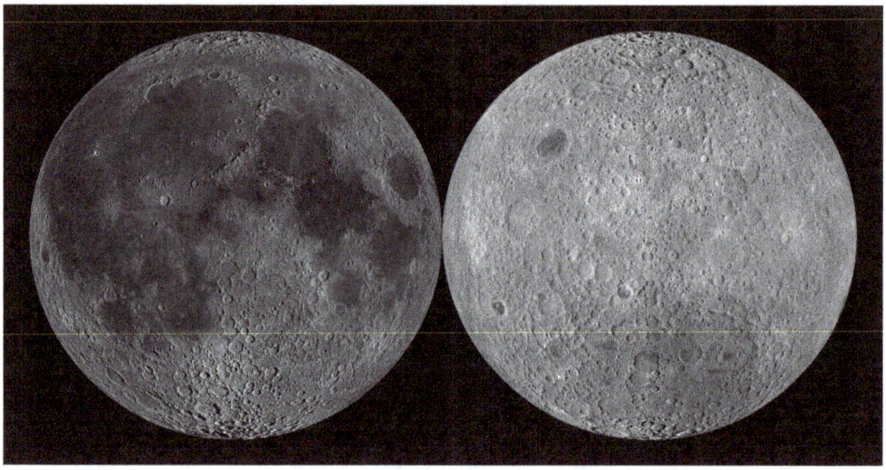

FIGURE 8.16 Composite photo of the near side (left) and far side (right) of the Moon captured by the Lunar Reconnaissance Orbiter spacecraft. NASA

success.[71] Data collected showed that water ice is present in large quantities at both of the lunar poles (Figure 8.15).[72]

71 The Lunar Prospector probe was designed to go into a low polar orbit around the moon and search for water ice and other minerals in the dark areas of craters. During its 19 month mission, Lunar Prospector completed a map of the moon's surface composition.

72 The data indicated that a large quantity of water ice (300 million metric tons) was mixed

The Lunar Reconnaissance Orbiter (LRO), launched in 18 June 2009, was the first in NASA's Robotic Lunar Exploration Program designed to map the surface of the moon and characterize future landing sites.[73] In December 2010 the most accurate topographic map of the moon surface based on LRO images was released. Seven LRO instruments were able to collect the most comprehensive lunar set of data to date.[74] A global map of the moon was also produced with a resolution of 100 m per pixel of the Lunar Reconnaissance Orbiter Camera (Figure 8.16).

4 Asian Missions to the Moon: Japan, China and India

4.1 *Japanese Missions*

The SELenological and ENgineering Explorer (SELENE), also known in Japan by its nickname *Kaguya*, was the second Japanese lunar orbiter spacecraft following the Hiten probe (launched on January 24, 1990).[75] The Selene spacecraft (launched on September 14, 2007) was a joint project of the Japan Aerospace Exploration Agency (JAXA)[76], the Institute of Space and Astronautical Science (ISAS) and the National Space Development Agency (NASDA). Selene (the largest lunar mission since the Apollo program) consisted of a main orbiter at 100 km altitude and two small satellites in polar orbit. The scientific instruments on board the main orbiter were used for global mapping of the lunar surface magnetic and gravity field. The main orbiter impacted on the lunar surface near the crater Gill[77] on June 10, 2009, after successfully orbiting the moon for a year and eight months. In September 2016, JAXA announced the full release of the SELENE (KAYUGA) High Definition Television (HDTV) data on its website, where all images can be viewed online (Figure 8.17).[78] Many of these images

into the regolith at each of the poles. This was the first direct evidence of the presence of water ice at the moon's poles.

73 The Lunar Crater Observation and Sensing Satellites (LCROSS) were launched along with the LRO spacecraft. LCROSS was mainly designed to search for water ice on the moon's surface.

74 More than 192 terabytes of data were collected, the equivalent of all other planetary missions combined.

75 <http://www.isas.jaxa.jp/e/enterp/missions/hiten.shtml>.

76 <http://www.kaguya.jaxa.jp/index_e.htm>.

77 Gill: 63.9°S, 75.9°E.

78 <http://global.jaxa.jp/projects/sat/selene/topics.html#topics8544> (last accessed Juny 9, 2017).

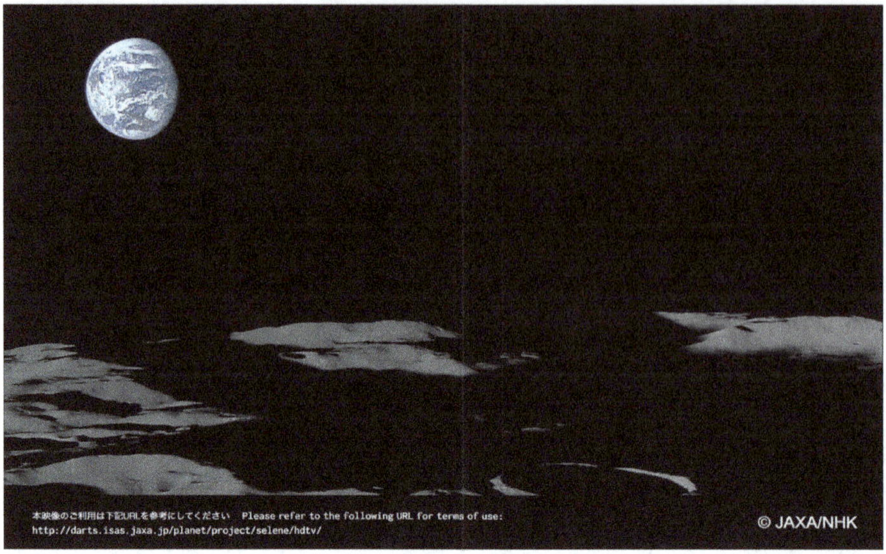

FIGURE 8.17 Earth Set above the South Pole, SELENE (KAYUGA), 6th November 2007, JAXA

were used to complete the Google Earth-Moon.[79] It is also possible to view online the impressive 6 last images taken from SELENE before it impacted on lunar soil.[80]

4.2 *Chinese Missions*

The Chinese Lunar Exploration Program (CLEP), also known as *Chang'e Program* and managed by the China National Space Administration (CNSA), launched its first spacecraft *Chang'e 1* on October 24, 2007. *Chang'e 1* mapped the moon from orbit. After completing its mission, *Chang'e 1* crashed into the lunar surface in 2009 as planned. A second orbiter, *Chang'e 2,* was launch on the 1 October 2010. In 2013, *Chang'e 3* with the *Yutu* rover became the first Chinese spacecraft to land on the moon.[81] There are two planned missions, *Chang'e 5* (Figure 8.18) and a second one which intends to land on the moon, which have caused quite a fierce controversy, as leading lunar scientist Paul D. Spu-

79 <https://www.google.com/earth/explore/showcase/moon.html> (last accessed Juny 9, 2017).

80 <http://www.kaguya.jaxa.jp/gallery/en/observation_mission/hdtv/hdtv_077.html> (last accessed Juny 9, 2017).

81 For the latest news about China's space program, see: <http://www.space.com/topics/latest-news-china-space-program/>.

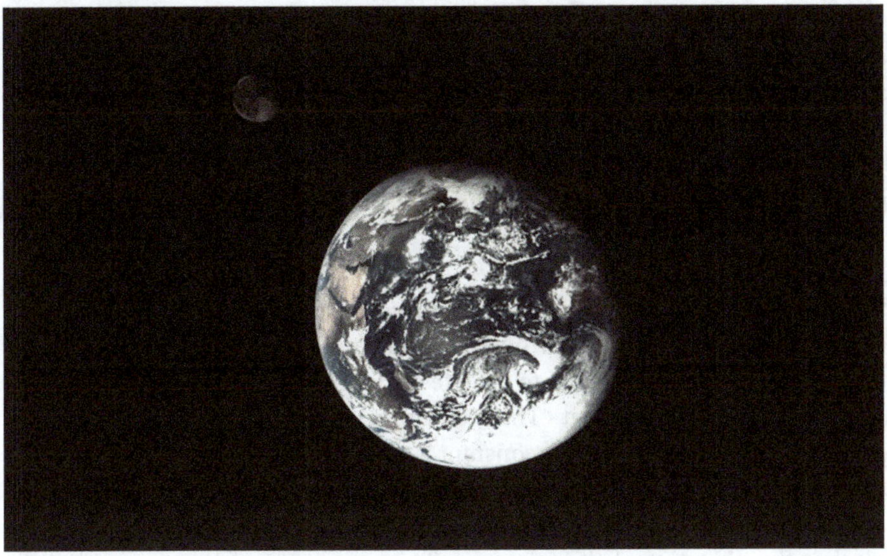

FIGURE 8.18 Earth and Moon, Chang'e 5, November 2015, CNSA

dis[82] warns in a recent published paper of dangerous intentions beyond science in the Chinese space program, whose ambitions were made public in a paper released by the CNSA in December 2016.[83]

4.3 *Indian Missions*

On the 22 October 2008 India launched *Chandrayaan-1*.[84] It was launched by the Indian Space Research Organisation (ISRO). On 14 November 2008, the Moon Impact Probe separated from the Chandrayaan orbiter and struck the south pole in a controlled manner. India became the second country to place its flag on the moon, but not a physical flag as the Americans did, but four small ones painted on the exterior of the spacecraft. The probe impacted near the Shackleton crater, ejecting sub-surface soil that could be analyzed for the presence of lunar water ice. *Chandrayaan-2*, the second lunar exploration mission after *Chandrayaan-1*, is planned to be launched by the beginning of 2018.

82 Paul D. Spudis, "China's Moon Missions Are Anything But Pointless. The value of the nation's lunar missions goes beyond science," in *Air & Space, Smithsonian*, 3rd January 2017: <http://www.airspacemag.com/daily-planet/chinas-moon-missions-are-anything-pointless-180961633/>.

83 You can read this paper in the following link: <http://www.spaceref.com/news/viewsr.html?pid = 49722>.

84 <http://isro.gov.in/>.

The United States, European countries (ESA), Russia, Japan, India and China are planning future manned lunar exploration missions or lunar outpost construction on the Moon between 2018 and 2025.

5 Epilogue – Amateur Astronomers: High Resolution Astrophotography

But did space era photography fully replace lunar photography from Earth? Did the new generation of astronaut-photographers end up with the old astronomer-photographer glued to his telescope and camera?

André Couder wrote that "The worst part of the telescope is the atmosphere"[85]. The atmosphere is a constant source of perturbation exhibiting inhomogeneity, due to winds, eddies and differences in temperature. The air refractive index varies with temperature causing an irregular wave front that reaches the objective of the telescope. The magnitude of these irregularities determines the quality of astronomical seeing. Seeing is usually measured taking into account the diameter of a star's Airy disk.[86] With small and medium-size telescopes (10 to 50 cm aperture) the seeing disk of star at the focal plane of a telescope varies from 0.3" to 3". Seeing varies with the wavelength of light and distance to the zenith;[87] the best seeing conditions are usually found at mountain top observatories.

Jean Texerau[88] describes in detail different types of turbulence that can affect astronomical seeing:

1. *High-level atmospheric turbulence*: high frequency temperature fluctuations of the atmosphere and the mixing of air "parcels" of different temperatures/densities;
2. *Local turbulence*: caused by the immediate environment related to the diurnal heating of the telescope followed by night time radiation;
3. *Instrumental turbulence*: produced inside the telescope tube.

---

85 André Danjon, and André Couder, *Lunettes et Telescopes* (Paris: Editions Revue d'Optique, 1935).

86 The Airy disk and Airy pattern are descriptions of the best focused spot of light that a perfect telescope lens can produce, limited by the diffraction of light.

87 Short wavelengths are more affected by seeing. Atmospheric thickness also plays an important role as far as dispersion is concerned. Dispersion is the 'smearing out' of light of different colors due to differential refraction as it passes through our atmosphere.

88 Jean Texerau, *How to make a Telescope* (Virginia: Willmann-Bell, 1984).

Many astronomical seeing scales have been described over the years. Eugène M. Antoniadi and William H. Pickering's scales are familiar to many visual observers, however with the advent of CCDs and webcams other seeing scales have been described (*e.g.* Damian Peach's *Modern scale of astronomical seeing for imagers*[89]):

1. *Excellent seeing*: A solid stable star disk with good contrast. Minor lunar and planetary details are held for long periods. No significant undulation or fuzziness.
2. *Good seeing*: A mostly solid stable star disk with good contrast. Minor details are frequently seen though not held for long periods.
3. *Fair seeing*: Slight or moderate undulation or fuzziness. Reasonable contrast. Minor lunar and planetary details occasionally seen.
4. *Poor to very poor seeing*: Severe undulations or fuzziness. Poor contrast. Large scale detail poorly defined. Minor details invisible.
5. *Extremely poor seeing*: Severe undulations or fuzziness. Very poor contrast. Little detail visible.

The first high-resolution astrophotographs of the moon were obtained from 1890 to 1900 with two different telescopes: i) 91 cm Lick Observatory and ii) 60 cm Paris observatory refractors.[90] These first photographs had a low resolution.[91] High resolution images obtained with the Mount Wilson Hooker 2.5 m telescope in the 1920s were not much better[92]. Planetary images obtained with the 60 cm Clark refractor of the Lowell observatory had a resolution of 0.6", decidedly a big step forward (1905 to 1930). Similar resolutions (0.4" to 0.6") in planetary images were recorded by several French astronomers working at Pic du Midi observatory (1940s). The best lunar and planetary images from this period had an average resolution of 0.5".

With the introduction in the 1990s of Kodak Technical Pan (TP-2415), amateur astronomers were able to produce lunar and planetary images with details as small as 0.3", a resolution obtained only by the best professional observatories. The extremely fine grain and high resolving power, coupled with high-contrast development of this film, permitted reproduction of very fine detail on objects in the solar system.[93]

89 <http://www.damianpeach.com/seeingscale.htm>.

90 Gérard de Vaucoulers, *La photographie astronomique* (Paris: Albin Michel, 1958).

91 The finest details recorded could be easily observed with a 10 cm telescope.

92 The smallest lunar features were easily visible in an amateur-sized 20 cm telescope.

93 TP-2415 was a "miracle film" for many astrophotographers because it combined speed, high contrast, resolving power and fine grain.

TP-2415 is the derivative of Technical Pan SO-115, developed by Kodak for solar patrol photography. Hypered[94] Tech-Pan was universally accepted as the film of choice for astrophotography (both in high resolution and deep-sky photography).

In 1969, Willard Sterling Boyle and George Elwood Smith of Bell Laboratories invented the *charge-coupled device* (CCD), a solid-state semiconductor. Boyle and Smith were investigating new ways of imaging with solid-state, silicon methods in an effort to develop the Picture phone. This method turned out to be an incredible innovation that changed imaging away from vidicon-type TV tubes and from emulsion based photography.[95]

The first Boyle & Smith paper[96] began with the following words:

> A new conductor device concept has been devised which shows promise of having wide applications (...)

A CCD array consists of a thin wafer of pure silicon subdivided into a grid of picture elements or pixels.[97] The pixels act as light sensors like the tiles in a mosaic. As photons strike a given pixel they generate an electrical charge with a magnitude proportional to the intensity of the incident light. Each picture element acts as a tiny capacitor accumulating a charge for a given time interval or integration time.

CCDs have many advantages over other electronic and photographic imaging devices. CCDs are remarkably efficient light detectors[98], linear in their response to light, and exhibit an excellent sensitivity over a wide range of wavelengths.

Amateur astronomers basically use three different types of digital cameras:

1. Digital SLR cameras (DSLR)
2. Dedicated, cooled astronomical CCD cameras (CCD)
3. Webcams and astronomical digital video cameras

---

94 Film hypersensitazion (or hypering) is the process of making film more sensitive by baking it in an atmosphere of 8% hydrogen and 92% nitrogen forming gas. Hypersensitazion can also be achieved by freezing the film to near-cryogenic temperatures.

95 The 2009 Nobel Prize in Physics was awarded to W.S. Boyle and G.E. Smith *"for the invention of an imaging semiconductor circuit – the CCD"*.

96 W.S. Boyle, G.E. Smith, "Charge Couple semiconductor devices", *The Bell System Technical Journal*, 49 (1970): 40: 587-593.

97 "Pixel" is the contraction of PICture ELement.

98 The sensitivity or quantum efficiency of a CCD can be as high as 90%. By comparison, the fastest photographic emulsions are 10 times less efficient. The detector quantum efficiency of the human eye is only 5%.

DSLR and cooled astronomical CCD cameras are mainly used for shooting deep-sky objects. Webcams and specialized high-end digital video cameras are better suited to imaging the moon, sun and planets.

The recent advent of CCD cameras that can be operated in a video mode, taking 10 or more images per second for periods of up to a few minutes, can be used with excellent results for high-resolution imaging.[99]

Webcams[100] and astronomical digital video cameras are equipped with a color or a black & white CCD or CMOS.[101] They can be operated with different interfaces: USB 1.0, UBS 2.0, USB 3.0, FireWire and GigE. These cameras can capture several hundred to thousands of individual images (frames) in rapid succession and store them in popular video formats.[102] This video file includes frames seriously degraded by seeing and others that are less affected. Image processing software can then be used to stack and align the best frames in order to produce a low noise high-resolution image.

Individual frames of the video are analyzed and the sharpest frames selected using dedicated software. These good frames (less affected by seeing) are stacked together to produce a high resolution image in the form of a single still picture. The composite images have less noise, higher contrast and better resolution than a single exposure.

The pioneers in composite image stacking were two Lowell Observatory astronomers: Earl C. Slipher and Harold L. Johnson. These astronomers developed and improved astronomical photographic techniques that became standard and are now used in the digital age. This technique was known as integration printing. The method consisted of capturing many high-resolution photographs of a planet and then aligning and exposing these plates on a single print. The end result was a photographic print with less film grain than a single negative could produce (better signal-to-noise ratio). This method of integration printing has evolved into what we now know as staked-video imaging.

Amateur astronomers today regularly capture images of the sun, the moon and planets that rival those taken by professional astronomers. These images

---

99 Some high-end video cameras have high-speed data transfer of up to 120/s (USB 3.0, GigE and FireWire interfaces).

100 The first webcams were mainly used as video conferencing devices.

101 Complementary metal–oxide–semiconductor (CMOS). In the recent past CMOS technology proved inferior to CCD for capturing high resolution images of the sun, moon and planets. Modern CMOS detectors have a better overall quality and are considerably cheaper to produce.

102 8-bit avi files, 12-bit ser files.

often constitute valuable scientific contributions.[103] Specialized software[104] aligns, sorts and stacks hundreds to thousands of images, automatically producing a low noise composite image.[105] These images can be processed using aggressive image processing tools to bring out hidden detail.[106] The best amateur high-resolution images of the moon have resolutions of 300-500 m. These images are far better than some professional Earth-based telescopic atlases.

In 2009, several well-known astrophotographers[107] created an image that has eclipsed any other so far taken of the moon by ground-based astronomers.[108] This world record image was part of an ambitious project to create the largest ground-based mosaic image of the moon.[109] The moon was imaged using high-end astronomical video cameras coupled to different large amateur telescopes. Each telescope targeted a small portion of the lunar surface. These high-resolution images were assembled in a giant mosaic of the gibbous moon (9-day old).

Almost 1000 different areas were imaged, corresponding to 1.2 million captured video frames and 1.1 terabytes of data. The 288 individual image panes were selected based on their overall quality and assembled in a high resolution 87.4 megapixel mosaic.[110]

6 Conclusion

During the last 40 years of the 20th century, robotic lunar probes and the Apollo missions taught us much more about the moon (and especially about its hidden half) than had been learnt during the previous four centuries of lunar research. The act of photography was redefined as the spatial scenario changed,

103 Sun activity (e.g. flares), moon transient phenomenon (video impact monitoring), Atmospheric activity of Jupiter, Saturn and Mars.

104 Registax – <http://www.astronomie.be/registax/>, Autostakkert – <http://www.autostakkert.com/>, Avistack – <http://www.avistack.de/>.

105 Images are aligned using hundredths to thousands of reference points. The best resolution images are then staked producing a high signal to noise ratio final composite image.

106 Usually wavelet-based image restoration algorithms.

107 Ten different astrophotographers used a range of 9.25-inch to 14-inch Schmidt–Cassegrain telescopes working between 5.5 and 7.8 meter focal length and high frame rate digital astronomical video cameras to create the world record lunar mosaic image.

108 <http://www.lunarworldrecord.com/image.php>.

109 The mosaic Moon image was submitted to the 2010 Guinness World of Records, <http://www.lunarworldrecord.com/>.

110 The resolution of the mosaic image is better than 1 km on the moon surface.

from pioneer astronomer-photographers working from leading observatories around the world (or more precisely, in America, Europe and Australia) to astronaut-photographers working from space or from the lunar soil. Pioneer astronaut-photographers on the moon operated very successfully with the space Hasselblad camera, performing the first documentary photographic reports, not only of lunar soil, but also of the first scientific experiments that took part outside Earth, in another celestial body. Even if professional astrophotographers (astronaut-photographers) took the lead in lunar photography, amateur astronomers performing their work from the Earth kept producing images that equaled in quality those taken from a closer distance and better perspective from the moon.

CHRONOLOGY

History of Moon-photography: Timeline

Pedro Ré and Carmen Pérez González

Note: The italics indicate Asian contributions

1608	The Dutch astronomer Hans Lippershey constructs a telescope.
1609	Thomas Harriot makes the first known telescopic sketch of the moon on 5 August 1609.
1610	Galileo Galilei constructs a telescope and uses it to make his first telescopic drawings of the moon.
1611	The oldest preserved source that most likely refers to the phenomenon of libration is a note made by Thomas Harriot on Dec. 1611 in the course of one of the three lunar observations he made that month. He noticed that two features of the Moon's surface (today called Mare Frigoris and Sinus Moris) appeared closer to the edge of the lunar disk than was the case on the map he had drawn sometime earlier.
1613	*First telescope arrives in Japan*
1613	*First telescope arrives in Siam*
1614	Moon drawing by Christopher Scheiner
1615	*Sir Thomas Roe, British Ambassador, brought a telescope to the Mughal court of Jahangir*
1619	Moon drawing by Charles Malapert
1619	*A "Keplerian" astronomic telescope arrives in China with Jesuit mission*
1620	Moon drawing by Giuseppe Biancani
1627	Moon drawing by Christopher Borri
1629-30	Moon drawing by Francesco Fontana
1635-37	Remarkable drawings of three lunar phases by Claude Mellan
1645	The first real map of the moon by Michiel van Langren
1647	Between fall 1643 and spring 1646, Johannes Hevelius systematically observed the lunar surface with a telescope, publishing his results in a book entitled *Selenographia: sive, Lunae descriptio* (1647), which consisted of 40 prints showing the moon in different chronological phases, starting from when the crescent becomes visible after the new moon, and three large maps. These are the first complete renditions of the regions close to the edge of the lunar disk that periodically appear and disappear due to the joint effect of the principal types of optical libration.

 | DOI 10.1163/9789004298873_011

1651 P. Giovanni Riccioli publishes the *Almagestum Novum,* which includes a section on the moon with two new maps

1660s First telescope arrives in Iran: the French Capuchin Raphael du Mans, is apparently responsible for the introduction of the telescope at his mission in Isfahan.

1671 The French Capuchin Chérubin d'Orleans publishes a large book on optics, *La dioptrique oculaire,* and a lunar map with topography.

1730s Maharajah Jai Singh publishes his astronomical tables, Zij-i Muhammad Shahi, with the only known Indian telescopic drawing of the moon.

1779 First Japanese drawing of the moon by Asada.

1793 Lunar sketches of the moon by Zenbei Iwashi.

1796 Copperplate engraving "Getsurin Shinkei" no zu (*True Moon*) *by Shiba Kokan.*

1813 First lunar map by Masatami Hotta.

1816 Joseph Nicéphore Niépce combines the camera obscura with photosensitive paper.

1825 First surviving photograph taken by Nicéphore Niépce.

1826 Joseph Niépce produces the first permanent image (heliograph) using a camera obscura and white bitumen.

1829 Niépce and Louis Daguerre sign a ten year agreement to work in partnership in the development of their new recording medium.

1834 Henry Fox Talbot creates permanent (negative) images using paper soaked in silver chloride and fixed with a salt solution. Talbot created positive images by contact printing onto another sheet of paper. Talbot's *The Pencil of Nature,* published in six steps between 1844 and 1846 was the first book to be illustrated entirely with photographs.

1834 First telescope produced in Japan (*Gregorian reflector telescope*) *by Tobei Kunitomo.*

1836 Kunimoto produces the first sketches of the lunar disk with his telescope.

1837 Louis Daguerre creates images on silver-plated copper, coated with silver iodide and "developed" with warmed mercury (first Daguerreotypes).

1838 Charles Wheatstone wrote the first paper on the stereoscopic effect, and suggested that the libration of the moon should be used to obtain the necessary baseline to produce stereoscopic photographs of that celestial object.

1839 Louis Daguerre patents the Daguerreotype. The Daguerreotype process is released for general use in return for annual state pensions given to Daguerre and Isidore Niépce (Louis Daguerre's son): 6000 and 4000 francs respectively.

1839 John Frederick William Herschel uses the term *Photography* ("writing with light") for the first time.

1839	First unsuccessful Daguerreotype of the moon obtained by Daguerre (blurred image long exposure).
1839	François Jean Dominique Arago announces the Daguerreotype process at the French Academy of Sciences (January 7 and August 19). Arago predicts the future use of photographic technique in the fields of selenography, photometry and spectroscopy.
1838-39	Frau Hofrätin Wilhelmine von Witte of Hanover constructs a lunar globe using wax and mastic (26 inches high, 12 inches in diameter).
1840	John William Draper obtains the first successful (correctly exposed) Daguerreotype of the moon using a 13 cm reflector with a long focal length (20 min exposure).
1841	Henry Talbot patents his process under the name "Calotype".
1842	Austrian astronomer Gian Alessandro Majocchi obtains the first photograph of the partial phase of a solar eclipse on a Daguerreotype on July 8 (2 min exposure).
1843	William Parsons 3rd Earl of Rosse completes the 72-inch reflector "*Leviathan of Parsonstwon*" at Birr Castle (Ireland).
1847	First light of the Harvard College Observatory 15-inch (38 cm) refractor made by Merz & Mahler (Munich, Germany).
1849/1852	William Cranch Bond and John Adams Whipple obtain a series of lunar Daguerreotypes with the 38 cm Harvard refractor (40 s exposures).
1851	Frederick Scott Archer improves photographic resolution by spreading a mixture of collodion (nitrated cotton dissolved in ether and alcohol) and chemicals on sheets of glass. Wet plate collodion photography was much cheaper than Daguerreotype, and the negative/positive process permitted unlimited reproduction. The process was published but not patented.
1851	*Introduction of modern astronomy in Iran through Dar al-Funun (polytechnic institute) in Tehran.*
1852	First wet plate Collodion images of the moon obtained by Warren de la Rue using a 13-inch (33 cm) reflector with 3.05 m focal length. Mount without clock drive.
c. 1853	Joseph Sidebotham takes collodion photographs of James Nasmyth's Moon models, and they are published in 1855 in *The Strines Journal: A Monthly Magazine of Literature, Science and Art.*
1853	J. Phillips photographs the moon with a 159 mm refractor (3.35 m focal length) (60 s exposures).
1854	Julius Schmidt and Thomas Dickert together build a relief model of the moon at Bonn Observatory.
1855	Warren de la Rue publishes "*A series of twelve photographs of the Moon.*"

1855 The dry Collodium-Albumen process is introduced by Jean Marie Taupenot.

1856/1858 Lewis Morris Rutherford photographs the moon and the sun using an achromatic refractor of 285 mm aperture.

1856 Collotype process introduced by Louis-Alphonse Poitevin.

1858/1859 Warren de la Rue publishes the first stereographs of the moon by obtaining images at different librations, as he explained in detail in his article "The Present State of Celestial Photography in England" published in 1859 in *The Report of the 29th Meeting of the British Association for the Advancement of Science.*

1859 Abdull Ghaffar Najm al-Dawleh, court astronomer to the Qajar king Naser al-Din Shah, starts teaching modern astronomy in Dar al-Funun and writes several textbooks on modern science for students at Dar al-Funun

1858-1865 European travels of the Iranian student Mahmod Khan Ghomi to study modern astronomy in Paris (5 years, Sorbonne University and Paris Observatory) and Brussels (2 years)

1860 Warren de la Rue photographs (wet Collodion) the total eclipse of the sun in Spain (July, 18) with the Kew photoheliograph (60 s exposures). Angelo Secchi also obtains excellent photographs of the same eclipse (Spain).

1861 James Clerk Maxwell demonstrates a color photography system involving three black and white photographs, each taken through a red, green and blue filter.

1864 The Collodio-Bromide ("*dry collodion*") process described by William Blanchard Bolton and Benjamin Jones Sayce.

1864 First known lunar photograph taken by an Iranian photographer, the court photographer Reza Akkasbashi

1863/1865 Henry Draper images the moon using a 40 cm reflector built by himself.

1865 Lewis Morris Rutherfurd obtains excellent moon images using a specially corrected (photographic) 290 mm lens. He also obtains hundreds of Collodion images of stars and star clusters.

Abdull Ghaffar Najm al-Dawleh publishes the Qanun-i-Nasiri, the first treatise on modern astronomy written in Iran

1866 Pietro Angelo Secchi establishes the "*Secchi*" three different classes of stellar spectra.

1866 Pierre Jules César Janssen maintains that the "*telluric lines*" in the sun's spectra originate from the presence of water vapor in the earth's atmosphere.

1868 Pietro Angelo Secchi described a fourth class of stellar spectra (carbon stars).

1868	Helium is discovered by Pierre Jules Janssen and Joseph Normam Lockyer: total solar eclipse (August 18).
1869	John Adams Whipple photographs the solar corona during a total eclipse (August 12).
1869	Pierre Jules Janssen describes the spectroheliograph.
1871	German astronomer Hermann Carl Vogel obtains excellent photographs of the sun using a 294 mm refractor equipped with an electrical shutter (1/5000 to 1/8000 s exposures). The period of rotation of the sun is also measured by measuring the Doppler shift in the absorption lines of its spectrum.
1871	Richard Leach Maddox, proposes the use of an emulsion of gelatin and silver bromide on a glass plate, the "dry plate" process.
1873	Hermann Wilhelm Vogel describes the process of dye sensitization.
1874	James Nasmyth and James Carpenter publish *The Moon: Considered as a Planet, a World, a Satellite* (John Murray), where they published photographs of plaster and wood moon models produced by Nasmyth.
1875	*First known Japanese moon photograph.*
1878	Charles Harper Bennett describes a method of increasing the sensitivity of gelatin-silver bromide emulsions.
1878	William Huggins and Margaret Lindsay Murray obtain a first photographic spectrum of the moon and Jupiter.
1879	George Eastman starts mass production of dry photographic plates.
1879/1882	Henry Draper photographs the spectra of the sun, moon, Mars, Jupiter, Comet C/1881 K1, several stars and the Orion nebula.
1882	Edward Charles Pickering starts a program of photographic astronomical spectroscopy at Harvard Observatory using objective prisms. This setup enabled Pickering to obtain several spectra on a single plate.
1884	George Eastman describes the flexible film made by coating paper in a photographic emulsion.
1885	David Gill and Jacobus Cornelius Kapetyn start work on the Cape Photographic Durchmusterung (CPD).
1886	In 1886, the librarian of the l'Observatoire Royal de Bruxelles, Albert Benoit Marie Lancaster, publishes his book *Liste Générale des Observatoires et des Astronomes des Sociétés et des Revues Astronomiques.*
1889	George Eastman introduces flexible plastic film.
1889	Albert Taylor, Andrew Ainslie Common's assistant, photographs planet Uranus with the 60-inch reflector. This photograph shows a distinct disc.
1892	The Crocker Astrograph is built using a 6-inch portrait lens.
1894–1910	Moritz Loewy and Pierre-Henri Puiseux obtain 6000 photographs (during 500 nights) of the Moon using the 60 cm Paris observatory Coudé refractor.

The *Atlas Photographique de la Lune* was edited from 1896–1910 by the Paris observatory.

1897 Edward Singleton Holden and Ladislaus Weinek publish the Lick Observatory Atlas of the Moon (19 photographs with a low resolution and poor quality).

1900 Ladislaus Weinek publishes the *Photographischer Mond-Atlas*.

1901 William Henry Pickering starts his Photographic Atlas of the Moon (published in 1903), Mandeville, Jamaica.

1901/1902 Karl Friedrich Küstner took nearly 100 plates of the moon with the double refractor at Bonn Observatory.

1902 Georges Méliès produces his film *Le Voyage dans la Lune*.

1904 George Ellery Hale installs the Snow Solar Telescope at Mount Wilson Observatory.

1906 Charles Edward Kenneth Mees invents the panchromatic photographic plate.

1917 Completion of the 100-inch (2.54 m) Mount Wilson telescope, California.

1928 Kodak research laboratory introduces the 103a series of photographic plates and emulsions.

1930 Bernhard Voldemar Schmidt builds the Schmidt camera (wide-field astrophotography).

1948 Edwin Powell Hubble uses the 200-inch (5.08 m) Hale telescope for the first time (Mount Palomar, California).

1955 Gerard Kuiper and colleagues start work on a large photographic lunar atlas. Photographs are obtained in five different observatories (Mount Wilson, Lick, Pic du Midi, McDonald, and Yerkes). The atlas is published in 1960.

1959 The first photographs of the *far side* of the Moon were obtained by the Soviet Luna 3 probe, in October 1959.

1964 Ranger 7 transmitted to Earth several high-resolution images of the area 100 km south-west of Fra Mauro, July 31, 1964.

1965 Ranger 8 obtains high-resolution images of Mare Tranquilitatis in February.

Ranger 9 returns images of the crater Alphonsus.

1966 The first panoramic photographs of the moon surface are obtained by the Soviet Luna 9 in February.

Luna 13 transmits high-resolution panoramic images of the Moon in December.

1969 The Charge-Coupled Device (CCD) is invented at AT&T Bell Labs by Willard Sterling Boyle and George Elwood Smith.

Apollo 11 is the first manned spacecraft to land on the Moon. The *Eagle* Lunar Module lands in Mar Tranquilitatis on July 20. Several Hasseblad cameras are used during *extravehicular activities.*

1970 The lunar rover (Lunokhod 1) taken in Luna 17 mission, took 20,000 photographs and 200 panoramic photos.

1971 *The Lunar Orbiter Photographic Atlas of the Moon (LOPAM) is published.*

1973 First commercial CCD camera is produced by Fairchild Imaging (100 × 100 pixels).

1974 First CCD image of an astronomical object (moon). Jim Janesick uses a 100 × 100 Fairchild CCD coupled to an 8-inch telescope.

1979 First professional CCD camera is used at Kitt Peak observatory (320 × 512 pixels) (0.9 m telescope).

1990 Hubble Space Telescope launched (April 24).

1994 *Clementine* (Deep Space Program Science Experiment) was a joint US Department of Defense/NASA mission. More than 2 million images of the Moon are obtained from an eccentric moon orbit between February and May 1994.

1996 SBIG Santa Barbara Imaging Group and Celestron Corporation introduce the Pixel 255 CCD based of the TC-255 chip (320 × 240 pixels).

1998 SBIG and Celestron launch the 357 CCD imaging system that later becomes the SBIG ST-237 (640 × 480 pixels).

The *Lunar Prospector* probe complements *Clementine* data and improves understanding of the origin, evolution and current state of the moon.

2004 SBIG introduces the STL11000 research CCD camera (4008 × 2672 pixels).

2007 SELENE spacecraft (launched September 14, 2007) is a joint project of Japan Aerospace Exploration Agency (JAXA), Institute of Space and Astronautical Science (ISAS) and National Space Development Agency (NASDA). The main orbiter impacts on the lunar surface on June 10, 2009, after successfully orbiting the Moon for a year and eight months.

China launches its Chang'e first lunar explorer on October 24.

2008 India launches Chandrayaan-1 on October 22.

2009 The Lunar Reconnaissance Orbiter (LRO), launched on June 18, 2009, is the first NASA Robotic Lunar Exploration Program designed to map the surface of the moon and locate future landing sites.

2010 The most accurate topographic map of the moon's surface based on LRO images is released in December.

Bibliography

Aït-Touati, Frédérique. *Fictions of the Cosmos*. Chicago: Chicago University Press, 2011.

Alexander, Amir R. "Lunar maps and coastal outlines: Thomas Hariot's [sic] mapping of the moon." *Studies in the History and Philosophy of Science* 29 (1998): 345-368.

Andrews, William J. H., ed. *The Quest for Longitude*. Cambridge: Collection of Historical Scientific Instruments, Harvard University, 1996.

Anon. (probably Heinrich Johann Mädler), "Ueber die Erfindung des Hrn. Daguerre," *Königlich privilegirte Zeitung von Staats- und gelehrten Sache* (*Vossische Zeitung*), 16 (19 January 1839), 1-2. Reprinted in *Neues Licht. Daguerre, Talbot und die Veröffentlichung der Fotografie im Jahr 1839*, edited by Steffan Siegel, 57-60. Paderborn: Wilhelm Fink, 2014.

[Anoymous], *Prospectus of Messrs. De Schlagintwetis' Collection of Ethnographical Heads from India and High Asia*, 2nd edition. Leipzig: J. A. Barth, 1859.

[Anonymous], Review of "The Moon: Considered as a Planet, a World, and a Satellite, by James Nasmyth, C.E., and James Carpenter, FRAS." *The Edinburgh Review or Critical Journal* 140 (1874): 78.

Ansari, Razaulah S.M. "The Observatories Movement in India During the 17-18th Centuries." *Vistas in Astronomy* 28 (1985)*:* 179-385.

Ansari, Razaulah S.M. *Introduction of Modern Astronomy in India during 18th-19th centuries.* New Delhi: Institute for the History of Medicine and Medical Research, 1985.

Ansari, Razaulah S.M., ed. "Modern Astronomy in Indo-Persian Writings." In *History of Oriental Astronomy*. Proceedings of the Joint Discussion-17 at the 23rd General Assembly of the International Astronomical Union, organised by the Commission 41 (History of Astronomy), held in Kyoto, August 25-26, 1997, edited by S. M. Razaulah Ansari, 133-144. Dortrecht: Kluwer Academic Publishers, 2002.

Ansari, Razaulah S.M. "The First Comprehensive Book in Indo-Persian on Modern Astronomy." *Proceedings of the Fifth International Conference on Oriental Astronomy,* held in Chiang Mai, Thailand, 4-8 Oct. 2004, edited by K.Y. Chen, W. Orchiston, B. Soonthenthum and R. Stroom, 81-91. Chiang Mai : Chiang Mai University, 2006.

Apollo 11 Lunar Photography. Nasa NSSDC Report 70-06. Washington D.C: NASA, 1970.

Arago, François. "Fixation des images qui se forment au foyer d'une chambre obscure." *Comptes Rendus* 8 (1839): 4-7.

Arago, François. "Le daguerréotype." *Comptes Rendus Hebdomadaires des Séances de l'Académie des Sciences* 9 (19 August 1839).

Arjomand, Kamram. "Transfer of Knowledge to Iran in the Safavid Era: A Letter on Modern Astronomy Explaining the Tychonic Model." *Majjalah-i tarikh-i Ulum* 10 (2012): 1-26 (in Persian).

Armitage, Geoff. "The Schlagintweit Collections." *Indian Journal of History of Science* 24 (1989): 67-83.

Armstrong, Carol. *Scenes in a Library: Reading the Photograph in the Book, 1843-1875*. Cambridge: MIT Press, 1998.

Armstrong, Carol. "Der Mond als Fotografie." In *Diskurse der Fotografie. Fotokritik am Ende des fotografisches Zeitalters,* edited by Herta Wolf, 359-383. Frankfurt am Main: Suhrkamp Verlag, 2003.

Arthur, D.W. G., and Whitaker, Ewen A. *Orthographic Atlas of the Moon-Supplement Number One to the Photographic Lunar Atlas*. Tucson: University of Arizona Press, 1960.

Ashtiyani, Abbas Iqbal. "Awwalin durbin-i nujum-i jadid dar Iran." ("The First Modern Telescope in Iran") *Yadegar* 10 (1945): 33-36.

Ashworth, Jr., William B. *The Face of the Moon. Galileo to Apollo*. Kansas City: Linda Hall Library, 1989.

AstroArts. *Stellar Navigator,* ver. 6.0. Tokyo: Astro Arts Inc, 1998.

Atai, Fahrhad. "The Sending of Iranian Students to Europe, 1811-1906." PhD diss., University of California, Berkeley, 1992.

Atami, Atafallah. "Introduction of modern astronomy in Qajar Era." *Nojum* 10 (2009): 28-31. (In Persian).

Bajac, Quentin. *Dans le champ des étoiles. Les photographes et le ciel, 1850-2000*. Paris: Editions de la Réunion des Musées Nationaux, 2000.

Ball, Sir Robert S. *The Story of the Heavens*. London: Cassell and Co., 1897.

Barger, M. Susan, and William B. White. *The Daguerreotype. Nineteenth-Century Technology and Modern Science*. Baltimore and London: The Johns Hopkins University Press, 1991.

Barger, Susan M. *The Daguerreotype: Nineteenth-Century Technology and Modern Science.* Washington and London: Smithsonian Institution Press, 1991.

Barnard, Edward E. "Some of the results of astronomical photography pertaining specially to the work with a portrait lens." *Proceedings of the American Philosophical Society* 46/187 (1907): 422.

Bartholomew, C. E. "Discovery of the Solar Granulation." *Quarterly Journal of the Royal Astronomical Society* 17 (1976): 263-289.

Batchen, Geoffrey. "Desiring Production." In *Each Wild Idea. Writing, Photography, History,* edited by Geoffrey Batchen, 3-24. Cambridge, Mass. and London: The MIT Press, 2001.

Beamish, Richard. *The Psychonomy of the Hand,* 2nd edition. London: Frederick Pitman, 1865.

Beer, Wilhelm, and Johann H. von Mädler. *Mappa selenographica totam lunare hemisphaeram visibilem complectens*. Berlin: Schropp, 1834.

Bennett, James A. *The Mathematical Science of Sir Christopher Wren*. Cambridge: Cambridge University Press, 2002.

Bennett, James A. "A Double Anniversary for John Russell, R. A." *Sphaera* 2 (1995):2. Online: <http://www.mhs.ox.ac.uk/about/sphaera/sphaera-issue-no-2/a-double-anniversary-for-john-russell-r-a/>. Accessed April 16, 2016.

Bennett, Jim. "Hooke's instruments." In *London's Leonardo: The Life and Work of Robert Hooke,* edited by Jim Bennett, Michael Cooper, Michael Hunter, and Lisa Jardine. Oxford: Oxford University Press, 2003.

Bigg, Charlotte. "Photography and the Labour History of Astronomy and Astrophysics." *The Role of Visual Representations in Astronomy: History and Research practice,* special issue edited by Klaus Hentschel and Axel Wittmann, *Acta Historica Astronomiae,* 9 (2000): 90-106.

Bigg, Charlotte. "In weiter Ferne so nah. Bilder des Titan." *Bildwelten des Wissens,* 5 (2007): 9-19.

Bigg, Charlotte. "The panorama, or la nature à coup d'œil." In *Observing Nature - Representing Experience. The Osmotic Dynamics of Romanticism 1800-1850*, edited by Erna Fiorentini, 73-95. Berlin : Reimer Verlag, 2007.

Blanchard, Guillaume, "Chérubin d'Orleans and his binocular microscope." In *Treasury of Optics. The Collections of the Optisches Museum Jena,* edited by Optisches Museum Jena, 127-144. Jena: Optisches Museum of Ernst Abbe Foundation, 2013.

Blunck, Juergen. "Deutsche Pionierarbeiten der Monddarstellung mit erhabenem Relief." *Der Globusfreund* 47/48 (2000): 293-303.

Blunt, Henry. "On a Model of the Moon's Surface." *Notices and Abstracts of Communication to the British Association for the Advancement of Science, at the Birmingham Meeting, September 1849* . London: John Murray, 1850.

Boyle, Colleen. "You Saw the Whole of the Moon: The Role of the Imagination in the Perceptual Construction of the Moon." *Leonardo* 46/3 (2013): 246-252.

Boyle, Willard, and George E. Smith. "Charge Couple semiconductor devices." *The Bell System Technical Journal,* 49 (1970): 587-593.

Brandstetter, Thomas. "Time Machines: Model Experiments in Geology." *Centaurus* 53 (2011): 135-145.

Briefel, Aviva. *The Racial Hand in the Victorian Imagination*. Cambridge: Cambridge University Press, 2015.

Brock, William H. *William Crookes (1832-1919) and the Commercialization of Science*. Hampshire: Asghate, 2008.

Brunet, François. *La naissance de l'idée de photographie.* Paris : Presses Universitaires de France, 2000.

Brusius, Mirjam. "Impreciseness in Julia Margaret Cameron's Portrait Photographs." *History of Photography* 34 (2010): 342-355.

Brusius, Mirjam, Katrina Dean, and Chitra Ramalingam, eds. *William Henry Fox Talbot. Beyond Photography.* New Haven: Yale University Press, 2013.

Brusius, Mirjam, *Fotografie und museales Wissen. William Henry Fox Talbot, das Altertum und die Absenz der Fotografie*. Berlin, Boston: De Gruyter, 2015.

Capuano, Peter J. *Changing Hands: Industry, Evolution, and the Reconfiguration of the Victorian Body*. Ann Arbor: University of Michigan Press, 2015.

Carder, Robert W. *Mapping of the Moon: Past and Present*. Dordrecht: Springer Science+Business Media, 1974.

Carlier, André H. *La photographie aérienne pendant la guerre*. Paris: Delagrave, 1921.

Carus, C. G. *Ueber Grund und Bedeutung der venchiedenen Formen der Hand in verschiedencn Personen*. Stuttgart, 1846.

Carver, Ben. "'A Gleaming and Glorious Star': Rethinking History in the Plurality-of-Worlds Debate." *Journal of Victorian Culture* 18 (2013): 429-452.

Ceragioli, Roger, and Walter Stephani. "Optiken für die Hamburger Sternwarte. Bernhard Schmidt und Richard Schorr – Zeugnisse und Ergebnisse einer Zusammenarbeit." In *Kometen, Sterne, Galaxien – Astronomie in der Hamburger Sternwarte',* Zum 100jährigen Jubiläum der Hamburger Sternwarte in Bergedorf, edited by : Wolfschmidt, Gudrun. Hamburg: tredition (Nuncius Hamburgensis; Band 24), 2014.

Chambers, Robert. "Sky-Hieroglyphics." *Chamber's Journal of Popular Literature, Science, and Arts*, 180 (1857): 372-374.

Chapman, Allan. "A World in the Moon - Wilkins and his Lunar Voyage of 1640." *Quarterly Journal of the Royal Astronomical Society* 32 (1991): 121-132.

Chapman, Allan. *England's Leonardo: Robert Hooke and the Seventeenth Century Scientific Revolution*. Boca Raton: CRC Press, 2004.

Chapman, Alan. *The Victorian Amateur Astronomer: Independent Astronomical Research in Britain, 1820-1920*. Chichester: Wiley, 1998.

Chéreau, Clément. "Le jeu des amateurs." In *L'Art de la Photographie, des Origines à nos Jours*, edited by André Gunthert and Michel Poivert, 256-271. Paris: Citadelles & Mazenod, 2007.

Cole, Sir Henry. "Mr James Nasmyth, C.E., on Freehand Drawing." *Journal of the Society of Arts* 28 (1880): 871-72.

Cortright, Edgar M. *Apollo expeditions to the Moon* (Scientific and Technical Information Office National Aeronautics and Space Administration Washington, DC. SP-350. Washington: NASA 1975: <http://www.hq.nasa.gov/office/pao/History/SP-350/toc.html>. Accessed March 6, 2017.

Crampton, Josiah. *The Lunar World: Its Scenery, Motions, Etc., Considered with a view to Design*. Edinburgh: Adam and Charles Black, 1863.

Crawford, William. *The Keepers of Light. A History of Working Guide to Early Photographic Processes.* New York: Dobbs Ferry Morgan, 1979.

Cressy, David. “Early Modern Space Travel and the English Man in the Moon.” *The American Historical Review* 111 (2006): 961-982.

Crookes, William. “On the Photography of the Moon.” *Proceedings of the Royal Society of London* 8 (1856-1857): 363-371.

Crowe, Michael J. *The Extraterrestrial Life Debate, 1750-1900: The Idea of a Plurality of Worlds from Kant to Lowell.* Cambridge: Cambridge University Press, 1988.

Crowe, Michael J., ed. *The Extraterrestrial Life Debate: Antiquity to 1915. A Source Book.* Paris: University of Notre Dame Press, 2008.

Cunningham, Clifford. *Discovery of the First Asteroid, Ceres: Historical Studies in Asteroid Research.* Dordrecht: Springer, 2016.

Danjon, André, and André Couder. *Lunettes et Telescopes*. Paris: Editions Revue d’Optique, 1935

Darius, Jon. *Beyond Vision.* Oxford: Oxford University Press, 1984.

D’Arpentigny, Casimir S. *La chirognomie, ou l’art de reconnaitre les tendances de l’intelligence d’apres les formes de la main.* Paris: C. le Clere, 1843.

Daston, Lorraine, and Peter Galison, eds. *Objectivity.* New York: Zone Books, 2010.

Davies, Merton E., and Bruce C. Murray. *The view from space. Photographic exploration of the planets.* New York and London: Columbia University Press, 1971.

De la Rue, Warren. “The Present State of Celestial Photography in England.” *The Report of the 29^{t}h Meeting of the British Association for the Advancement of Science* (1859): 130.

De la Rue, Warren. “Report on Celestial Photography in England.” *Reports of the British Association for the Advancement of Science for 1859* (London: Taylor and Francis, 1860), 131- 153.

Della Valle, Pietro. *Viaggi di Pietro Della Valle il pellegrino, con minuto ragguaglio di tutte le cose notabili osservate in essi: descritti da lui medesimo in 54 lettere familiari all’erudito suo amico Mario Schipano, divisi in tre parti cioè: la Turchia, la Persia e l’India,* Volume 1. Torino: 1843.

Desbarrolles, Adolphe. *Les mystéres de la main révélés et expliqués*. Paris: E. Dentu, Libraire- Éditeur, 1859.

Duerbeck, Hilmar W. “The German Transit of Venus expeditions of 1874 and 1882: organization, methods, stations, results.” *Journal of Astronomical History and Heritage* 7/1 (2004): 8-17.

Du Mans, Raphaël. *Estat de la Perse en 1660,* per Le P. Raphaël du Mans, Supérieur de la Mission des Capuchins d’Isfahan. Publié avec notes et appendice per Ch. Schefer, membre de l’Institut. Paris: Ernest Leroux Editeur, 1890.

Dunn, Richard, and Rebekah Higgitt. *Finding Longitude: How ships, clocks and stars helped solve the longitude problem.* London: Collins and Royal Museums Greenwich, 2014.

Eelsalu, Eino, Jürgen Hamel and Dieter Herrmann. *Johann Heinrich Mädler, 1794-1874. Eine Dokumentarische Biographie*. Berlin: Akademie-Verlag, 1985.

Ehktiar, Maryam. "The Dar al-Funun: Educational Reform and Cultural Development in Qajar Iran." PhD diss., New York University, 1994.

Ehktiar, Maryam. "Nasir al-Din Shah and the Dar al-Funun: The Evolution of an Institution." *Iranian Studies* 34/1-4, (2001): 153-164.

Farrington, Oliver C. "The Schmidt-Dickert Moon Model." *Science* 9:210 (1899): 35-36.

Faye, Hervé. "Sur l'état de la photographie astronomique en France." *Comptes Rendus Hebdomadaires des Séances de l'Académie des Sciences* 50 (1860): 965-967.

Faye, Hervé. "Sur les photographies de la lune de M. Lewis Rutherfurd." *CRAS* 75 (1872), 1072.

Ferguson, Eugene. *Engineering and the Mind's Eye*. Cambridge: MIT Press, 1994.

Feuvrier, Jean-Baptiste. *Trois ans à la cour de Perse*. Paris: F. Juven, 1899.

Finkelstein, Gabriel. "'Conquerors of the Kuenluen'? The Schlagintweit Mission to High Asia, 1854-57." *History of Science* 38 (2000): 179-218.

Forbes, Eric Gray. "Tobias Mayer (1723-62): A Case of Forgotten Genius." *British Journal for the History of Science* 5 (1970): 1-20.

Franz, Julius. "Neue Berechung von Hartwig's Beobachtungen der physischen Libration des Mondes." *Astronomische Nachrichten* 116 (1886): 1-12.

Franz, Julius. "Aufforderung zu Meridianbeobachtungen des Mondkraters Mösting A." *Astronomische Nachrichten* 122 (1889a): 237-240.

Franz, Julius. "Die Konstanten der physischen Libration des Mondes abgeleitet aus Schlüters Königsberger Heliometer-Beobachtungen." *Astronomische Beobachtungen in Königsberg* 38/1 (1889b): 1-47.

Franz, Julius. "Die Figur des Mondes." *Astronomische Beobachtungen in Königsberg* 38/5 (1899): 1-35.

Franz, Julius. "Ortsbestimmung von 150 Mondkratern." *Astronomische Mitteilungen der Koeniglichen Universitaets-Sternwarte zu Breslau* 1 (1901): I-VI, 1-51.

Franz, Julius. *Der Mond*. Leipzig: Teubner Verlag, 1912.

Franz, Julius. „Die Randlandschaften des Mondes." *NOVA ACTA, Abh. Der Kaiserl. Leop.-Carol. Deutschen Akademie der Naturforscher*, Band IC, Nr.1 (1913): 1-95.

Friendly, Michael, Pedro Valero-Mora, and Joaquín Ibáñez Ulargui. "The First (Known) Statistical Graph: Michael Florent van Langren and the "Secret" of Longitude." *The American Statistician* 64 (2010): 174-184.

Funke, Otto. *Bericht ueber die Schlagintweit'schen Reliefe des Monte-Rosa in der Schweiz und der Gruppe der Zugspitze und des Wettersteines in den bayerischen Voralpen*. Leipzig: J. A. Barth, 1855.

Galilei, Galileo. *Siderevs// Nvncivs : Magna, Longeqve Admirabilia Spectacula pandens, suspiciendáque proponens vnicuique, praesertim verò Philosophis, atq[ue] Astronomis*. Venetiis : Baglionus, 1610.

Galilei, Galileo, and Albert van Helden, ed./transl. *Sidereus Nuncius or the Sidereal Messenger*. Chicago, London: The University of Chicago Press, 1989.

Gaucheraud, Hippolyte. "Fine Arts—The Daguerotype." *The Literary Gazette; and Journal of the Belles Lettres, Arts, Sciences, &c.*, January 19 (1839): 43, 44.

Geffert, Michael; Martin W. Kuemmel; and Hans Schmidt. "Internal motions in the Pleiades cluster." *Astron. Astrophys. Suppl.* 112 (1995): 229-234.

Geimer, Peter. *Theorien der Photografie zur Einführung*. Hamburg: Junius, 2009.

Geimer, Peter. *Bilder aus Versehen. Eine Geschichte photographischer Erscheinungen*. Hamburg: Philo Fine arts, 2010.

Ghaffar Najm al-Dawleh, Abdull. *Qanun-i-Nasiri* (Tehran, 1865).

Gillespie, Sarah K. "John William Draper and the Reception of Early Scientific Photography." *History of Photography* 36 (2012): 241-254.

Godwin, Robert, *Apollo 11. The NASA Mission Reports*, Volume 1. Burlington: Apogee Books Space Series, 1999.

Godwin, Robert, *Apollo 11. The NASA Mission Reports*, Volume 2. Burlington: Apogee Books Space Series, 1999.

Godwin, Robert, *Apollo 11. The NASA Mission Reports*, Volume 3. Burlington: Apogee Books Space Series, 1999.

Godwin, Robert, *Apollo 12. The NASA Mission Reports*, Volume 1. Burlington: Apogee Books Space Series, 1999.

Godwin, Robert, *Apollo 12. The NASA Mission Reports*, Volume 2. Burlington: Apogee Books Space Series, 1999.

Godwin, Robert, *Apollo 14. The NASA Mission Reports*. Burlington: Apogee Books Space Series, 2001.

Godwin, Robert, *Apollo 15. The NASA Mission Reports*, Volume 1. Burlington: Apogee Books Space Series, 2001.

Godwin, Robert, *Apollo 16. The NASA Mission Reports*. Burlington: Apogee Books Space Series, 2002.

Godwin, Robert, *Apollo 17. The NASA Mission Reports*, Volume 1. Burlington: Apogee Books Space Series, 2002.

Goodings, L., S.D. Brown, and M. Parker. "Organizing images of futures-past: Remembering the Apollo Moon Landings." *International Journal of Management Concepts and Philosophy* 7/3-4 (2013): 268-288.

Graney, Chistopher M. *Setting Aside All Authority: Giovanni Battista Riccioli and the Science Against Copernicus in the Age of Galileo*. Paris: University of Notre Dame Press, 2015.

Great exhibition of the works of industry of all nations, 1851: official descriptive and illustrated catalogue, vol. 1. London: Spicer Brothers and W. Clowes, 1851.

Greenslade Jr., Thomas B. "The First Stereoscopic Pictures of the Moon." *Am. J. Phys.* 40 (1972): 536-40.

Grindlay, Jonathan, Sumin Tang, Edward Los, and Mathieu Servillat. "Opening the 100-Year Window for Time Domain Astronomy." *New Horizons in Time-Domain Astronomy: Proceedings IAU Symposium* 285 (2012): 29-34.

Guignard, Laurence. „Le `tourment lunaire' de Jules Pierrot-Desseilligny. Pratiques amateur d'observations de la lune." *Romantisme* 166 (2014): 65-78.

Gunthert, André. "La rétine du savant. La fonction heuristique de la photographie." *Etudes Photographiques* 7 (2000): 28-48.

Gunthert, André. "Spectres de la photographie. Arago et la divulgation du daguerréotype." In *Les Arago, acteurs de leur temps,* 441-451. Perpignan: Archives départementales des Pyrénées-orientales, 2010.

Gurney, John D. "Pietro della Valle: The Limits of Perception." *Bulletin of the School of Oriental and African Studies* 49 (1986): 103-11.

Hagen, Hans J., Detlef Groote, Dieter Engels, and Dieter Reimers. "The Hamburg Quasar Survey. I. Schmidt observations and plate digitization." A&AS 111 (1995): 195

Hall, R. Cargil, and Paul Dickison. *Lunar Impact: A History of Project Ranger. NASA SP-4210*. Washington, D.C.: NASA, 1977: <http://history.nasa.gov/SP-4210/pages/TOC.htm>. Accessed March 6, 2017.

Hallett, Michael. "The Strines Journal and the Nasmyth Steam Hammer." *History of Photography: An International Quaterly*, 13 (1989): 221-222.

Harland, David M. *Paving the Way for Apollo 11*. Dordrecht: Springer/Praxis Publishing, 2009.

Harley, B., and Paul Lexton, eds. *The New Nature of Maps: Essays in the History of Cartography*. Baltimore: Johns Hopkins University, 2002.

Hasselblad Press Info: *Hasselblad – 30 Years on the moon!* Göteborg, July 2, 1999.

Hasselblad Press Release: *Hasselblad 10 years in space. A unique photographic anniversary 1962-1972*. Göteborg, November 1972.

Heacock, Raymond L. "Lunar Photography: techniques and results." *Space Science Reviews* 8 (1968), 214-257.

Head, James W., and Douglas Lloyd. "Appendix: Near-Terminator and Earthshine Photography." In *Apollo 17. Preliminary Science Report,* 4-33 to 4-39. Washington D.C.: Lyndon B. Johnson Space Center, 1973.

Heidarzadeh, Tofigh. *From the Maragha School to the Dar al-Funun. A Historical Review of Astronomy in Iran from the 13th to the 19th Century,* Paper presented at the Symposium on Science and Technology in the Turkish and the Islamic World, Istanbul, 3-5 June, 1994. (unpublished).

Hentschel, Klaus. *Mapping the Spectrum. Techniques of Visual Representation in Research and Teaching.* Oxford and New York: Oxford University Press, 2002.

Hentschel, Klaus. "Why not one more Imponderable?: John William Draper and his `Tithonic rays'." *Foundations of Chemistry* 4 (2002): 5-59.

Hentschel, Klaus. *Visual Cultures in Science and Technology: A Comparative History.* Oxford: Oxford University Press, 2014.

Hermann, Dieter B. "An Exponential Law for the Establishment of Observatories in the Nineteenth-Century." *Journal for the History of Astronomy* 4 (1973): 57-58.

Herschel to de la Rue (Letter from), 10 October 1858 (Royal Society HS.6.D.143)

Herschel, John F. to William H. Fox Talbot, 4 December 1839, *The Correspondence of William Henry Fox Talbot,* collection number: 1937-4854. <http://foxtalbot.dmu.ac.uk/letters>. Accessed November 15, 2015.

Herschel, John. "Description of model of the surface of the moon by Frau H. Witte." *Monthly Notices of the Royal Astronomical Society* 7 (1845): 15-16.

Hingley, Peter and Françoise Launay. "Expéditions scientifiques: éclipses et passages, 1860-19010." In *Dans le champ des étoiles. Les photographes et le ciel, 1850-2000,* edited by Quentin Bajac, 117-124. Paris: Editions de la Réunion des Musées Nationaux, 2000.

Hockey, Thomas. *Biographical Encyclopedia of Astronomers.* New York: Springer Publishing, 2009.

Hoffmann, Martin, and Michael Geffert. "Photographische Astrometrie am Observatorium Hoher List." *Veröff. Univ. Bonn,* Nr. 89 (1979): 1-12.

Holden Edward S. ed. *Memorials of William Cranch Bond and of his son George Phillips Bond.* San Francisco: C.A. Murdock and Co./ New York City: Lehmke and Buechner, 1897.

Hopmann, Josef. "Nekrolog Karl Friedrich Küstner." *Vierteljahresschrift Astron. Ges.* 72 (1937): 24-34.

Hudec, Rene. "Astronomical Photographic Plate Archives." *Astroplate 2014* proceedings, edited by Linda Mišková, Stanislav Vítek. Prague: Institute of Chemical Technology, 2014.

Huff, Toby E. *Intellectual Curiosity and the Scientific Revolution. A Global Perspective.* Cambridge: Cambridge University Press, 2011.

Hughes, David E., and J. Kendrick, Bowker. *Lunar Orbiter Photographic Atlas of the Moon.* Washington D.C.: NASA 1970. SP-206. <http://www.lpi.usra.edu/resources/lunar_orbiter/book/lopam.pdf>. Accessed March 6, 2017.

Hukui, Kyuzo. 1937. *Shodaimyo no Bungei to Gakujutsu* (Literature and Scholastic Achievements by Daimyo Warlords, in Japanese). Tokyo: unknown publisher, 1937. Reprinted by Hara Shobo Inc. (1976).

Humphrey, Samuel D. "Lunar Daguerreotypes." *Daguerreian journal* 1 (1 nov. 1850)

Hunter, Michael. "Hooke the Natural Philosopher." In *London's Leonardo: The Life and Work of Robert Hooke,* edited by Jim Bennett, Michael Cooper, Michael Hunter, and Lisa Jardine, 105-62. Oxford: Oxford University Press, 2003.

Hurd, David, and John Matelok. *Getting a Feel for Lunar Craters.* Collingdale: Diane Publishing Co., 2011.

Ikeda, Kazuyuki. "Bunka Nenkan" "*Taiin no Zu*" (The lunar map drawn in 1813, in Japanese). *Tenkai* 123 (1931): 337-340.

Iwahashi, Zenbei. *Heitengi Zukai* (Illustrated Explanation of Planisphere and Introduction to Astronomy, in Japanese). Osaka: publisher unknown, 1801.

Janssen, Pierre Jules César. "Sur les spectres de l'oxigène." *CRAS* 106 (1888): 1118-19.

Janssen, Jules. "Mission du Japon pour l'observation du passage de Vénus." *l'Annuaire du Bureau des Longitudes de 1876* (1876): 572-588.

Jardine, Boris. "Made Real: Artifice and Accuracy in Nineteenth-Century Scientific Illustration." *Science Museum Group Journal* 1 (2014). <http://dx.doi.org/10.15180/140208>. Accessed November 24, 2015.

Jardine, Lisa. *On a Grander Scale: the Outstanding career of Sir Christopher Wren*. London: Harper Collins Publishers, 2002.

Kage, Toshio. *Collected Works of Asada Goryu,* Letter No.10 (Ed. T. Kage, in Japanese). Oita: Historical Documents Office, 1999.

Keil, Inge. *Augustanus Opticus: Johann Wiesel (1583-1662) und 200 Jahre optisches Handwerk in Augusburg.* Berlin: Akademie Verlag, 2000.

Kemp, Martin. *Visualizations: The Nature Book of Art and Science*. Berkeley: University of California Press, 2000.

Kemp, Martin. *Seen/Unseen: Art, Science, and Intuition from Leonardo to the Hubble Telescope*. Oxford: Oxford University Press, 2006.

Kenaan, Hagi. "Photography and Its Shadow." *Critical Inquiry* 41 (2015): 541-72.

Kenaan, Jan. "Counting on the Unexpected: Aime Civiale's Mountain Photography." *Science in Context* 22 (2009): 409-437.

Kessler, Elizabeth. *Picturing the Cosmos. Hubble Space Telescope Images and the Astronomical Sublime*. Minneapolis: University of Minnesota Press, 2012.

Keunig, Johannes. "The Van Langren Family." *Imago Mundi* 13 (1956): 101-109.

King, David A. *Astronomy in the Service of Islam*. Aldershot: Variorum, 1993.

King, Henry C. *The History of the Telescope*. New York: Dover Publications, 1979.

Kircher, Athanacius. *Mundus Subterraneus, quo Universae Denique Naturae Divitiae* (Underground World, in Latin). Amsterdam: Janssonio Waesbergiana, 1664-1678.

Kobe City Museum. *Shiba Kokan Hyakka Jiten* (Encyclopedia of Shiba Kokan, in Japanese). Kobe: Kobe City Museum, 1998.

Koeberl, Christian. "Craters on the Moon from Galileo to Wegener: A short history of the impact hypothesis, and implications for the study of terrestrial impact craters." In *Earth-Moon Relationships*, edited by Cesare Barbieri, and Francesca Rampazzi, 209-24. New York: Springer, 2001.

Kuiper, Gerard P., David W.G. Arthur, E. Moore, J.W. Tapscott, and Whitaker, Ewen A. *Photographic Lunar Atlas Based on Photographs Taken at the Mount Wilson, Lick, Pic du Midi, McDonald and Yerkes Observatories*. Chicago: University of Chicago Press, Nat. Sci. Foundation, 1960.

Kuiper, G.P., R.G. Strom, R. G, Ewen A. Whitaker, J.W. Fountain, and S.M. Larson. *Consolidated lunar atlas. Supplement no. 3 & 4 to the USAF Photographic lunar atlas.* Air Force Cambridge Research Laboratory (Office of Aerospace Research) Lunar and Planetary Laboratory. Tucson: University of Arizona, 1967.

Kunitzsch, Paul. "Sufi Latinus." *Zeitschrift der Deutschen Morgenlandischen Gesellschaft: ZDMG* 115 (1965): 66-74.

Kunitzsch, Paul. "Abd-al-Rahman B. 'Omar Sufi." *Encyclopaedia Iranica,* I/2 (1982): 148-149.

Kunitzsch, Paul. "The Astronomer Abu 'l Husayn al-Sufi and his Book on the Constellations." *Zeitschrift für Geschichte der arabische islamischen Wissenschaften: ZGAW* 3 (1986): 56-81.

Kunitzsch, Paul. "The Constellations." In *Images of Islamic Science. Illustrated manuscripts from the Iranian world,* 1, edited by Ziva Vesel and Azedine Beschaouch, 104-105. Paris: Institut Français de Recherches en Iran, 2009.

Küstner, Friedrich. "Jahresbericht der Sternwarten für 1900: Bonn." *Vierteljahresschrift Astron. Ges.* 36 (1901): 83-86.

Küstner, Friedrich. "Jahresbericht der Sternwarten für 1901: Bonn." *Vierteljahresschrift Astron. Ges.* 37 (1902): 84-89.

Küstner, Friedrich. "Jahresbericht der Sternwarten für 1902: Bonn." *Vierteljahresschrift Astron. Ges.* 38 (1903): 86-88;

Küstner, Friedrich. "Jahresbericht der Sternwarten für 1903: Bonn." *Vierteljahresschrift Astron. Ges.* 39 (1904): 63-65.

Küstner, Friedrich. "Eine spektrographische Bestimmung der Sonnenparallaxe." *Astronomische Nachrichten* 169 (1905): 241-264.

Küstner, Friedrich. "Radial velocities of 99 stars of the second and third spectral classes observed at Bonn." *Astrophysical Journal* 27 (1908): 301-324.

Küstner, Friedrich. "Radialgeschwindigkeiten von 227 Sternen des Spektraltypus F bis M beobachtet 1908 bis 1913 am Bonner 30 cm-Refraktor." *Astronomische Nachrichten* 198 (1914): 409-448.

Küstner, Friedrich. "Der kugelförmige Sternhaufen Messier 56." *Veröff. Univ. Bonn* 14 (1920): 1-47.

Lacy, Jodi, and Marvin Bolt. *Mapping the Universe.* Chicago: Adler Planetarium & Astronomy Museum, 2007.

Lancaster, Albert. *Liste Générale des Observatoires et des Astronomes des Sociétés et des Revues Astronomiques,* 3rd edition. Harvard: Harvard University, 1890.

Lane, K. Maria D. *Geographies of Mars; Seeing and Knowing the Red Planet.* Chicago: University of Chicago Press, 2011.

Lankford, John. "The impact of photography on astronomy." In *Astrophysics and twentieth-century astronomy to 1950,* part A, edited by Owen Gingerich, 16-39. Cambridge: Cambridge University Press, 1984.

Le Conte, David. "Warren De La Rue – Pioneer astronomical photographer." *The Antiquarian Astronomer* 5 (February 2011): 14-35.

Lankford, John. "Photography and the nineteenth-century transits of Venus." *Technology and Culture* 28 (1987): 648-657.

Launay, Françoise. "Jules Janssen et la photographie." In *Dans le champ des étoiles. Les photographes et le ciel, 1850-2000,* edited by Quentin Bajac, 22-31. Paris: Editions de la Réunion des Musées Nationaux, 2000.

Lavater, Johann Casper. *Essai sur la physiognomonie, destiné à faire connoître l'homme et à le faire aimer*. The Hague : Jaques Van Karnebeek, 1781.

Legros, Victor. *Sommaire de Photogrammétrie. Application élémentaire de la photographie à l'architecture, à la topographie, aux observations scientifiques et aux opérations militaires*. Paris: Sociétés d'éditions scientifiques, 1891.

Lequeaux, James. *Le Verrier – Magnificent and Detestable Astronome.* Dorndrecht: Springer, 2013.

Levitt, Theresa. "Biot's Paper and Arago's Plates: Photographic Practice and the Transparency of Representation." *Isis* 94 (2003): 456-476.

Levitt, Theresa. *The Shadow of Enlightenment. Optical and Political Transparency in France, 1789-1848*. Oxford: Oxford University Press, 2009.

Levitt, Theresa. "I thought this might be of interest... The Observatory as a Public Enterprise." In *The Heavens on Earth. Observatories and Astronomy in Nineteenth-Century Science and Culture*, edited by David Aubin, Charlotte Bigg and H. Otto Sibum, 285-304. Durham, NC: Duke University Press, 2010.

Loewy, Maurice and Henri Puiseux. "Sur les photographies de la Lune obtenues au grand équatorial Coudé de l'Observatoire de Paris." *Comptes Rendus Hebdomadaires de l'Académie des Sciences* 119 (1894): 254-259.

Loewy, Maurice and Pierre-Henri Puiseux. *Atlas photographique de la lune publié par l'observatoire de Paris, 1er fascicule*. Paris: Imprimerie Nationale, 1896.

Loewy, Maurice and Pierre-Henri Puiseux. *Atlas photographique de la lune publié par l'observatoire de Paris, 2e fascicule*. Paris: Imprimerie Nationale, 1898.

Louwman, Peter J.K. and Huib J. Zuidervaart. *A Certain Instrument for Seeing Far.* Wassenaar: Louwman Collection of Historical Telescopes, 2013.

Maegraith, Michael, ed. *Mondlandung. Dokumentation der Weltraumfahrt USA und UdSSR*. Stuttgart: Besler Verlag, 3rd edition, September 1969.

Mason, Richard, and John Caiger. *A History of Japan,* revised edition. Tokyo: Tuttle Publications, 1997.

Mathur, M. N. "Qanun-i Nasiri. A Persian treatise on modern astronomy." *Studies in History of Medicine & Science* IX/3-4 (1985): 153-159.

Matthee, Rudi. "Christians in Safavid Iran: Hospitality and Harassment." *Studies on Persianate Societies* 3 (2005): 3-43.

Matthee, Rudi. "The Safavid Under Western Eyes: Seventeenth-Century European Travellers in Iran." *Journal of Modern History* 13 (2009): 137-171.

Matthee, Rudi. "Safavid Iran Through the Eyes of European Travellers." *Harvard Library Bulletin* 23/1-2 (2012): 10-24.

Mayall, Nichollas U. "Bernhard Schmidt and his coma free reflector." PASP 58 (1946): 282.

McCauley, Anne Elizabeth. *Industrial Madness. Commercial Photography in Paris, 1848-1871.* New Haven and London: Yale University Press, 1994.

McEwen, M. C. and Uel S. Clanton. "Photographic Summary." In *Apollo 17. Preliminary Science Report,* 4-1 to 4-3. Washington D.C.: Lyndon B. Johnson Space Center, 1973.

Mercer, R.D., L. Dunkelman, R.E. Evans. "Part A: Zodiacal Light photography"; and R. M. MacQueen; C. L. Ross; and R. E. Evans, "Part B – Solar Corona Photography." In *Apollo 17. Preliminary Science Report.* Washington D.C.: Lyndon B. Johnson Space Center, 1973.

Mitchell, Colin. "Les voyageurs français en Perse de 1600 à 1730." *Eurasian Studies* 4/1 (2005): 41-7.

Montgomery, Scott L. *The Moon and the Western Imagination*. Tucson: University of Arizona Press, 1999.

Morgan, Hal and Andreas Brown. *Prairie Fires and Paper Moons. The American Photographic Postcard: 1900-1920.* Boston: David R. Godine Publisher, 1981.

MS.42502 (438 B1); MS.3241, 205, v.229; MS.3242, 10-12, v.38, 176, James Nasmyth Archive, National Library of Scotland, Edinburgh.

Myers, Natasha. *Rendering Life Molecular: Models, Modellers, and Excitable Matter.* Durham/London: Duke University Press, 2015.

Nabavi, Negin. "Dar al-Funun: The First Modern College in Iran." Master Thesis, St. Anthony's College, Oxford, 1990.

Nakamura, Tsuko. "Telescope making and its Western influence in the modern history of Japan: Kunitomo's reflector telescopes." *Proceedings of the Third International Conference on Oriental Astronomy* (2000): 135-142. Fukuoka: Fukuoka University of Education, Ed. M. Hirai

Nakamura, Tsuko. "The Earliest Telescope Preserved in Japan." *Journal of Astronomical History and Heritage*, 11/3 (2008): 203-212.

Nakamura, Tsuko. "A Lunar Map Made in 1813 by a Japanese Feudal Warlord." In *Mapping the Oriental Sky. Proceedings of the Seventh International Conference on Oriental Astronomy- ICOA,* edited by Tsuko Nakamura, Wayne Orchiston, Mitsuru Soma and Richard Strom, 82-87. Tokyo: National Astronomical Observatory of Japan, 2011.

Nakayama, Shigeru. *A History of Japanese Astronomy*. Cambridge: Harvard University, 1969.

Nakayama, Shigeru. "Asada Goryu." *Dictionary of Scientific Biography* 1. New York: Charles Scribner's Sons, 1970.

Nall, Joshua. *News from Mars. Transatlantic Mass Media and the Practice of New Astronomy, 1870-1910*. Ph.D. diss., University of Cambridge, Cambridge, 2013.

Nasim, Omar W. *Observing by Hand: Sketching the Nebulae in the Nineteenth Century.* Chicago: Chicago University Press, 2013.

Nasim, Omar W. "Observation, Working Images and Procedure: the "Great Spiral" in Lord Rosse's Astronomical Record Books and Beyond." *British Journal for the History of Science* 43 (2010): 353-389.

Nasim, Omar W. "Handling the Heavens: Things and Photo-Objects of Astronomy." In *Photo-Objects: On the Materiality of Photographs and Photo-Archives in the Humanities and Sciences*, edited by Julia Bärnighausen, and Costanza Caraffa (2018).

Nasmyth, James. "Appendix B. Remarks on the Introduction of the Slide Principle in Tools and Machines Employed in the Production of Machinery". In *Practical Essays on Mill Work and other Machinery with notes and additional articles by Thomas Tredgold and revised into a third edition with additions by George Rennie.* London: John Weale, 1841.

Nasmyth, James. "Some Remarks on the Telescopic Appearance of the Moon." *Monthly Notices of the Royal Astronomical Society* 6 (1844): 79-82.

Nasmyth, James. "New Theory of the Nature and Source of the Sun's Light and Heat." In *The Year-Book of Facts in Science and Art*, edited by John Times, 152-54. London: David Bogue, 1852.

Nasmyth, James. "Submarine Mortar Frigate." *Scientific American* 8 (1853): 178.

Nasmyth, James. "Proposed Theory of the Origin of the Asteroids." In *Notices and Abstracts of Miscellaneous Communications to the Sections*. London: John Murray, 1853.

Nasmyth, James. "Scheme Book, c1838-1856" Institution of Mechanical Engineers, <http://www.imeche.org/flipbook/ils/ims_98/>. Accessed December 17, 2015.

Nasmyth, James. "The Moon and its Surface Inequalities." *The Strines Journal: A Monthly Magazine of Literature, Science and Art* 4 (1855): 110-115.

Nasmyth, James. "Suggestions respecting the Origin of the Rotatory Movements of the Celestial Bodies and the Spiral Forms of the Nebulae as seen in Lord Rosse's Telescopes." *Monthly Notices of the Royal Astronomical Society* 15 (1855): 220-21.

Nasmyth, James. "On the Plastic Origins of the Cuneiform Characters and its Relation to our own Alphabet." In *Notices and Abstracts of Communication to the British Association for the Advancement of Science, at Cheltenham in August 1856*. London: John Murray, 1857, 118-19.

Nasmyth, James. "On the Form of Lightning." In *Notices and Abstracts of Miscellaneous Communications to the Sections of BAAS*. London: John Murray, 1857, 14.

Nasmyth, James. "On the structure of the luminous envelope of the Sun." *Memoirs of the Literary and Philosophical Society of Manchester*, 1 (1862): 407-411.

Nasmyth, James. "On the Physical Aspects of the Moon's Surface." *The Quarterly Journal of Science* 1 (1864): 395-401

Nasmyth, James. "On the Physical Aspects of the Moon's Surface." *Notices of the Proceedings at the Meeting of the Members of the Royal Institution of Great Britain* 4 (1866): 395-6.

Nasmyth, James. "On War Rockets." *Memoirs of the Literary and Philosophical Society of Manchester* 4 (1871): 206-10.

Nasmyth, James and James Carpenter. *The Moon, considered as a Planet, a World, a Satellite*. London: John Murray, 1874.

Nasmyth, James. "Hints on the Education of the Eye and Fingers." *James Murray's Magazine: A Home and Colonial Periodical for the General Reader*, May (1887): 312.

Nasmyth, James. *James Nasmyth: Engineer, An Autobiography*, edited by Samuel Smiles. London: John Murray, 1912.

National Geographic, *First Explorers on the Moon. The Incredible Story of Apollo 11 in Five Parts* 136/6, December 1969.

Neison, Edmund. *The Moon and the Condition and Configurations of its Surface*. London: Longman, Green and co., 1876.

Neugebauer, Paul V. "Anzeige des Todes von Julius Franz." *Astronomische Nachrichten* 193 (1913): 443.

Newhall, Beaumount, and Louis Jacques Mandé Daguerre. *Daguerre* (*An historical and descriptive account of the various processes of the Daguerreotype and the Diorama, by Daguerre*). New York: Winter House, 1971.

Nezami, Pouria. "German Venus Transit Expedition to Isfahan." *Nojum* 3 (2012): 42-45. (In Persian).

Nicolson, Marjorie Hope. *Voyages to the Moon.* New York: The Macmillan Company, 1948.

Norman, Daniel. "The Development of Astronomical Photography." *Osiris* 5 (1938): 560-594.

Nogéus, Gunilla. "Hasselblad Space Cameras – A Brief History." *Hasselblad Magazine* 1970: 3, 28.

Nozette, S., *et al.* The Clementine mission to the Moon: Scientific overview. *Science*, 266, No. 5192, 1835-1839, Dec. 1994.

Odenkirchen, Michael, Peter Brosche, Michael Geffert, and Hans-Joachim Tucholke. "Globular cluster orbits based on Hipparcos proper motions." *New Astronomy* 2, 5 (1997): 477-499.

Ogilvie, Marilyn Bailey. "Obligatory Amateurs: Annie Maunder (1868-1947) and other British Women Astronomers at the Dawn of Professional Astronomy." *The British Journal for the History of Science* 33 (2000): 67-84.

Oreskes, N. "From Scaling to Simulation: Changing Meanings and Ambitions of Models in Geology." In *Science without Laws: Model Systems, Cases, Exemplary Narratives*, edited by Angela N. H. Creager, Elizabeth Lunbeck, M. Norton Wise, 93-124. Durham/ London: Duke University Press, 2007.

Payervand Sabet, Fariba. "Abdull Ghaffar Najm al-Dawleh." *Nojum* 10 (2009): 54-55. (In Persian).

Payervand Sabet, Fariba. "Mahmod Khan Ghomi." *Nojum* 7 (2009): 53-55. (In Persian).

Pérez González, Carmen. "From the Observatory to the Classroom: space images in the Keystone *600 SET* and *1200 SET.*" *Stereo & Immersive Media Journal* 1/1 (2017): 40-63.

Pérez González, Carmen. *Through the Lens of the 19th Century Iranian Photographers.* Leiden: Leiden University Press, 2012.

Pérez González, Carmen, and Reza Sheikh. "From the Inner Sanctum - Men who were Trusted by the Kings". In *The Eye of the Shah: Qajar Photography and the Persian Past*, edited by Jennifer Y. Chi, 132-157. New York: ISAW and Princeton University Press, 2015.

Phillips, John. "Notes on the Drawing of "Copernicus" presented to the Royal Society by P.A. Secchi." *Proceedings of the Royal Society of London* 8 (1856-57): 73-75.

Phillips, John. "Suggestions for the Attainment of a Systematic Representation of the Physical Aspect of the Moon." *Proceedings of the Royal Society of London* 12 (1862-63): 31-32.

Phillips, John. "Notices of Some Parts of the Surface of the Moon." *Philosophical Transactions of the Royal Society of London* 158 (1868): 333-345.

Polak, Dr. Jakob Eduard. *Persien. Das land und Ihre Bewohner.* Leipzig: J. M. Brodhaus, 1865.

Pumfrey, Stephen. "Harriot's Maps of the Moon: New Interpretations." *Notes and Records of the Royal Society of London* 63 (2009): 163-168.

Prodger, Phillip. *Darwin's Camera: Art and Photography in the Theory of Evolution.* Oxford: Oxford University Press, 2009

Radau, Rodolphe. "Les Applications scientifiques de la photographie." *Revue des Deux Mondes* 25 (1878) : 13.

Ratcliff, Jessica. *The Transit of Venus Enterprise in Victorian Britain.* London: Pickering & Chatto, 2008.

Ratcliff, Jessica. "The East India Company, the Company's Museum, and the Political Economy of Natural History in the Early Nineteenth Century." *Isis* 107 (2016): 495-517.

Rao, Kameswara N. "Father J. Richard and Early Telescope Observations in India." *Bulletin of the Astronomical Society of India* 12 (1984): 81-85.

Reaves, Gibson, and Carlo Pedretti. "Leonardo da Vinci's Drawings of the Surface Features of the Moon." *Journal for the History of Astronomy* xviii (1987): 55-58.

Rheita, Anton Maria Schyteus. *Occulus enoch & Elliae.* Antwerp: ex officina typographica Hieronymi Verdussij,1645.

Richard, Francis. *Raphael du Mans, missionnaire en Perse au XVIIe s.*, Vol. I: *Biographie. Correspondance*, Vol.II: *Estats et Mémoire;* Moyen Orient & Océan Indien XVe-XIX^{e}s; Société d'Histoire de l'Orient. Paris: Editions L'Harmattan, 1995.

Richard, Francis. "Raphaël du Mans". *Encyclopaedia Iranica* VII/6 (2004): 571-572. Online version: <http://www.iranicaonline.org/articles/du-mans>. Accessed September 3, 2016.

Richard, Francis. "Capuchins in Persia". *Encyclopaedia Iranica* IV/7 (2007): 786-788. Online version: <http://www.iranicaonline.org/articles/capuchins-in-persia>. Accessed February 18, 2016.

Robertson, Frances. "Science and Fiction: James Nasmyth's Photographic Images of the Moon." *Victorian Studies* 48 (2006): 595-623.

Rothermel, Holly. "Images of the Sun: Warren De la Rue, George Biddell Airy and Celestial Photography." *The British Journal for the History of Science* 26 (1993): 137-169.

Rudwick, Martin J. "Poulett Scrope on the Volcanoes of Auvergne: Lyellian Time and Political Economy." *British Journal for the History of Science* 7 (1974): 205-42.

Ruiz-Castell, Pedro. *Astronomy and Astrophysics in Spain (1850-1914)*. Cambridge: Cambridge Scholars Publishing, 2008.

Ryan, Peter. *The Invasion of the Moon 1969. The Story of Apollo 11.* Harmondsworth: Penguin, 1969.

Saito, Kuniji and Shinozawa Shizuyo. "On the observations of the transit of Venus over the Sun with particular emphasis on the December 9, 1847 event observed in Japan: Part I." *Tokyo Astronomical Observatory Report*, 16/1 (1972): 72-162; Ibid. Part II, 16/2 (1973): 259-385.

Saliba, George. "Persian scientists in the Islamic world: astronomy from Maragha to Samarqand." In *The Persian Presence in the Islamic World,* edited by Richard G. Hovannisian, and George Sabagh, 126-146. Cambridge: Cambridge University Press, 1998.

Sanner, Jörg, Martin Altmann, Jens Bunzendorf, and Michael Geffert."Photometric and kinematic studies of open star clusters. II. NGC 1960 (M 36) and NGC 2194." *Astron. Astrophys.* 357 (2000): 471-483.

Satow, M. Earnest. *The Voyage of Captain John Saris to Japan*. London: The Hakluyt Society, 1900.

Savage-Smith, Emily. *Islamicate Celestial Globes: Their History, Construction, and Use.* Washington, D.C.: Smithsonian Institution Press, 1985.

Sayili, Aydin. "An early seventeenth century Persian manuscript on the Tychonic system." *Anatolia* 3 (1958): 84-87.

Schaaf, Larry. "Piazzi Smyth at Teneriffe: Part 1 The Expedition and the Resulting Book." *History of Photography* 4 (1980): 289-307.

Schaaf, Larry. *Out of the shadows: Herschel, Talbot and the invention of photography*. New Haven and London: Yale University Press, 1992.

Schaaf, Larry J. *Selected Correspondence of William Henry Fox Talbot, 1823-1874*. Bradford: Science Museum and National Museum of Photography, Film & Television 1994.

Schaefer, Bradley E. "The transit of Venus and the notorious black drop effect." *Journal for the History of Astronomy* 32 (2001): 325-336.

Schaefer, Bradley E. "The Epoch for the Constellations on the Farnese Atlas and Their Origin in Hipparchus's Lost Catalogue." *Journal for the History of Astronomy* xxxvi (2005): 167-196.

Scheiner, Julius, *Photographie der Gestirne*. Leipzig: W. Engelmann, 1897.

Schlagintweit, Adolphe and Hermann. *Épreuves de cartes géographiques produites par la photographie d'après les reliefs du Mont Rose et de la Zugspitze*. Leipzig: J. A. Barth, 1854.

Schmidt, Johann FJ. *Das Relief der sichtbaren Halbkugel des Mondes*. Bonn: Carl Georgi, 1854.

Schmidt, Johann F.J. *Charte der gebirge des mondes nach eigenen beobachtungen in den jahren 1840-1874 entworfen*. Berlin: Dietrich Reimer, 1878.

Schorr, Richard. "Die Hamburgische Sonnenfinsternis-Expedition nach Souk-Ahras (Algerien)." In *Jahrbuch der Hamburgischen Wissenschaftlichen Anstalten XXII 1904*. Hamburg: Hamburger Sternwarte 1905.

Schorr, Richard. "Anzeige des Todes von Bernhard Schmidt." AN 258 (1936): 45.

Schramm, Jochen. "Reisen zur verfinsterten Sonne." *Sterne Weltraum, Special* 4 (*Sonne*) 1999: 28.

Schramm, Jochen. "Der Große Hamburger Schmidtspiegel". In *Kometen, Sterne Galaxien - Astronomie in der Hamburger Sternwarte,* Zum 100-jährigen Jubiläum der Hamburger Sternwarte in Bergedorf , edited by Gudrun Wolfschmidt. Hamburg: tredition (Nuncius Hamburgensis; Band 24), 2014.

Scrope, George Julius P. *On the Mode of Formation of Volcanic Cones and Craters*. London: Taylor and Francis, 1859.

Scrope, George Julius P. "Review of The Moon." *Geological Magazine* 1 (1874): 272-79.

Secchi, Angelo Pietro. "Lunar Photography." *The Daguerrian Journal* 8/23 (1 April 1857): 354.

Sekidou, Naruo. *Tenmon Shoukei* (Introduction of Astronomy, in Japanese). Tokyo: Ministry of Education, 1875.

Shah of Persia, A *Diary Kept by His Majesty the Shah of Persia, During his Journey to Europe in 1878,* translated by Albert Houtum Schindler and Baron Louis de Norman. London: Richard Bentley and Son, 1879.

Sharpe, Tom and P.J. McCartney. *The Papers of H. T. De la Beche (1796-1855) in the National Museum of Wales*. Cardiff: Llyfrau Amgueddfa Cymru/ National Museum Wales Books, 1998.

Sheehan, Tanya and Andrés M. Zervigón, eds. *Photography and Its Origin.* New York and London: Routledge, 2015.

Sheehan, William, and Richard Baum. "Observation and inference: Johann Hieronymous Schröter, 1745-1816." *Journal of the British Astronomical Association* 105 (1995): 171-175.

Sheehan, William P., and Thomas A. Dobbin. *Epic Moon - A History of Lunar Exploration in the Age of the Telescope*. Richmond: Willman-Bell, 2001.

Sheikh, Reza and Carmen Pérez González, eds. "The First Hundred Years of Iranian Photography." *History of Photography* 37/1 (2013).

Shiba, Kokan. *Shiba Kokan Zenshu* (Collected Works of Shiba Kokan, in Japanese, four volumes, reproduced version). Kyoto: Yasaka Shobo Inc, 1992-1993.

Shiba, Kokan. *Tenkyu Zenzu* (Figure of the Celestial Sphere, in Japanese), 12 plates. Tokyo: Shunparo, c.1796.

Sicard, Monique. "Passage de Vénus: Le révolver photographique de Jules Janssen." *Etudes Photographiques* 4 (1998): 45-63.

Siegel, Steffan, ed. *Neues Licht. Daguerre, Talbot und die Veröffentlichung der Fotografie im Jahr 1839*. Paderborn: Wilhelm Fink, 2014.

Sluiter, Engel. "The First Known Telescopes carried to America, Asia and the Arctic, 1614-39." *Journal for the History of Photography* 28 (1997): 141-145.

Smiles, Samuel. *Industrial Biography: Iron-Workers and Tool-Makers*. Boston: Ticknor and Fields, 1864.

Smyth, Charles P. "Sketches Chiefly in Scotland 1849-", vol. 2, 77. Edinburgh: Royal Observatory of Edinburgh: December 1850.

Smyth, Charles P. "On Certain Connecting Points between Lunar and Terrestrial Volcanoes." *Monthly Notices of the Royal Astronomical Society* 18 (1858): 156-162.

Smyth, Charles P. *Report on the Teneriffe Astronomical Experiment of 1856, Addressed to the Lords Commissioners of the Admiralty*. London and Edinburgh: Richard Taylor and William Francis, 1858.

Snyder, Joel. "Res Ipsa Loquitur." In *Things That Talk. Object Lessons from Art and Science,* edited by Lorraine Daston, 195-221. New York : Zone Books, 1998.

Sohrabi, Naghmeh. *Taken for Wonder. Nineteenth-Century Travel Accounts from Iran to Europe*. Oxford: Oxford University Press, 2012.

Soojung-Kim Pang, Alex. "'Stars should henceforth register themselves'. Astrophotography at the Early Lick Observatory." *British Journal for the History of Science* 30 (1997): 177- 202.

Soojung-Kim Pang, Alex. *Empire and the Sun. Victorian Solar Eclipse Expeditions.* Stanford: Stanford University Press, 2002.

Sopwith,Thomas. *Account of the Museum of Economic Geology, and Mining Records Office*. London: John Murray, 1843.

Sorensen, T. C., and Paul D. Spudis, "The Clementine mission -- A 10-year perspective." *J. Earth Syst. Sci.* 114/6 (2005): 645-668.

Spencer, Herbert. "Recent Astronomy, and the Nebular Hypothesis." *The Westminster Review* 14 (1858): 185-225.

Spudis, Paul D. "China's Moon Missions Are Anything But Pointless. The value of the nation's lunar missions goes beyond science", in *Air & Space, Smithsonian,* 3rd January 2017: <http://www.airspacemag.com/daily-planet/chinas-moon-missions-are-anything-pointless-180961633/>. Accessed March 6, 2017.

Staubermann, Klaus B. "Making stars: projection culture in nineteenth-century German astronomy." *British Journal for the History of Science* 34 (2001): 439-51.

Stephen Pinson. *Speculating Daguerre. Art and Enterprise in the work of L.J.M. Daguerre.* Chicago: University of Chicago Press, 2012.

Stevenson, Sara. *The Personal Art of David Octavius Hill.* New Haven: Yale University Press, 2002.

Storey, G. A. "Philip Hermogenes Calderon, A. R. A. (1833-98)." *Magazine of Art* 22 (1897-1898): 446-52.

Sumira, Sylvia. *Globes: 400 Years of Exploration, Navigation and Power.* Chicago, London: The University of Chicago Press, 2014.

Sutton, Thomas. "Photographic Exhibition of the Manchester Photographic Society." *Photographic Notes: Journal of the Photographic Society of Scotland and of the Manchester Photographic Society* 1 (1856): 204.

Tahmasbpour, Mohammad Reza. *Aqa Reza, the Court Photographer.* Tehran: Akshane-ye Shar Museum and Tehran Municipality Department of Art and Culture, 2006. (In Persian).

Talbot to Herschel (letter from), London Royal Society, Herschel Collection, HS 17:299, December 7, 1839.

Talbot, William H. Fox to J. F. Herschel, 15 November 1839, *The Correspondence of William Henry Fox Talbot,* collection number: HS 17:298, <http://foxtalbot.dmu.ac.uk>. Accessed November 15, 2015.

Talbot, William H. Fox. *The Pencil of Nature.* London: Longman, Brown, Green, & Longman, 1844-6.

Tani, Tateki. *Tani Jinzan's Jinki Roku* II (Tani Jinzan's Dialogues with Shibukawa Harumi, II). Tokyo: publisher unknown, 1910. (In Japanese).

Terpak, Frances. "Objects and Contexts." In *Devices of Wonder: From the World in a Box to Images on a Screen,* edited by Barbara Maria Stafford, and Frances Terpak, 143-364. Los Angeles: Getty Research Institute, 2001.

Texerau, Jean. *How to make a Telescope.* Virginia: Willmann-Bell, 1984.

Thomas, Ann. *Beauty of Another Order: Photography in Science.* Yale: Yale University Press, 1997.

Thomas, Ann. "Capturing the Light: Photographing the Universe." In *Beauty of Another Order. Photography in Science,* edited by Ann Thomas and Marta Braun, 186-219. New Haven: Yale University Press, 1997.

Tomita, Yoshio, and Tsuko Nakamura. "Investigation of the Gregorian telescopes made by the gunsmith Tobei Kunitomo in the 1830s." *Kokuritsu Tenmondaihou* (National Astronomical Observatory Report) 4 (1998): 9-41, in Japanese (abstract in English).

Trachtenberg, Alan. "Likeness as Identity: Reflections on the Daguerrean Mystique." In *The Portrait in Photography*, edited by Graham Clarke, 173-192. London: Reaktion Books, 1992.

Tresch, John. *The Romantic Machine: Utopian Science and Technology after Napoleon.* Chicago & London: The University of Chicago Press, 2012.

Tucker, Jennifer. *Nature Exposed: Photography as Eyewitness in Victorian Science.* Baltimore and London: Johns Hopkins, 2005.

Van Helden, Albert. "Telescopes and Authority from Galileo to Casssini." *Osiris* 9 (1994): 8-29.

Van Helden, Albert. "Moon maps in History (review of E. Whitaker's Mapping and Naming the Moon." *Journal for the History of Astronomy* 33 (2002): 285-289.

Van Helden, Albert, Sven Dupré, Rob Van Gent, and Huib Zuidervaart, eds. *The origins of the telescope.* Amsterdam: KNAW Press, 2011.

Vaucoulers, Gérard de. *La photographie astronomique.* Paris: Albin Michel, 1958.

Vertesi, Janet. "Picturing the moon: Hevelius's and Riccioli's visual debate." *Studies in History and Philosophy of Science* 38 (2007): 401-421.

Vertesi, Janet. "Instrumental images: the visual rethoric of self-presentation in Hevelius's *Machina Coelestis.*" *British Journal for the History of Science* 43 (2010): 209-243.

Von Brevern, Jan. "Resemblance after photography." *Representations* 123 (2013): 1-22.

Von Humboldt, Alexander. "Ideen zu einer Physiognomik der Gewächse." In *Ansichten der Natur* 1. Tübingen: J. G. Cotta, 1808.

Von Humboldt, Alexander. *Views of Nature: or Contemplations on the sublime Phenomena of Creation*, trans. Elise C. Otté and Henry G. Bohn. London: Henry G. Bohn, 1850.

Von Humboldt, Alexander to François Arago, February 1839. In *Correspondance d'Alexandre de Humboldt avec François Arago (1809-1853),* preface E.-T. Hamy. Paris: Guilmoto, 1908.

Warner, Deborah Jean. "The Draper family material: National Museum of American History." *History of Photography* 24 (2000), 16-23.

Wertz, Myriam, Dieter Horns, Detlef Groote, Taavi Tuvikene, Stefan Czesla, and Schmitt JHMM. "Hamburger Sternwarte plate archives: Historic long-term variability study of active galaxies based on digitized photographic plates." *Astron. Nachr.*/AN (2016);00:1-6. DOI:10.1002/asna.201613201

Whitaker, Ewen A., G.P. Kuiper, W.K. Hartmann, and L.H. Spradley. *Rectified Lunar Atlas - Supplement Number Two to the Photographic Lunar Atlas.* Tucson: Arizona University Press, 1963.

Whitaker, Ewen A. "Selenography in the seventeenth century." In *The General History of Astronomy, Vol. 2: Planetary Astronomy from the Renaissance to the rise of astrophysics, Part A: Tycho Brahe to Newton,* edited by René Taton, Curtis Wilson, and Michael Hoskin, 119-143. Cambridge: Cambridge University Press, 1989.

Whitaker, Ewen A. *Mapping and Naming the Moon. A History of Lunar Cartography and Nomenclature.* Cambridge: Cambridge University Press, 1999.

Wilder, Kelley. "Photography and the Art of Science." *Visual Studies,* 24/2 (2009): 163-168.

Wilder, Kelley. *Photography and Science.* London: Reaktion Books, 2009.

Wilder, Kelley. "Visualizing Radiation: The Photographs of Henri Becquerel." In *Histories of Scientific Observation,* edited by Lorraine Daston and Elizabeth Lunbeck, 349-368. Chicago: University of Chicago Press, 2011.

Wilder, Kelley. "A Note on the Invention of Photography: Reconsidering the Invention Story." In *Photography and its Origins*, edited by Tanya Sheehan and Andrés Zervigon, 208-220. New York: Routledge, 2015.

Winker, Mary G., and Albert van Helden. "Representing the Heavens: Galileo and Visual Astronomy." *Isis* 83 (1992): 195-217.

Wisotzki, Lutz, T. Köhler, Detlef Groote, and Dieter Reimers. "The Hamburg/ESO survey for bright QSOs I. Survey design and candidate selection procedure."*az* 115 (1996)*:* 227

Wlodarczyk, Jaroslav. "Libration of the Moon, Hevelius Theory and its early Reception in England." *Journal for the history of astronomy* 42 (2011): 495-519.

Wolfschmidt, Gudrun ed. *Sonne, Mond und Sterne Meilensteine der Astronomiegeschichte.* Zum 100jährigen Jubiläum der Hamburger Sternwarte in Bergedorf. Hamburg: tredition (Nuncius Hamburgensis, Band 29), 2013.

Wolfschmidt, Gudrun ed. *Kometen, Sterne, Galaxien - Astronomie in der Hamburger Sternwarte.* Zum 100jährigen Jubiläum der Hamburger Sternwarte in Bergedorf. Hamburg: tredition (Nuncius Hamburgensis; Band 24), 2014.

Wu, Sherman S. C., Francis J. Schafer, Gary M. Nakata, and Raymond Jordan, "Repeatability of elevation measurements – Apollo photography." In *Apollo 17. Preliminary Science Report,* 33-35 to 33-40. Washington D.C: Lyndon B. Johnson Space Center, 1973.

Wynter, Harriet and Anthony Turner. *Scientific Instruments.* London: Studio Vista, 1975.

Yamamoto, Issei. "Kunitomo and his astronomical activities in the pre-Meizi era." *Isis,* 26/2 (1937): 330-335.

Yazdi, Hamid-Reza Giahi. "Nasir al-Din Tusi on lunar visibility and an Analysis with Modern Altitude-Azimuth Criteria." *Suhayl* 3 (2002-3): 231-243.

Yoichi, Isahaya. "Gravity of Modernity: reactions to the "New Astronomy" in Iran and Japan." In *Translation, History and Arts: New Horizons in Asian Interdisciplinary Humanities Research,* edited by Ji, Meng and Ukai Atsuko, 27-47. Newcastle upon Tyne: Cambridge Scholars Publishing, 2013.

Yunli, Shi. "The First Telescopic Map of the Moon's Surface Drawn by a Chinese Astronomer." *China Historical Materials of Science and Technology* 12/4, 1991: 88-92. (In Chinese).

Zimmerman, A. "Die Gipsmasken der Brüder Schlagintweit: Verkörperung kolonialer Macht". In *Über den Himalaya: Die Expedition der Brüder Schlagintweit nach Indien und Zentralasien 1854-1858,* edited by Moritz von Brescius, Friederick Kaiser, and Stephanie Kleidt, 241-49. Cologne: Böhlau, 2015.

Zoka, Yayar. *The History of Photography and Pioneer Photographers in Iran.* Tehran: Sherkate Entesharat Elemi wa Fahrhangui, 1997. (In Persian).

Zoomers, Henk. "The Netherlands, Siam and the Telescope. The first Asian encounter with a Dutch invention." In *The Origins of the Telescope,* edited by Albert Van Helden, Sven Dupré, Rob Van Gent, Huib J. Zuidervaart, 301-320. Amsterdam: Knaw Press, 2010.

Zuidervaart, Huib J. "The reflecting telescope in the Netherlands." *Annals of Science* 61 (2004): 407-452.

Index

www.ingramcontent.com/pod-product-compliance
Lightning Source LLC
Jackson TN
JSHW070954230826
107171JS00003B/61
* 9 7 8 9 0 0 4 2 9 8 8 6 6 *